T0180191

# Waste Management, Processing and Valorisation

Abu Zahrim Yaser · Husnul Azan Tajarudin ·
Asha Embrandiri
Editors

# Waste Management, Processing and Valorisation

*Editors*
Abu Zahrim Yaser
Chemical Engineering Programme
Faculty of Engineering
Universiti Malaysia Sabah
Jalan UMS
Kota Kinabalu, Sabah, Malaysia

Husnul Azan Tajarudin
Division of Bioprocess Technology
Jalan Transkrian
Universiti Sains Malaysia
Gelugor, Malaysia

Asha Embrandiri
Department of Environmental Health
College of Medicine and Health Sciences
(CMHS)
Wollo University
Dessie, Ethiopia

ISBN 978-981-16-7655-0 ISBN 978-981-16-7653-6 (eBook)
https://doi.org/10.1007/978-981-16-7653-6



This Springer imprint is published by the registered company Springer Nature Singapore Pte Ltd.
The registered company address is: 152 Beach Road, #21-01/04 Gateway East, Singapore 189721, Singapore

*Naimmaton Sa'adiah Mustain: for love that ease the long journey*
*...Abu Zahrim Yaser*

*To Him who gives my breath. My parents, which always pray for me as fast as windy, my wife and children as companions like sunshine in the darkroom. My friends are always behind me and to all of my readers. Thank you for your tear of joy.*
*...Husnul Azan Tajarudin*

*To my parents Easwaran and Jayalakshmi Embrandiri*
*...Asha Embrandiri*

# Foreword

It gives me a great honour to write the foreword for this important book as a scholar with deep passion for environment-related niches. Waste management and valourization are trending topics of modern-day waste management practices and as part of the sustainable development goals of United Nations.

Dr. Abu Zahrim Yaser, whom I had the honour of working with, has many years of experience in this field of waste management with keen interest in composting towards zero food waste and technologies. This time, he is teaming up with Dr. Husnul Azan Tajarudin (Universiti Sains Malaysia) and Dr. Asha Embrandiri (Wollo University, Ethiopia), who have experience from other environmental fields to co-edit this book. I have no doubt that this publication with evidence-based practices from countries, technical information and reviews will be of great use to the scientific community. I believe this book will be of great benefit to developing countries and rural communities from such comparisons to promote green affordable technologies to enhance circular economies. Topics ranging from agro-industrial wastes to compost designs and plastic wastes from case studies in Asia and Africa have been compiled and brought together in this book.

In addition, upcoming global concerns such as alternative renewable energy from microbial fuel cells, life cycle analysis and waste to wealth/energy have also been addressed. In this COVID-19 era, while the world is on a reverse tangent, waste experts have great concerns for the inevitable use of PPEs the world left behind and its impact on the future generations. This book covered knowledge related to Sustainable Development Goals (SDGs), particularly Goal 6 (clean water and sanitation), Goal 7 (affordable and clean energy), Goal 11 (sustainable cities and communities) and Goal 15 (life on land).

The uniqueness of this book is the case studies of African and Asian waste management-related practices and challenges and how solutions are being experimented. The necessity for environmentally friendly technologies in future will require the expertise of engineers and environment experts to incorporate sustainability into products, processes, technology systems and services, which involves the design for energy efficiency, mass efficiency and low environmental emissions.

I am confident that this book will enlighten and empower academic community across the globe and masses to initiate the need for sustainability for the future yet unborn.

Editor-in-Chief | *DESALINATION*
The International Journal on the Science and Technology of Desalting and Water Purification

Nidal Hilal
D.Sc., Ph.D., CEng, FIChemE, FLSW
Professor of Engineering
Director | NYUAD Water Research Center
New York University Abu Dhabi
Abu Dhabi, UAE

Global Network Professor of Chemical Engineering
Tandon School of Engineering
New York University
New York, USA

Professor in Water Process Engineering Emeritus
Swansea University
Swansea, UK

# Preface

Challenges in the management of municipal (including sewage), agricultural as well as industrial wastes are being covered in this book. Themes on waste characterization, processing and life cycle assessment are also reviewed and discussed. Due to land scarcity, waste disposal through landfilling should be minimized. Therefore, other than landfill, this book explained the possibility of future processing techniques as an alternative: thermal (pyrolysis and incineration) as well as biological means (solid state fermentation, composting and microbial fuel cell). Authors from various countries, i.e. Ethiopia, Nigeria, Malaysia, India, Thailand, Indonesia and Spain, involved in this publication.

Owing to wide application of plastic, plastic waste has become a cause for major health and environmental concerns worldwide. Embrandiri *et al.* stated that the successful management of plastic wastes requires not only a good infrastructure and operation skill but also full support from the community. Plastic waste processing to any of commercial value, e.g. Waste Plastic Oil (WPO), could motivate the community in assisting the collecting and disposal of plastic waste. In this regard, Mistoh *et al.* investigated the production of Waste Plastic Oil (WPO) from the pyrolysis of plastics.

Since solid wastes generated from municipal and industrial sectors consist of mixture of plastic, food waste, metal, yard waste, etc., waste auditing is important prior to waste processing. Hussain *et al.* presented a summary of the most common components of Municipal Solid Waste (MSW) generation trends and their percentage-wise distribution in landfills. Waste processing at source should be carried out by the community and industry to prolong the landfill lifespan. For rural areas, centralized landfill is seemed to be unpractical due to transportation problem. Sarbatly and Sariau suggested that the small-scale incinerators can be applied at the source as a method of municipal and agricultural solid waste management in rural areas.

For organic waste, composting can be an alternative for the treatment. Composting can turn waste into soil conditioner and passive (natural) aeration normally chosen due to lower cost compared to forced aeration. Composting at source can minimize the waste to landfill, and the compost produce can be a source of income. Chu *et al.*

recommended inclined roof version to be the most promising for large-scale natural aeration systems based on CFD analysis.

Another useful application for solid waste is by turning them into adsorbent. For this purpose, raw waste (without processing) or after processing, e.g. biochar, can be used. Chong *et al.* show some physicochemical characterization of raw palm oil mill solid wastes-based adsorbent and its capability as heavy metal adsorbent. In addition, Ismail *et al.* investigated the comparison between fresh and degraded bamboo biochars towards ammonium ion removal.

Rapid urbanization and population growth increase the generation of sewage. Current research development is towards reusing treated water, and its sludge is becoming vital. Mokhtar *et al.* discussed the possible health risk or hazard associated with sewage sludge in more detail. Hussain *et al.* reported that sewage treatment system integrated with microalgae is having excellent nutrient, heavy metal and emerging pollutant removal capacity.

Due to high demand from industries such as food and beverage, and biofuel, the global enzyme market size was expected to grow at a Compound Annual Growth Rate (CAGR) of 7.1% from 2020 to 2027 (source: https://www.grandviewresearch.com). Salikin and Makhtar explore the potential of using waste as a cheaper fermentation media for the production of pectinases, while Ja'afar and Shitu explain the production of enzyme using fungal. Besides that, Zulfigar and Ahmad review the potential of seafood waste to be converted to amino acids and other high-value product.

Waste to energy is another area that is critical to be developed in future. Over the years, anaerobic digestion of waste has been implemented for the production of biogas for electricity generation. Currently, research also geared towards strengthening microbial fuel cell as well as bioethanol as bioenergy. Shamsudin *et al.* describe the theory and current developments in biological means to generate electricity. Watcharawipas *et al.* developed superior yeast strains for the production of bioethanol from lignocellulosic biomass.

Finally, since all processing techniques have their own advantages and disadvantages, proper evaluation for each technique should be carried out properly to ensure the sustainability of the project. Saburi stated that the application of life cycle assessment to waste management practices favours a circular economy and waste valourization.

Kota Kinabalu, Malaysia

Abu Zahrim Yaser
zahrim@ums.edu.my

Acknowledgements

The editors gratefully acknowledge the following individuals for their time and efforts in assisting the editors with the reviewing of manuscript. This book would not have been possible without the commitment of the reviewers.

ALIYU SALIHU
ALLWAR
AYALEW ASTATKIE
AZWAN AWANG
EE LING YONG
FARED MURSHED
HAFIZ KASSIM
HAFIZI ABU BAKAR
HAFIZUL SELAMAT
HUSZALINA HUSSIN
IJANU EMANUEL
JONO SUHORTO
K. A. BAWA-ALLAH
KHAIRUL FIKRI TAMRIN
LAW MING CHIAT
MANOJ EMBRANDIRI
MARDIANA IDAYU
MOHAMADU BOYIE JALLOH
MOHD TAUFIQ BIN MAT JALIL
NAHRUL HAYAWIN ZAINAL
NAPISAH BINTI HUSSIN
NAZREIN ADRIAN AMALUDIN
NOVI CAROKO
NUR AAINAA HASBULLAH
NUR HANIE BINTI MOHD LATIFF
NUR HIDAYATUL LAILI
OO CHUAN WEI

PARVEEN RUPANI
RACHMAWATI SUGIHHARTATI DJ
ROKIAH HAHIM
RUBIA IDRIS
SAQIB HASSAN
SYARIFAH BT AB RASHID
TAN JHU SHUN
UTTARIYA ROY
WOLYU KORMA
YUSOF AB
ZAINUL AKMAR ZAKARIA

Special thanks to the Symposium on Advanced Coagulation-Adsorption Processes (ACAD 2021) Committee for suggesting few papers from the symposium to be included in this book.

# Contents

# About the Editors

**Abu Zahrim Yaser** is Associate Professor in Waste Processing Technology at the Faculty of Engineering, Universiti Malaysia Sabah, Malaysia. He obtained his Ph.D. from Swansea University, UK. He has published 7 books, 19 chapters and over 90 other refereed technical papers. He is a recognized reviewer for several impact factor journals and was a guest editor for Environmental Science and Pollution Research (Springer) and International Journal of Chemical Engineering (Hindawi) special issues. His inventions on waste processing system have been adapted by Tongod and Nabawan district in Sabah. He is Visiting Scientist at the University of Hull and Member of Institutions of Chemical Engineers, UK, Board of Engineers (Malaysia) and MyBIOGAS.

**Husnul Azan** is Programme Chairman and Associate Professor at Division of Bioprocess, Universiti Sains Malaysia. He received Ph.D. under field of Bioprocess from Swansea University. He obtained his degree and MSc related to Environmental Engineering from Universiti Tun Hussein Onn Malaysia and Universiti Sains Malaysia. He published more than 70 manuscripts indexed by ISI and Scopus, more than 10 chapters in books and 4 books. He is active researcher received more than 20 grants as a leader or member. His recent research is related to sustainable waste treatment or conversion waste to wealth via biological process. He is also interested to explore research on development of new material or technology via biological process.

**Asha Embrandiri** is Assistant Professor, Department of Environmental Health, College of Medicine and Health Sciences (CMHS), Wollo University, Ethiopia. She got her B.Sc. in Environmental Management and Toxicology (UNAAB, Nigeria); M.Sc. in Ecology, Environmental Sciences at Pondicherry University, India; Ph.D. in Environmental Technology from Universiti Sains Malaysia; and a Post-Doctorate from UMK, Malaysia. She has some years of work experience in different capacities during and after her Ph.D., as Assistant Professor, Department of Environmental Studies, Kannur University, India, and other research positions. She has 6 chapters and more than 25 other refereed technical papers.

# The Menace of Single Use Plastics: Management and Challenges in the African Context

**Asha Embrandiri, Genanew Mulugeta Kassaw, Abebe Kasssa Geto, Belachew T/yohannes Wogayehu, and Manoj Embrandiri**

**Abstract** The persistent nature of single use plastics has become a cause for major health and environmental concerns worldwide as a result of commercialization. Africa as a continent generates considerable amounts of single use plastic wastes from the total plastic wastes generated. The prolonged use and mismanagement of existing wastes poses environmental and public health risks. Current practices involve managing single use plastic wastes in combination with other municipal and medical wastes. As many countries in Africa have put into law some legislation on the use of single use plastics, the level of enforcement and implementation is questionable provided that no tangible measures have been put in practice. This is evidenced by the evcr-present single use plastics strewn around. Although infrastructural and skill-related challenges are pertinent to the management of single use plastics across the continent, single use plastic waste reduction, community-based indigenous model practices which are cost effective should be encouraged. This waste reduction and recycling practice in the long run make a positive impact on people's lives by achieving sustainable development goals.

**Keywords** Single use plastics waste · Africa · Ban · Levy · African alliance · Packaging

A. Embrandiri (✉) · G. M. Kassaw · A. K. Geto · B. T. Wogayehu
Department of Environmental Health, College of Medicine and Health Sciences, Wollo University, P.O Box 1145, Dessie, Amhara, Ethiopia
e-mail: asha.embrandiri@wu.edu.et

M. Embrandiri
Barium Selat Technologies, No. 6-1 Jalan Tasik Utama 7, Medan Niaga Tasik Damai Sungai Besi, 57000 Kuala Lumpur, Malaysia



A. Z. Yaser et al. (eds.), *Waste Management, Processing and Valorisation*,
https://doi.org/10.1007/978-981-16-7653-6_1

# 1 Introduction

## *1.1 Plastics: Production and Utilization*

The use of plastics has become a routine part of human existence. *Bakelite*, the first synthetic plastic, produced in twentieth century marked the beginning of the plastic era [1]. Notwithstanding, fast growth in plastic production worldwide began only after the 1950s. In the subsequent years, production of plastics rose to almost 200-folds resulting in about 381 million tons by 2018 which is nearly about two-thirds of the world population [2]. Plastics are the polymerization products of low-molecular-weight monomers' organic materials derived majorly from petroleum. Plastics or "*plastikos*" in Greek owes the name for its molding property which allows it to be shaped, pressed, or designed into a variety of shapes and sizes [3]. Due to desirable properties such as lightweight and cheapness, the use of plastics and its products have immensely risen and thus replaced other materials and traditional items such as glass, wood, clay, and cloth.

Estimation the global production and management profile of plastic waste might not be enough to appraise localized implications/impact of such waste on the ecosystem for remedial/mitigation measures. As population is projected to rise to more than 9.5 billion by 2025 especially in Asia and Africa [4], it will become extremely tedious to quantify and estimate the generation rates at country levels. Broadly, plastics constitute three portions: plastics in use, post-consumer managed plastic waste, and a mismanaged plastic waste (MPW) fraction which include urban litter [5]. According to the winning International Statistic of 2018, un-recycled plastic wastes accounts for 90.5% of the total waste produced. Studies have estimated that out of the 6300 million metric tons of plastic wastes produced, 12% of it is incinerated with close to 79% left lying around in the environment or sent to landfills [5, 6].

Plastics used for packaging as the name implies have most peculiarly a very short use period such as wrapping materials, covers, and holders and thereby contribute largely to municipal wastes and mismanaged wastes [7]. Mismanaged waste also comprises of improperly contained wastes such as open dumps, insecure landfills, and pits which are liable to spread by rodents, scavengers, and rag pickers or by wind and rain [8]. Figure 1 illustrates the global plastic production by various sectors as at 2015 revealing 146 million tons for the packaging industry alone. This spells doom for the upcoming generations as these will all end up in some corner of the earth.

The production, utilization of single use plastic, and generation profile of the respective waste product in Africa did not well addressed. However, people use SUP in their routine activities since plastic are easy to handle and convenient for utilization. This leads to the higher production and consumption where relevant statistical profile among some African countries has presented in the following chart (Fig. 2).

The chart notifies the production of plastic waste in each indicated nation increase from time to time. Moreover, among those above nations, South Africa has found to

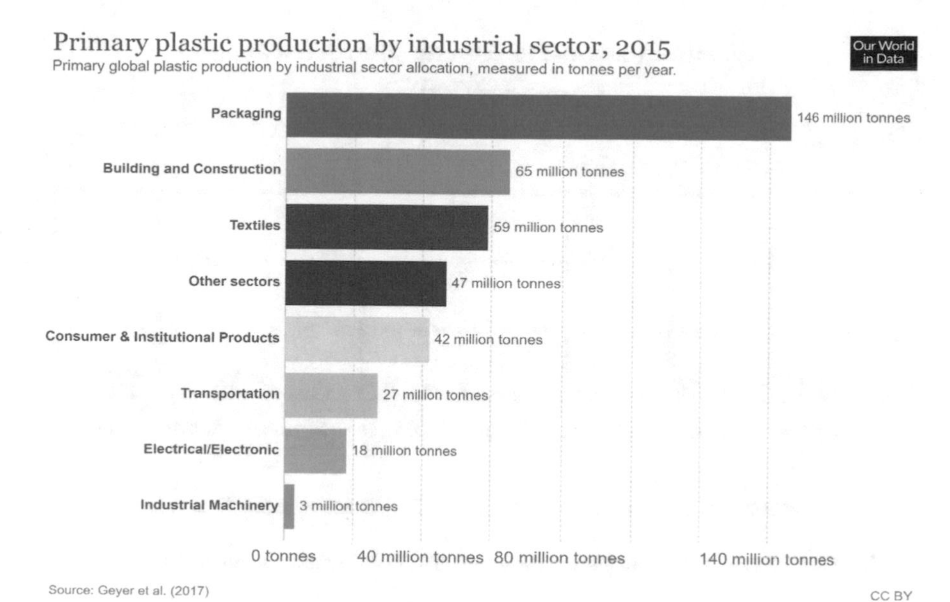


Fig. 1 Global plastic production by industrial sector, 2015. Reprinted from Ricthie and Max [1, 5] with permission

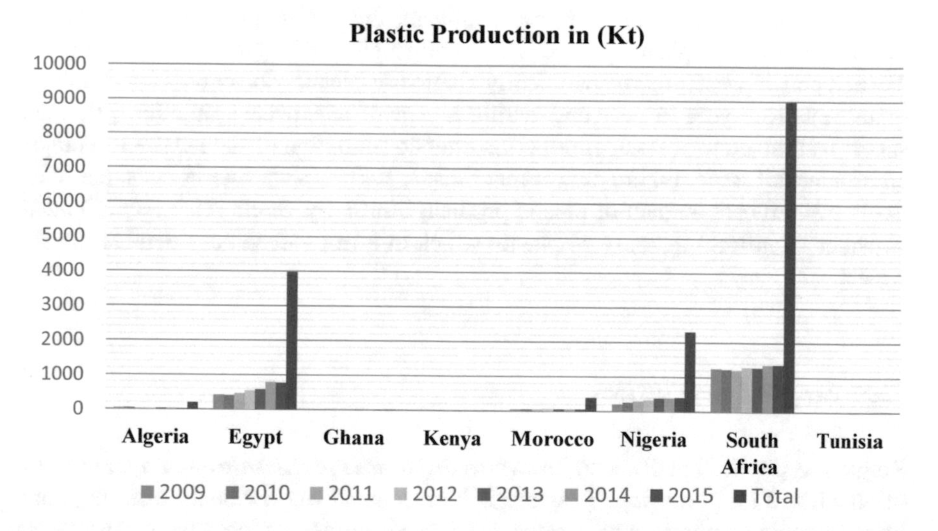


Fig. 2 Production of plastic waste among some African countries in thousand ton. *Source* Ref. [1]

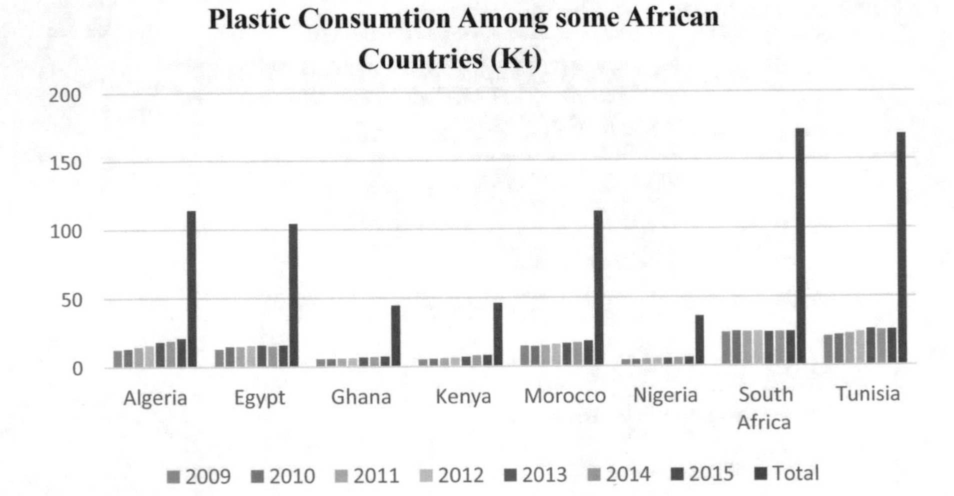


**Fig. 3** Plastic consumption among some African countries in kiloton. *Source* Ref. [1]

produce the highest tons of plastic in the continent. This much production of plastic signifies highest production of plastic waste in the continent (Fig. 3).

According to some studies in the continent, the consumption of plastic waste more importantly single use plastic is becoming higher and higher. The consumption of such plastic waste, more specifically single use plastic, does not show dramatic increment from year to year and within a year of unexpected situation [2]. This shows that the issue of waste production after consumption of such waste is a major environmental issue. Figure shows that Ghana, Kenya, and Nigeria consume relatively insignificant amount of plastic product. Contrarily, South Africa and Tunisia consume significant quantity of plastic which in turn produce considerable plastic waste.

## *1.2 Single Use Plastics*

Single use plastics (SUP), also known as ***disposable plastics/throwaway***, are most often utilized for packaging. They are in high demand worldwide used as shopping bags, wrapping, cutlery, cups, straws, food packs, bottles, sanitary pads, cosmetics, and of recent in personal protective equipment. They are materials that are meant for one time use before they are thrown away or sent for recycling [9]. Which gained tremendous popularity since the 70's [10]. Traditional materials such as cloth, glass, or paper were replaced with lighter or more durable and affordable plastic alternatives. Many significant uses for plastics have been in existence for instance surgical gloves, masks, and straws for disabled and so on which take up only a minor fraction of single use plastics. SUP polymers are the largest segment of general plastics

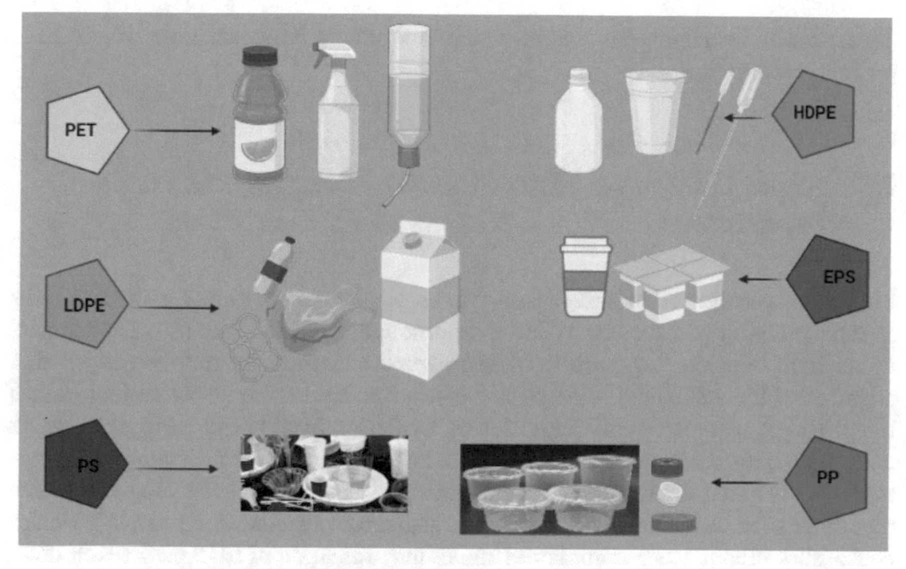


Polyethylene Terephthalate (PET), Low Density Polyethylene (LDPE), Polystyrene (PS), High Density Polyethylene(HDPE), Expanded polystyrene (EPS), Polyvinyl-chloride (PVC), Polycarbonate, Polypropylene (PP)

**Fig. 4** Main polymers used in single use plastic items and most common applications

fabricated around the world where they accounted for half of the total amounts of plastics produced with SUP packaging having the highest portion (40%) [11]. Both thermoset and thermoplastics are used in the production of SUP. Figure 4 illustrates the common household items we use and the main polymers that are present in them.

SUP products such as low-density polyethylene (LDPE) are mainly made for films, bags, housewares, etc.; polypropylene (PP) for food packaging, microwave containers, pipes, etc.; polyethylene terephthalate (PET) for water, drink bottles, dispensing bottles; polystyrene (PS) for cutlery items, and high-density polyethylene (HDPE) for coffee cups, straws, droppers, etc. [7, 12]. Another major group of single use plastics are personal-hygiene-related products such as sanitary pads, diapers, and wet wipes/baby wipes. These are one time use products which eventually find their way into our municipal waste system.

Common sources of single use plastics include: food and drink service establishments (cafeterias, groceries, hotels, and night clubs), mini and supermarkets, tourist attraction sites, beaches and recreational areas, public institutions like universities, prisons, colleges, governmental offices and healthcare facilities, electronics, residential areas, agricultural sector, manufacturing industry, transport industry, etc.

The statistics on production and consumption of single use plastic varies from nation to nation and among companies within each country. Though global companies are rated in terms of the production and consumption rate, the statistics of single waste generation and consumption is not well reported among African nations which needs further investigation.

## 1.3 Single Use Plastic Waste Pollution, Dangers, and Health Implications

Single use plastic waste and plastic wastes in general account for distinct risks to human health and the ecosystem at almost every phase of its use. The dangers range from both the exposure to plastic particles themselves and their associated chemicals [13, 14]. Single use plastic enters the human body via various routes like direct exposures through ingestion or inhalation from burning fumes by waste collectors. This leads to an array of health impacts, including inflammation, genotoxicity, oxidative stress, apoptosis, and necrosis which are linked to negative health outcomes such as cancer, skin infections, and cardiovascular diseases [15]. Health risk is attributed to the accumulation and/or magnification of SUP particles through ingestion and/or inhalation, and such micro plastic might affect the cellular function of the body by imposing oxidative stress and making the cellular structure break apart [3]. Figure 5 illustrates the routes of exposure from the waste SUP's to humans and the environmental implications.

Left alone, single use plastics do not break down on their own, they disintegrate over time by climatic factors (sun and heat) or mechanical (modern technology) slowly degrading the plastics into tiny fragments until they eventually become what are known as microplastics [16]. These eventually find their way to water bodies, eaten by wildlife, and inside our bodies via the food. They can be particularly dangerous when eaten by animals, e.g., goats, cows, birds, and even fish in the sea, and accumulate in the animal's body causing fatal consequences like punctured organs or fatal intestinal blockages and eventual death [10]. Studies from University of Exeter report that sharks contained huge quantities of microplastics and other human-made fibers [17]. Natsha Daly reports in her National Geographic article about how a plastic straw was extracted from a sea turtle's nostril in Costa Rica and flesh-footed shearwaters off the coasts of Australia and New Zealand eat more plastic than any other marine animal. Similarly anchovies were found to eat plastic because of food remains on them, and some studies estimated that about 700 species of marine animals have eaten or become entangled in most especially single use plastics [18].

Man's dependence on single use plastics implies that huge quantities of the wastes are projected in the nearest future. We produce 300 million tons of plastics each year worldwide, half of which is for single-use items [1, 19]. Reducing plastic use by taking reusable bags and containers is one the most effective means of avoiding this waste (and the impacts linked to plastic production and use) in our daily lives. Recycling more plastic, more frequently, reduces its footprint. PET bottles is most often recycled into different products from polyester fabric, artifacts, and planting

**Fig. 5** Routes of exposure from the waste SUP's to humans, animals, and their environmental implications

materials to automotive parts [20, 21]. Items like straws, forks, spoons, knives, and carry bags are most often harder to recycle because they get stuck in the crevices of machines for recycling/sorters and consequently are frequently rejected by the recycling centers [9]. Although single use plastic pollution accumulates most visibly on our streets, our water suffers even more. Litter can be the first stage in a waste stream that enters waterways as plastics tossed on the street are washed away via storm drains into rivers and streams [16]. Approximately between 4.8 million and 12.7 million metric tons of global plastic per year was estimated to find their ways to the oceans from residents of the coastlines [10]. The majority of this pollution dominated by single use plastic waste comes from countries lacking infrastructure to properly manage waste, particularly in Africa and Asia. India, for example, generates 25,940 tons of plastic waste every day but collects only 60 percent of it [22]. Laura Parker's reports, [23] in her National Geography column, described how single use plastic bottles have gone from once revered miracle container to a hated trash. She also described how these miracle bottles/bags/ containers which once fast tracked modern existence, has begun to choke our waterways. The perils of this wastes are more long term, and therefore, a collective effort needs to be followed. Governmental, non-governmental organizations, movements, international corporations, and local bodies have been rallying together toward the prevention of the infiltration of plastics (particularly SUP's) into the environment [16]. Phasing out the use of single use plastics and reduction of plastic footprint, raising awareness, and effecting changes in attitudes seems to have brought some good in the recent past [22].

## 2 Africa Single Use Plastic Waste Generation Profile

Global estimates of plastic wastes generated annually were 300 million tons in 2015 [6], and 50% of it accounts for single use plastic waste [4]. Babayemi [24] reported that Africa imported around 72 metric tons of polymers and plastics estimated at $285 billion between 1990 and 2017. It is, however, unknown the quantity of wastes this had attributed to, but it is easy to predict that over 80% of it is still lying off somewhere. A Nigerian analysis indicated that the per capita consumption of plastics was 6.5 kg in 2017 and was estimated to be 7.5 kg by 2020 [25] and about four-fifths of this ends up in dumpsites across the country.

Plastic waste generation rate is high in Tunisia (0.358 kg/person/day) as compared to other African countries. This is followed by Togo which is 0.325 kg/person/day, Algeria (0.015 kg/capita/day), and Angola 0.016 kg/capita/day, respectively [1, 26]. In addition to this, Nigeria was ranked the highest in mismanaged plastic wastes with between 2.91 and 2.98 MMT/yr being produced in 2017, with Egypt, Ethiopia, and Morocco following behind with 1.02–1.68, 0.90–1.00, and 0.90–0.93 MMT/yr, respectively [2].

According to reports the total amount of waste imported, generated, and utilized in Africa is about 120 million tons where plastic waste constitute 16.25 million tons (13%) [27]. Out of this, 7% (8 million tons) is estimated as single use plastics [6, 9] as seen from Fig. 6.

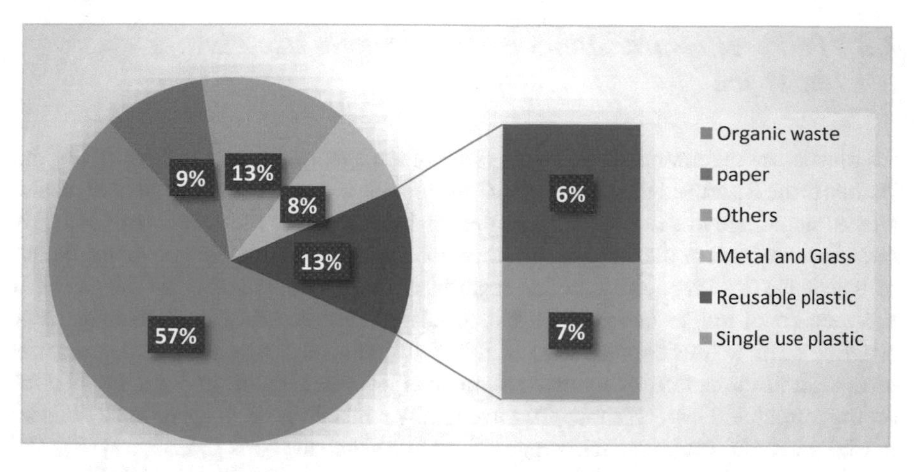


**Fig. 6** Proportion of single use plastic waste from the overall municipal wastes in Africa. *Source* Ref. [27]

## 2.1 *Beating Plastic Pollution: Management Practices in Africa*

In line with the UN "World Environment Day 2018" theme: *Beat Plastic Pollution,* countries across the globe Africa inclusive had pledged toward environmentally friendly practices *by* awareness creation [25]. The management of single use plastics is dependent on its plastic constituents, available resources, motivation, and infrastructures of the country. In Africa, about 80% of plastic wastes generated are disposed off at open dump sites and landfills sited across the country, or left mismanaged, while the rest of which (4%) is taken for recycling [25]. Africa faces the challenges of management of plastic wastes including single use plastics due to lack of resources, skilled man power, and priority. Despite these setbacks, many nations in Africa have made major strides in curbing the use of single plastics in the form of banning/ placing levies on their use and policies put in place, as well as the formation of recycling alliances that have come together with the goal of sustainable management of these single use plastics.

Numerous studies [13, 28–30] have brought to light the environmental health impacts of single use plastic wastes which have paved the way for private organizations in tackling the problems of single use plastics in the continent including the recycling of SUP waste to other valuable product [29].

## 2.2 Policies Toward Management of Single Use Plastics in Africa

As plastic waste remains the topmost issue of the agenda of the Conferences of the Parties to the Basel and the Stockholm Conventions, concerns on the impact of plastic waste, single use plastics, and microplastics have come to the fore front [24]. The need for reduction or curbing the consumption and ensuring sustainable management of waste plastics is required. In this regards, African nations strive not only to curb the menace of single use plastics but to ensure sustainability in its management practices. Thirty four countries out of fifty four in the continent have imposed some laws with the intention of being implemented or even already implemented based on the formulated law. Of those, 16 have totally banned plastic bags or have done so partially without yet introducing regulations to enforce the bans [24, 31, 32] (Fig. 7). Some countries have banned the production, importing, and utilization of single use plastics, while others impose levies on the consumers and sellers and the rest manage the wastes through certain level of known management practices such as landfilling, incinerating, and recycling. Countries in the forefront in the fight against single use plastics are South Africa [33–35] Ghana and Nigeria [36], and Kenya [37, 38]. This is due to the joint involvement of the government, non-governmental organizations, and private sectors. BBC Journalist; *Sulley Lansa*, reported tangible evidence about the appreciable minds of African youths/entrepreneurs on the management of single plastic waste [21]. Countries imposed fee up on those who violate the formulated law of banning; South Africa and Botswana's activities is the principal example of such stringent measure.

## 2.3 Africa's Challenges to Single Use Plastics Waste Management

Although environmentally sustainable plastic waste management has been documented in Africa [21, 39] the trimming from plastic wastes and plastic usage, bulk plastic waste, un-standardized waste management infrastructures including inconsistent plastic waste recycling and disposal sites have resulted in varying health and environment problems for the continent. In addition to loss of biodiversity, plastics that end up in the oceans result in severe disruption in the aquatic diversity, Aesthetic quality, and other environmental problems [7, 15, 40].

In recent times, use of colorant on plastic products have become a trendy way of life. They are attractive and are used to create certain distinction as well as improving usability and quality with great risk to health [41].

The principal challenges facing African nations with regards to single use plastic waste managements are balancing between promoting recycling and protecting consumers against harmful chemical substance. In addition, there are inaccurate records of importation, production and utilization of single use plastic wastes; as well

**Fig. 7** Practices and legislative measures of African countries toward management of single use plastic wastes

as documents related to recycling, energy recovery, health implications of improper waste management [24, 25, 39]. In order to best manage these wastes, experiences from more developed nations should be emulated and introduced bans and regulations should be implemented [8, 9, 42].

## 2.4 The Ethiopian Scenario

Camilla Louise Bjerkli in her report on the cycle of plastic waste: An analysis on the informal plastic recovery system in Addis Ababa, Ethiopia, discovered that most of the low status residents in Addis Ababa depend on plastic recovery for their livelihoods [43]. In addition to that, it has been presented that this practice adds a lot of value to solid waste management via the informal sector which signifies the lack in cooperation between government sector and private sector [44]. The results indicated that the larger proportion (76.52%) of the population used plastic bags more frequently than any other plastic products regardless of their age, occupation, and economic and educational status [2–4, 45]. As different types of plastic materials are in use, zero attention is given to their environmental concern.

## 2.5 The Kenyan Experience

Lucy Luo Minghui's 2000 tons capacity plastic recycling factory in Nairobi is one of the slews of regional initiatives to tackle the world's spiraling plastic waste crisis. The intent was to produce low-cost textile fibers from plastic bottles [46]. However, after the ban on plastics by Chinese recycling plants, they run below their full capacity at just 650 tons. Like many developing nations, Kenya has no organized recycling system in place to collect the plastics and sort them from the huge municipal solid wastes. PETCO Kenya had recycled 2400 tons of plastic waste so far this year against its 2019 target of 6000 tons, a target it plans to increase in coming years [47]. Unfortunately, much more PET plastic waste (20,000 tonnes) is produced annually [37].

## 2.6 The Recycling Alliances

Consumer brands Coca-Cola, Unilever, and Nestle are among the world's biggest producers of plastic trash, according to an October 2018 Greenpeace report [31]. Coca-Cola, Nestle, Unilever, and Diageo formed an African plastic recycling alliance to build on national recycling initiatives, which member companies already collaborate with, in South Africa, Nigeria, and Ghana. South Africa is so far the only nation to have a successful recycling model as reports indicated that they were able to recycle almost two-thirds of its plastic waste in 2018. Kenya's Casper Durandt which had been taken to the PETCO (South African company) model pilot plants are being tested in Ethiopia, Tanzania, and other Asian countries [25, 46].

The West Africa ENRG Company Nigeria commissioned its first Materials Recovery Facility in 2015 which routed useful fraction of waste from dumpsites/landfills in Lagos State and returns for recycling into local industries such as paper mills, plastic molded and blown manufactures, and smelters throughout the country. Likewise, in 2018, a consortium of Nigerian beverage companies liaised with building giants LafargeHolcim, Ltd in utilizing the plastic bottles as fuel in their kilns [48].

The Ghana Recycling Initiative by Private Enterprises (GRIPE) formed by a coalition of companies (Nestlé Ghana Limited; Guinness Ghana Breweries Limited; Coca-Cola; Unilever; PZ Cussons; KGM Industries; Volti; Dow Fan Milk; etc.) to incorporate sustainable waste management solutions with particular emphasis on plastic wastes in Ghana. They established 4000 tons capacity/ annum plastic recycling plant, however, as at now they collect only 1200 tons annually [49]. Environmentalists urge that companies should be responsible for their waste in accordance to the polluter pays principle. If such regulations are not imposed, manufacturers will not take the initiatives, and the cycle will go on. Even as recycling or repurposing of SUPs is still in effect, there is no safeguarding the environment as plastic is leached out as microfibers which infiltrate into the ecosystem [46, 50]. Such alliances had faced various challenges such as resistance from industries and civil societies, leading to set backs.

## 3 Lessons from the Global Community

The Swedish recycling revolution has taken the country's recycling revolution to great heights from dumping rubbish in landfills, to recycling to reusing [42]. The country's waste management system takes into consideration of all kinds of waste being generated and implements strategies for the management of plastic waste including single use plastic [51]. These strategies are pivotal for the proper and effective management of single use plastic wastes ensuring sustainability. However, some countries in Africa had attempted a "copy and paste" approach in dealing with single use plastics but had failed to reach the desired goals due to a number of reasons such as funding, electricity, climate, etc. [48]. In the UK, the use of straws was banned from all public places and even for sale, while in India and USA, energy recovery from plastic wastes are being researched on [3–6]. Other countries like Germany, Japan, and South Africa practise the extended producer responsibility models which proved effective in reducing littering from PET bottles vis-a-vis enhancing the recycling sector. Table 1 provides a summary of some global practices/action plan toward combating single use plastics.

### *3.1 The Way Forward: Pledges by International Brands*

Some of the world's largest companies have pledged to eliminate/reduce the amount of plastic waste (single use and recyclable) they send to landfills or to use more recycled materials in their products and packaging, thereby creating a bigger market

**Table 1** Practices or policies for management of SUPs in various countries

| S/N | Countries | Practices | References |
|---|---|---|---|
| 1 | Ghana | Plasma pyrolysis technology (PPT)<br>Waste stock exchange (WSE)<br>Converting of plastic waste in to artefacts | [21, 39], |
| 2 | Nigeria | Stringent policy | [7] |
| 3 | Thailand | Material flow analysis | [52] |
| 4 | Indonesia | Recycling<br>Landfilling in secured area | [53] |
| 5 | India | Material recovery facilities (MRFs)<br>Plastic waste to fuel<br>Pyrolysis<br>Gasification<br>Plastic waste for road construction | [22, 54] |
| 6 | USA | Implementation of integrated plastic waste management techniques (i.e., Source reduction, reuse and recycling)<br>Advanced recycling technologies like glycolysis and ammonolysis<br>Energy recovery | [55] |
| 7 | Sweden | Implementation of circular economy (using only fully recyclable items), i.e., cradle-to-cradle approach<br>Reformation of the tax system in order to encourage reuse. (cheaper repairs on used items)<br>The pant system- money back system for recycling bottles and cans | [42] |
| 8 | Canada | Single use plastic interventions (e.g., bans vs. levies) | [16] |
| 9 | Pakistan | Policy development to ban on SUP | [56] |
| 10 | China | Global ideas and actions for reducing quantities of non-recyclable plastic waste, redesigning products, and funding domestic plastic waste management | [57] |
| 11 | UK | Banning plastic straw | [58, 59] |
| 12 | Germany, Japan, and South Africa | Extended producer responsibility (EPR)<br>Deposit-return schemes | [9] |

for recycled content and incentivizing recyclers to expand their operations. Some of the commitments have been tabled in Table 2.

**Table 2** Pledges or action plans of different international companies toward reduction in their plastic footprints

| S/N | Company | Products | Action plan/pledges |
|---|---|---|---|
| 1 | Trex Company, Inc USA | Wood-alternative composites | Utilizes recycled grocery bags, bread bags, and dry cleaning bags to make environmentally responsible outdoor products |
| 2 | BD Recykleen™, US | Sharps collectors and accessories | Produces its sharps collector line with recycled materials |
| 3 | Coca-Cola, International | Bottled drinks | Pledged to collect and recycle the equivalent of every bottle or can it sells throughout the world by 2030 |
| 4 | PepsiCo, International | Bottled drinks | Designs its packaging to be recyclable, biodegradable by increasing use of recycled materials and decreasing its packaging's carbon impact |
| 5 | SC Johnson &Sons USA | Household cleaning supplies and other consumer chemicals | Pledges to increase reuse and recycle by avoiding sending of waste to landfills by 2021 |
| 6 | John Deere, USA | Machinery | Pledged to recycle 75% of its total waste by 2018 |
| 7 | The Ford Motor Company | Automobile parts | A five-year goal of reducing waste to landfill by 40% The use of recycled materials for upholstery fabrics |
| 8 | McDonald's McD | Fast foods | Pledged that its product packaging would come from renewable, recycled, or certified sources by 2025 |
| 9 | Ecover, Belgium | Eco-friendly cleaning products | All bottles would be made with 100% recycled plastic by 2020 |
| 10 | Amcor, Switzerland | Flexible packaging, rigid containers, cartons, cases for food, beverage, pharmaceutical, medical-device, home and personal-care products | Pledged to make all its product packaging recyclable or reusable by 2025 |

(continued)

**Table 2** (continued)

| S/N | Company | Products | Action plan/pledges |
|---|---|---|---|
| 11 | Danone, France | Food products | The water bottles used by the Evian brand will be made with 100% recycled plastic by 2025 |
| 12 | Unilever, International | Canned foods, energy drinks, baby food, soft drinks, cheese, ice cream, tea, etc. | Committed to making 100% of its plastic packaging recyclable by 2025 |
| 13 | Walmart, US | departmental stores | Pledged that 100% of its packaging for its private-brand products would be recyclable by 2025 |
| 14 | L'Oréal International | Cosmetics | By 2025, plastic packaging will all be rechargeable, refillable, recyclable or compostable |
| 15 | ANN INC | Clothing store | At its stores, provides shopping bags made with between 40 and 80% post-consumer waste |
| 16 | H&M | Clothing store | Circular strategy for packaging by 2025 by the reduction of packaging by 25% as well as designing reusable, recyclable or compostable packaging |
| 17 | Evian | Bottled drinking water | By 2025, they plan to produce their plastic bottles from 100 percent recycled plastic |
| 18 | Guinness | Beverages | Replacing the packaging with 100% biodegradable/recyclable cardboards |
| 19 | Sodexo | Cafeteria-style meals | Sustainability plan that eliminates 245 million single-use items at their locations. Plastic bags and stirrers are no longer available<br>By 2025, they plan to get rid of polystyrene foam items such as cups, lids, and food containers too |

(continued)

**Table 2** (continued)

| S/N | Company | Products | Action plan/pledges |
|---|---|---|---|
| 20 | Nestlé | Packaged food company | In 2019, eliminated all plastic straws in its products<br>Changed drinks from plastic to paper containers<br>Increase PET content in its bottles to 50 percent in the US<br>Make a product's packaging recyclable or reusable |
| 21 | Starbucks | Coffee drinks | By 2020, Starbucks pledged to eliminate all plastic straws from their 28,000 stores worldwide. Provide recyclable, straw-less plastic lids |
| 22 | IKEA | Home furnishing | In 2008, IKEA eliminated all plastic bags. Customers could purchase IKEA blue bags or bring their own |
| 23 | American Airlines | Tourism | Banned plastic straws and stirrers on their flights & planning to phase out single use plastic in their lounges around the world |
| 24 | The Walt Disney Company | Entertainment centers | Banned single use plastic straws and stirrers by 2019<br>Promised to cut down on single use plastic bags with reusable shopping bags |
| 25 | ARKET | Clothing brand | Offering textile recycling in-store, and selling swimsuits made from recycled plastic bottles |

## 3.2 *Biodegradable Plastics—Bane or Boon?*

As part of the drive toward beating plastic pollution, a number of countries have put into legislation to do away with conventional plastic items and promote or enforce the use and production of *biodegradable* products [60]. These so called biodegradable plastics are derived from renewable sources or from bacterial fermentation of sugar or lipids [61]. Bioplastics are often touted as being eco-friendly, but what is the reality? The reality is that not all materials under this category are the same which led to many misconceptions about the biodegradable plastics and their after-life processing, resulting in false advertising and consumer indifference [62]. For example, it is interesting to know that our so called biodegradable coffee cups are not entirely made from degradable material. In order for proper disposal, they have to be separated into parts; lid, inner lining, straw, and the cup itself which is more capital intensive than avoiding or using your own cup [61]. Correct segregation and disposal operations of these biodegradable plastics listed known, affecting their optimum degradation. They breakdown with the application of external heat and not naturally, thus mechanisms that produce heat such as incinerators and pyrolysis plants are required for faster breakdown and degradation into the environment. Therefore, specialized waste management systems are needed to be put in place in order to handle these wastes as well. Due to the environmental impacts of mismanaged single use plastic waste, countries are using biodegradable plastic which is environmentally sound. However, the cost of production for such biodegradable plastic is determined to be higher.

### *3.3 The Covid-19 Effect on Single Use Plastic Waste*

The Covid-19 world pandemic has resulted in an unimaginable surge of global plastics waste. From the array of discarded personal protective equipment (PPE) such as gloves, surgical masks, face shields, sanitizer bottles, medical gowns, aprons and food packages, it is very imminent that the world is at the threat from plastic pollution [30, 63]. The case is no different in Africa as studies report that over 12 billion medical and fabric face masks are discarded monthly [30] though country specific information is still lacking. As most countries have imposed the use of masks and other protective wears to fight against the virus, the projection is quite alarming, and it is therefore imperative that the governments of African nations also put into effect ways of disposal and handling these wastes in the near future.

### *3.4 Doing Your Bid-Avoid Single Use Plastics*

- Use recyclable shopping bags
- Avoid accumulating take out containers by taking your own.
- Try keeping metal or bamboo straws, cutlery, or chopsticks as you go.
- Use of cloth face masks whenever possible.
- Formulate and enforce relevant law
- Adopt model practices from model countries.

## 4 Conclusion

Plastic waste management strategies undertaken by a city or nation is dependent on the individual and collective choices of people. From this write up, it is evident that Africa has made commendable strides with regards to either banning/levying or some recycling initiatives; however, this is not sufficient to cater for the ever-growing demand for the use of single use plastics. It is imperative that all concerned, be it the manufacturers or the consumers are obliged to dispose of single use plastic wastes in the appropriate manner with proper consideration of sustainability. However, it is advisable to approach this problem with more environmentally sustainable actions. Alternatives for shopping bags, water bottles, cutlery, etc., will go a long way in reducing individual footprint. As African nations have formed various alliances toward expanding plastic waste management infrastructures in order to increase access to recycling and to energy recovery technologies, we hope that this would in the long run make a positive impact on our lives by the extension beyond a single use plastic while focusing on sustainable development.

## References

1. Ritchie, H., Roser, M.: *Plastic pollution.* Our World in Data 2018; Available from: https://www.ourworldindata.org/plastic-pollution
2. Cordier, M., et al.: Plastic pollution and economic growth: The influence of corruption and lack of education. Ecol. Econ. **182**, 106930 (2021)
3. Europe, P.: *Plastics—The facts. Plastic Europe* (2017) 20 Apr 2021. Available from: https://www.plasticseurope.org/en/resources/publications/274-plastics-facts-2017
4. Lebreton, L., Andrady, A.: Future scenarios of global plastic waste generation and disposal. Palgrave Commun. **5**(1), 6 (2019)
5. Geyer, R., Jambeck, J.R., Law, K.L.: Production, use, and fate of all plastics ever made. Sci. Adv. **3**(7), e1700782 (2017)
6. Jambeck, J.R., et al.: Plastic waste inputs from land into the ocean. Science **347**(6223), 768–771 (2015)
7. Rhodes, C.J.: Plastic pollution and potential solutions. Sci. Prog. **101**(3), 207–260 (2018)
8. Bashir, N.H.H.: Plastic problem in Africa. Jpn. J. Vet. Res. 2013-02. Available from: http://hdl.handle.net/2115/52347
9. (UNEP), U.N.E.P.: *Single-use plastics: a roadmap for sustainability.* Nairobi, United Nations Environment Programme (2018) [cited 90]
10. Courtney Lindwall: *Single-Use Plastics 101* (2020) [cited 2021]. Available from: https://www.nrdc.org/stories/single-use-plastics-101
11. Advisors, D., Wit, D.W., Hamilton, A., Scheer, R., Stakes, T., Allan, S.: *Solving Plastic Pollution Through Accountability* (2019) 1 August 2020. Available from: https://www.worldwildlife.org/publications/solving-plastic-pollution-through-accountability
12. Chen, Y., et al.: Single-use plastics: production, usage, disposal, and adverse impacts. Sci. Total Environ. **752**, 141772 (2020)
13. Manzoor, J., et al.: Plastic waste environmental and human health impacts. In: *Handbook of Research on Environmental and Human Health Impacts of Plastic Pollution.* IGI Global, pp. 29–37 (2020)
14. Alabi, O.A., et al.: Public and environmental health effects of plastic wastes disposal: a review. J. Toxicol. Risk Assess. **5**(021), 1–13 (2019)
15. Campanale, C., et al.: A detailed review study on potential effects of microplastics and additives of concern on human health. Int. J. Environ. Res. Public Health **17**(4), 1212 (2020)
16. Schnurr, R.E., et al.: Reducing marine pollution from single-use plastics (SUPs): a review. Mar. Pollut. Bull. **137**, 157–171 (2018)
17. University of Exeter: Plastics found in sea-bed sharks. *ScienceDaily* (2020) [29-05-2021]. Available from: https://www.sciencedaily.com/releases/2020/07/200722083807.htm

18. Daly, N.: *For Animals, Plastic Is Turning the Ocean Into a Minefield*. National Geographic Magazine, Planet Or Plastic? (2018) [cited 2021 27–05–2021]. Available from: https://www.nationalgeographic.com/magazine/article/plastic-planet-animals-wildlife-impact-waste-pollution
19. Lebreton, L., Andrady, A.: Future scenarios of global plastic waste generation and disposal. Palgrave Commun. **5**(1), 1–11 (2019)
20. Plastics Industry Association, E.: *The Purpose of Single-Use Plastics* (2021). Available from: https://thisisplastics.com/environment/the-purpose-of-single-use-plastics/
21. Lansah, S.: *Ghana businessman turns waste plastic into profit* (2018)
22. Vanapalli, K.R., et al.: *Challenges and strategies for effective plastic waste management during and post COVID-19 pandemic.* Sci. Total Environment. **750,** 141514 (2021)
23. Parker, L.: *We made plastic. We depend on it. Now we're drowning in it.* Planet Or Plastic? (2018). Available from: https://www.nationalgeographic.com/magazine/article/plastic-planet-waste-pollution-trash-crisis
24. Babayemi, J.O., et al.: Ensuring sustainability in plastics use in Africa: consumption, waste generation, and projections. Environ. Sci. Eur. **31**(1), 1–20 (2019)
25. Egun, N.K., Evbayiro, O.J.: Beat the plastic: an approach to polyethylene terephthalate (PET) bottle waste management in Nigeria. Waste Disposal Sustain. Energy **2**(4), 313–320 (2020)
26. Eriksen, M., et al.: Plastic pollution in the world's oceans: more than 5 trillion plastic pieces weighing over 250,000 tons afloat at sea. *PloS one* **9**(12), e111913 (2014)
27. Hoornweg, D., Bhada-Tata, P.: *What a Waste: A Global Review of Solid Waste Management*. Urban development series; knowledge papers 2012 [cited 15]. Available from: https://openknowledge.worldbank.org/handle/10986/17388
28. Roberts, B.D.: Plastic waste management in Africa. In: *Conference of the Society for Applied Anthropology at Albuquerque*, New Mexico. Accessed Apr 2014
29. Jiang, B., et al.: Health impacts of environmental contamination of micro-and nanoplastics: a review. Environ. Health Prev. Med. **25**(1), 1–15 (2020)
30. Benson, N.U., et al.: COVID-19 pandemic and emerging plastic-based personal protective equipment waste pollution and management in Africa. J. Environ. Chem. Eng. **9**(3), 105222 (2021)
31. Greenpeace Africa: *34 Plastic Bans in Africa | A Reality Check*
32. Chitotombe, J.W., Gukurume, S.: The plastic bag 'ban' controversy in Zimbabwe: an analysis of policy issues and local responses. Int. J. Dev. Sci. **3**(5), 1000–1012 (2014)
33. Godfrey, L.: Waste plastic, the challenge facing developing countries—ban it, change it, collect it? Recycling **4**(1), 3 (2019)
34. Hasson, R., Leiman, A., Visser, M.: The economics of plastic bag legislation in South Africa 1. S. Afri. J. Econ. **75**(1), 66–83 (2007)
35. Hanekom, A.: South African initiative to end plastic pollution in the environment. S. Afr. J. Sci. **116**(5–6), 1–2 (2020)
36. Adam, I., et al.: Policies to reduce single-use plastic marine pollution in West Africa. Marine Policy **116**, 103928 (2020)
37. Katunge, L.: *Prohibition of Plastics in Kenya: Moving Towards a Constitutionally Compliant Governance Regime for Single-use Plastics*. University of Nairobi (2019)
38. Horvath, B., Mallinguh, E., Fogarassy, C.: Designing business solutions for plastic waste management to enhance circular transitions in Kenya. Sustainability **10**(5), 1664 (2018)
39. Ampofo, S.K.: *The Options for the Effective Management of Plastic Waste in Ghana.* Report on Management of Plastic Waste in Ghana-21–328-STASWAPA.pdf (2015)
40. Embrandiri, A., et al.: "Microplastics": The Next Threat to Mankind? In: Affam, A.C., Henry, E.E. (eds.) *Handbook of Research on Resource Management for Pollution and Waste Treatment*, pp. 106–122. IGI Global (2020)
41. Abrams, R., et al.: Colouring plastics: fundamentals and trends. Plast. Addit. Compd. **3**(7–8), 18–25 (2001)
42. Ljungkvist Nordin, H., Westöö, A.-K.: *Plastic in Sweden Facts and Practical Advice: A Short Version of Kartläggning av plastflöden i Sverige (Mapping Plastic Flows in Sweden)* (2019)

43. Bjerkli, C.L.: *The Cycle of Plastic Waste: An Analysis on the Informal Plastic Recovery System in Addis Ababa, Ethiopia.* Geografisk institutt (2005)
44. O'Connor, D., et al.: *Universality, Integration, and Policy Coherence for Sustainable Development: Early SDG Implementation in Selected OECD Countries* (2016)
45. Adane, L., Muleta, D.: Survey on the usage of plastic bags, their disposal and adverse impacts on environment: A case study in Jimma City, Southwestern Ethiopia. J. Toxicol. Environ. Health Sci. **3**(8), 234–48 (2011)
46. Ndiso, J.: *Plastic, Plastic Everywhere But not for African Recyclers* (2019) [12 Apr 2021]. Available from: https://cn.reuters.com/article/africa-plastics-idINKCN1UZ0VG
47. Koech, M.K., Munene, K.J.: Circular economy in Kenya. In: Circular Economy: Global Perspective, pp. 223–239. Springer (2020)
48. Agbesola, Y.: *Sustainability of Municipal Solid Waste Management in Nigeria: A Case Study of Lagos* (2013)
49. Green, S.A.: *Advocating Policies that Advance the Growth and Development of Green Industries* (2020)
50. Ayeleru, O.O., et al.: Challenges of plastic waste generation and management in sub-Saharan Africa: a review. Waste Manage. **110**, 24–42 (2020)
51. Svenja, S.: Sweden: New sorting plant for plastic packaging. In: *Magazine for Business Opportunities & International Markets.* Global Recycling (2021)
52. Wichai-utcha, N., Chavalparit, O.: 3Rs Policy and plastic waste management in Thailand. J. Mater. Cycles Waste Manage. **21**(1), 10–22 (2019)
53. Putri, A.R., Fujimori, T., Takaoka, M.: Plastic waste management in Jakarta, Indonesia: evaluation of material flow and recycling scheme. J. Mater. Cycles Waste Manage. **20**(4), 2140–2149 (2018)
54. Bhattacharya, R., et al.: *Challenges and Opportunities: Plastic Waste Management in India* (2018)
55. Subramanian, P.: Plastics recycling and waste management in the US. Resour. Conserv. Recycl. **28**(3–4), 253–263 (2000)
56. Ali, S., et al.: Strategic analysis of single-use plastic ban policy for environmental sustainability: the case of Pakistan. In: *Clean Technologies and Environmental Policy*, pp. 1–7 (2021)
57. Brooks, A.L., Wang, S., Jambeck, J.R.: The Chinese import ban and its impact on global plastic waste trade. Sci. Adv. **4**(6), eaat0131 (2018)
58. Treasury, H.: *Tackling the Plastic Problem. Using the Tax System or Charges to Address Single-Use Plastic Waste.* HM Treasury, UK (2018). https://assets.publishing.service.gov.uk/government/uploads/system/uploads/attachment_data/file/690293/PU2154_Call_for_evidence_plastics_web.pdf
59. Romero Mosquera, M.: Banning plastic straws: the beginning of the war against plastics. Environ. Earth Law J. (EELJ) **9**(1), 1 (2019)
60. The Canadian Press: *Walmart Canada Plans to Stop Handing Out Free Plastic Bags.* (2016) [25 Jan 2016]. Available from: https://www.cbc.ca/news/canada/toronto/walmart-plastic-bags-1.3418598
61. Espinach, F.X., et al.: Study on the macro and micromechanics tensile strength properties of orange tree pruning fiber as sustainable reinforcement on bio-polyethylene compared to oil-derived polymers and its composites. Polymers **12**(10), 2206 (2020)
62. Nandakumar, A., Chuah, J.-A., Sudesh, K.: Bioplastics: a boon or bane? Renew. Sustain. Energy Rev. **147**, 111237 (2021)
63. Napper, I.E., Thompson, R.C.: Plastic debris in the marine environment: history and future challenges. Global Chall. **4**(6), 1900081 (2020)

# Upcycling of Plastic Waste

**Mohd Aizzan Mistoh, Andrea Galassi, Taufiq Yap Yun Hin, Nancy Julius Siambun, Jurry Foo, Coswald Stephen Sipaut, Jeffrey Seay, and Jidon Janaun**

**Abstract** The usage of plastic has increased exponentially over the past few years. This has led to an accumulation of plastic wastes in landfills and continuous depletion of fossil fuel resources. Mechanical recycling has been used to compensate for this problem; however, it is not sufficient as the separation process is costly due to feedstock requirements; hence, most plastic waste still ends up in landfills. Recently, there has been interest in the production of waste plastic oil (WPO) from the pyrolysis of plastics which is high in calorific value. Further research has been done in this area to develop the WPO to replace commercial fuel such as diesel. This chapter focuses on providing background information on the pyrolysis of plastics including the operating parameters and WPO characteristics from single and mixed plastic wastes. In addition, information on the performance analysis of WPO and the current progress on appropriate technology (AT) pyrolysis reactors is also provided.

**Keywords** Waste plastic oil (WPO) · Pyrolysis · Plastic wastes · Fuel · Appropriate technology (AT)

M. A. Mistoh · N. J. Siambun · C. S. Sipaut · J. Janaun (✉)
Faculty of Engineering, Universiti Malaysia Sabah, Jalan UMS, 88400 Kota Kinabalu, Sabah, Malaysia
e-mail: jidon@ums.edu.my

A. Galassi
Institute for Sustainability Science and Technology, Unversitat Polytèchnica de Catalunya, Barcelona, Spain

T. Y. Y. Hin
Faculty of Science and Natural Resources, Universiti Malaysia Sabah, Jalan UMS, 88400 Kota Kinabalu, Sabah, Malaysia

J. Foo
Faculty of Social Sciences and Humanities, Universiti Malaysia Sabah, Jalan UMS, 88400 Kota Kinabalu, Sabah, Malaysia

J. Seay
Faculty of Chemical Engineering, University of Kentucky, 211 Crounse Hall, 4810 Alben Barkley Drive, Paducah, KY 42002, USA



A. Z. Yaser et al. (eds.), *Waste Management, Processing and Valorisation*,
https://doi.org/10.1007/978-981-16-7653-6_2

# 1 Introduction

It has been determined that plastic waste is the third-largest contributor to municipal solid waste (MSW) globally, amounting for 20–30% of the total weight of MSW [1]. Plastic's resistance to degradation is detrimental toward the environment and public health as plastic persists in the environment. The usual method to deal with plastic waste is either to send the waste to landfills or perform mechanical recycling, reusing, incinerating, plastic upcycling, or energy recovery. The problem with landfilling is that plastic has a long lifetime and can leech toxins into the environment and subsequently can end up in the waterways [2]. Incineration methods are problematic because they directly convert plastic waste into toxic gases which can be inhaled when not performed in industrial facilities, therefore causing adverse health effects for onlookers [3].

Mechanical recycling has its drawbacks which include sensitivity to contamination. Contamination of plastics is common such as when food and debris are left on the plastics being sent to the recycling plant. A large share of plastics sent to recycling centers end up being rejected due to contamination and subsequently end up in landfills [4]. There is a strong case for mechanical recycling and energy recovery in European countries which have more systematic waste management systems. However, most developing countries lack infrastructure for properly disposing of plastic waste, and therefore, the locals usually resort to either burning their trash or engaging in public littering. Generally, these countries lack of specified waste collection hubs and subsequent waste collection systems result in plastic pollution persisting in the environment and ending up in the ocean [5].

As an alternative, there has been an increased interest by researchers in chemical recycling, specifically that of plastic pyrolysis. Plastic pyrolysis is a thermal degradation process that occurs in the absence of oxygen in which the solid hydrocarbon chains of plastics are broken down into shorter hydrocarbon chains producing desired products including char, petroleum liquids, and gases [6]. Pyrolysis is an attractive upcycling method for researchers because the process is less sensitive to contamination and can produce products that have similar or more value than the feedstock [3]. Recently, there has been a shift in research to focus on producing fuel from plastic waste specifically. It has been observed that diesel-like fuel from pyrolysis of plastic wastes is possible, and even waste plastic oil (WPO) blends can be used in conjunction with diesel fuel in a diesel engine [7, 8]. Lately, there has been interest in developing an appropriate or a simple technology pyrolysis reactor which can be used in developing communities which lack waste collection and recycling infrastructure. The advantages of appropriate technology (AT) include low cost and easy construction, and additionally AT can be operated by individuals with limited technical education which make AT suitable for implementation on a decentralized and local scale.

There have been various authors such as Anuar Sharuddin [6] and Wong et al. [9] discussing the prospect of plastic pyrolysis in producing useful chemicals. There have also been authors including Mangesh et al. [8] and Kalargaris et al. [10] studying the effect of replacing commercial diesel with WPO blends. This chapter aims to add to the existing knowledge by assessing the current progress of using WPO to replace commercial fuel in diesel engines while adding additional information regarding the progress of appropriate technology (AT) reactors in producing liquid oil similar to commercial fuel.

## 2 Feedstock for Plastic Pyrolysis

Plastics traditionally are derived from fossil fuel sources and are made out of polymer chains where the monomer is repeated creating the ongoing chain. There are many types of plastic feedstock available for pyrolysis, but the input streams mostly utilized for pyrolysis are polyethylene (PE), polypropylene (PP), and polystyrene (PS). These plastics have a high amount of stored energy and are also composed solely of hydrogen and carbon chains, therefore making them an ideal feedstock for an AT pyrolysis reactor [11]. The monomer chain for each mentioned plastics can be seen in Fig. 1. It can be seen that PS is derived from a monomer with a benzene aromatic ring which explains PS's differing properties from the other three plastics studied for pyrolysis [12] (Fig. 1).

The waste plastics, PE, PP, and PS can be identified by recycle number where high-density PE (HDPE), low-density PE (LDPE), PP, and PS have recycled number 2, 4, 5, and 6, respectively. Table 1 shows the description of the plastic wastes usually used in plastic pyrolysis.

### 2.1 *Polyethylene (PE)*

The main difference between HDPE and LDPE, structurally, is the amount of branching in each polymer. LDPE has branching chains while HDPE is a straight chain as shown in Fig. 2. HDPE has minimal branching and is made out of a long

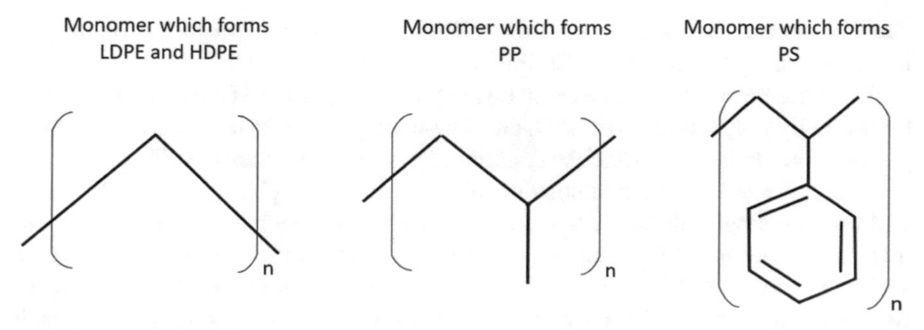


**Fig. 1** Monomer structure LDPE, HDPE, PP, and PS plastics [12–14]

**Table 1** Types of plastic suitable for pyrolysis [15–17]

| Recycle number | Plastic type | Degradation temperature | Description | Utilization |
|---|---|---|---|---|
| 2 | High-density polyethylene (HDPE) | 480 °C | Semi-rigid plastics, opaque | Milk jugs, plastic furniture, plastic shopping bags, shampoo and detergent bottles |
| 4 | Low-density polyethylene (LDPE) | 480 °C | Very flexible plastic, usually floats in water | Plastic bags, squeezable bottles, most wrappings, 6-pack rings, trays |
| 5 | Polypropylene (PP) | 470 °C | Semi-rigid, low gloss, usually floats in water | Bottle caps, yogurt containers, reusable food storage containers |
| 6 | Polystyrene (PS) | 420 °C | Dull appearance, Styrofoam floats in water | Styrofoam, plastic straws, disposable cutlery, drink lids, disposable coffee cups, packing foam |

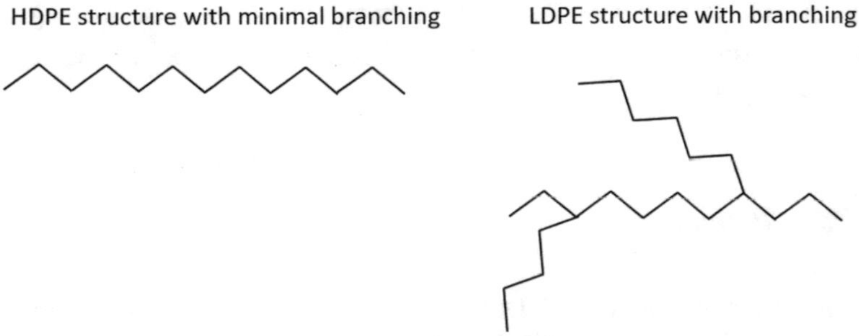

**Fig. 2** HDPE and LDPE structures

chain of the ethylene monomer, while LDPE has a branched structure attributing to its lower density and increased flexibility in comparison to HDEP.

There has been extensive research on the pyrolysis process of the two types of PE, high-density polyethylene (HDPE), and low-density polyethylene (LDPE) as these plastics contribute the most to the fraction of plastics in municipal solid waste.

Data collected on the composition of non-recyclable plastics within MSW in regions including Malaysia, USA, and UK, showed the global average of PE plastics was greater than 50% [18]. HDPE is made from a straight chain polyethylene structure, while LDPE has a branching structure. Physically HDPE has a higher rigidity compare to LDPE; hence, it is widely used as in the production of items such as shampoo bottles, milk jugs, and chemical containers [16]. Due to its flexibility,

LDPE makes a large portion of the grocery plastic bags that consumers use on a daily basis [17]. Through thermogravimetric analysis (TGA) while looking at the weight loss percentage of the individual PE, it was observed that the degradation temperature during the pyrolysis reaction for PE is around 480 °C [15].

### 2.2 *Polypropylene (PP)*

Polypropylene (PP) has similar properties to PE but it is more rigid and more heat resistant in comparison [17]. PP pyrolysis favors intramolecular radical transfer leading to the formation of lower molecular weight oligomer compounds [19]. Regarding the fuel's properties, the strong backbone structure allows the iso-alkanoic structure to be maintained during thermal decomposition. Despite having a higher melting point, the thermal degradation temperature for PP is lower than that of PE at 470 °C [15]. PP is usually used in the production of an assortment of materials including fibers and fabrics, carpets, rigid packaging, and car bumpers [16].

### 2.3 *Polystyrene (PS)*

Polystyrene (PS) has been extensively used in the food industry for disposable containers [16]. PS usually degrades at a lower temperature compared to other type of plastics. Through TGA methodology, the degradation temperature for PS is about 420 °C [15]. The main product of pyrolysis of PS consists of aromatics which is mainly due to the presence of a benzylic ring in the PS structure that stabilizes most of the produced oil into aromatics.

## 3 Operating Parameters for Pyrolysis

### 3.1 *Temperature*

Extensive studies have been done to determine the optimal temperature of the pyrolysis reaction with varying feedstock and reactor types. Most studies found that the optimum liquid yield was obtained for temperatures between 400–600 °C, where yields of over 80 wt.% were observed. The optimum reaction temperature for polystyrene in a semi-batch reactor was 450 °C yielding 80.8 wt.% of liquid product [20]. A similar study was performed with a blend of waste plastics and found that for a blend of LDPE, PP, PS, and HDPE the optimum reaction temperature observed was 475 °C with a liquid yield ranging from 66 to 90 wt.% [21]. The range in optimal reaction temperature can be caused by a difference in feedstock type since HDPE,

LDPE, and PP would require a reaction temperature at the higher end of the spectrum in comparison with PS due to its lower degradation temperature. It has also been found that virgin plastics have a higher activation energy due to being less affected by the environment [22]. The reason behind a lack of liquid yield at a lower temperature was because at lower temperatures incomplete decomposition would occur; hence, a higher char yield was observed [23]. According to Singh & Ruj [2], the production of gas was much higher during the pyrolysis of plastic wastes at 600 °C as oppose to 450 °C. This is because at higher temperature the liquid oil undergoes secondary reactions producing larger amounts of the gaseous fraction [11]. Therefore, depending on the production needs and feedstock type, varying temperatures would be needed to provide an optimum yield for the process.

## 3.2 Heating Rate

The heating rate of the pyrolysis reaction is also an important operating parameter to produce optimum WPO quality and yield. Previous research has studied a wide range of heating rates; in most cases, the range of 10–20 °C/min was used. This heating rate is coupled with an optimum reaction temperature between 400 and 600 °C to produce optimum liquid yield. A study of the effect of heating rates between 2 and 20 °C/min determined that there was an increase in liquid product yield for heating rates of 2 and 10 °C/min from 51.4%wt. to 53.5%wt. for a feedstock of oil shale mixed with plastics [24]. A comparison of the heating rate of 10 °C/min versus 20 °C/min was performed in which the lower heating rate oil yield contained lower range hydrocarbons when compared to the higher heating rate, this is partly because a lower heating rate would increase the residence time for the volatile substance which leads to further the cracking process [25]. A major difference in product yield was observed when comparing heating rates of 5 °C/min with 350 °C/min where a significant increase in gas production was seen with a decrease in liquid product at the higher heating rate with a final fixed temperature which can be explained by the rapid thermal cracking of primary pyrolysis product [26]. An exponentially higher heating rate coupled with a higher reaction temperature and longer residence time promotes the secondary cracking process. High heating rates would favor the promotion of gaseous yield. Comparison between lower heating ranges showed that the more prominent difference would be in the gas composition as opposed to the yield value, and lower heating rates favored the promotion of liquid yield.

## 3.3 Reaction and Residence Time

It has been observed that 75 min is the optimum reaction time for 1 kg of polystyrene at 450 °C, which can generate a yield of 80.8 wt.% of liquid oil, while any duration above the optimum time would not increase the yield significantly, but reaction time may vary depending on reactor configuration [20]. It has been found that a longer residence time leads to the further cracking of primary products [6]. This usually means an increased production of lower molecular weight hydrocarbons can lead to a higher production of non-condensable gases. Further experimental observation has shown an increase in temperature would reduce the residence time which would halt the process of cracking larger hydrocarbons at 600 °C, and this leads to the production of a wax-like substance which was attributed to heavier hydrocarbons [2]. The same observation could be seen for the hydrocarbon gases produced in which heavier gases were detected at this temperature. The correlation between temperature and residence time is that an increase in temperature would reduce the residence time for the process which would lead to a reduction in the cracking of heavier hydrocarbons into lighter hydrocarbons.

## *3.4 Catalyst*

Studies regarding catalysts focus on enhancing the quality of liquid oil produced in pyrolysis through catalytic reactions. Previous experiments have shown that acidity and surface area are the main driving forces that enhanced the cracking of the polymeric chains. Research has focused on testing different types of catalysts as well as the catalyst ratio to obtain optimum liquid oil quality and yield.

Heterogeneous catalysts are the catalysts most used to produce liquid oil from plastic fuel due to its economic and practical benefits. Examples of catalyst used for plastic pyrolysis include zeolite catalysts, spent fluid catalytic cracking (FCC) catalysts, and bentonite catalysts. The general trend showed a decrease in oil production for pyrolysis of PE, PP, and PS when a zeolite catalyst was introduced. Mesoporous catalyst with high acidity and high Brunauer–Emmett–Teller (BET) surface area favor higher gas production, whereas a microporous structure with low BET surface area and low acidity promoted higher liquid oil and char yields in comparison [27]. In a study of different catalyst types, it was observed the effect of acidity is greater compared to surface area when it comes to cracking intensity [28].

The main purpose of catalytic addition in the pyrolysis reaction is to help reduce the heavier hydrocarbon molecules within the liquid oil. The addition of zeolite HZSM-5 catalyst increased the amount of lower molecular weight hydrocarbons as well as the amount of aromatics [15]. The addition of spent FCC catalyst helps reduce the oxygenated content and increases the aromatic fraction within the liquid oil product [29]. It can be concluded that the addition of catalyst during pyrolysis increased the amount of lighter molecular weight hydrocarbons within the liquid oil product as well as increased the amount of aromatics produced.

### 3.5 *Carrier Gas*

Nitrogen has generally been used as the reaction medium for plastic pyrolysis; however, recently research has focused on carrying the reaction out under vacuum conditions. Using a vacuum would reduce the continuous need to provide nitrogen for the process which would reduce the cost of production, especially during the scale-up process. With increasing temperature, the liquid yield production for the vacuum scenario was found to be more stable since the vacuum process has the capability of reducing the further cracking of hydrocarbons, this also led to an increased production of diesel range hydrocarbons but also the production of wax [30]. Under vacuum conditions, the incidence of secondary reactions will be reduced thus favoring liquid production [19]. It was found that for the pyrolysis process under vacuum conditions, PP was the most viable to be used as a replacement for diesel engine fuel as the physical characteristics such as viscosity and cetane index were found to be the most similar [8]. These observations have shown that the pyrolysis reaction under vacuum conditions is a viable replacement for the pyrolysis process that utilizes $N_2$ gases, and this will help ease the scale up process.

## 4 Pyrolysis Products

Currently, much interest has been focused on creating a fuel substitute for diesel from plastic waste using pyrolysis. Therefore, this section will mainly focus on the viability of plastic pyrolysis to replace diesel fuel. This is done through comparisons of physicochemical characteristics and composition of the WPO produced from plastic and diesel fuel itself.

### 4.1 *Physicochemical Characteristics of Oil*

When it comes to comparing physicochemical characteristics of the WPO with diesel, the common characteristics usually include calorific value, density, viscosity, boiling point range, and cetane index. Typically when comparing calorific value which determines the amount of energy stored within the fuel, the calorific value of WPO from HDPE, LDPE, PP, and PS has been found to be above 40 MJ/kg, whereas usually the calorific value from diesel fuel ranges from 43 to 46 MJ/kg [16, 31, 32]. The calorific values for HDPE, LDPE, and PP were found to be 43.65 MJ/kg, 43.39 MJ/kg, and 43.37 MJ/kg, respectively [32]. This shows that the calorific value of WPO is very similar when compared to commercial diesel. As for PS, the presence of aromatic rings which has a lower combustion energy than aliphatic hydrocarbons makes the calorific value of PS lower from 37.8 to 41.6 MJ/kg when compared to the calorific value of HDPE, LDPE, and PP [20].

Density is considered an important aspect of petroleum products as it depends on the hydrocarbon blend of the fuel. The density of LDPE, HDPE, PP and PS was found to be 0.799 g/ml, 0.800 g/ml, 0.771 g/ml and 0.941 g/ml respectively in comparison with diesel fuel which has a density of 0.838 g/ml [8]. The density of WPO derived from LDPE, HDPE, and PP falls in the range of fuel oil, or diesel, whereas WPO derived from PS has a higher density compared to the that of commercial diesel.

Viscosity is a measure of the fluidity of a liquid compound and is another important physicochemical characteristic as the spray pattern and atomization of injected fuel in a combustion chamber is affected by the kinematic viscosity. Generally, a less viscous fluid represents a fluid with shorter carbon chain composition. The kinematic viscosity of WPO derived from PS, PP, and HDPE was found to be 1.461, 2.115, and 2.373 $mm^2/s$, respectively, whereas diesel has a kinematic viscosity of around 2.0–5.5 $mm^2/s$ [33]. A highly viscous WPO would lead to poor atomization, while a lower viscosity would lead to poor engine performance and severe leakage which would reduce the output of an engine caused by reduced fuel delivery [20, 34]. Compared to diesel, PS derived WPO has a lower kinematic viscosity which is related to the high amount of aromatics within the WPO produced by PS [33].

Boiling point range is an important aspect for the physical study of WPO because it gives an initial look into the hydrocarbon distribution of the WPO. Motor gasoline has a boiling point range of <190 °C, while diesel would have a boiling point range between 190 and 360 °C [35]. The boiling point range for WPO derived from PP, HDPE, and PS were found to be 148–355 °C, 119–364 °C, and 128–179 °C resulting in a broader range for PP and HDPE when compared to diesel [33].

The cetane index provides a better indication of the quality of diesel fuel during ignition where previous research has shown that a cetane index of >40 would be comparable to that of diesel fuel. The cetane index obtained for WPO derived from HDPE, LDPE, PP, and PS was 65, 68, 60, and 10, respectively, which was higher than the calculated diesel index of 52, except for PS [8]. Other studies have obtained a cetane index of 48 for the WPO produced from waste PP which is in a similar range to the diesel cetane index [34]. A higher cetane index can be attributed to the presence of alpha olefin and linear paraffin [36]. Since the pyrolysis products of polyolefin are usually high in the presence of paraffin and olefin, it is expected that the cetane index of HDPE, PP, and LDPE is comparable to that of diesel. As for PS, the cetane value is much lower than that of diesel which is attributed to the products of PS pyrolysis which are mainly aromatics, and hence, the olefin and paraffinic contents were reduced. Table 2 shows the physicochemical comparison between the WPO produced from plastic wastes.

### *4.2 Compositional Analysis of Oil*

The studies regarding composition of oil from single use plastics have been extensive, especially regarding HDPE, LDPE, PP, and PS where these four are usually categorically divided into polyolefin and polystyrene by itself. Thermal pyrolysis of

**Table 2** Physicochemical characteristics of waste plastic oil (WPO)

| | WPO derived from HDPE | WPO derived from LDPE | WPO derived from PP | WPO derived from PS | Diesel |
|---|---|---|---|---|---|
| Calorific value (MJ/kg) [20, 32] | 43.65 | 43.39 | 43.37 | 41.58 | 46.95 |
| Density at 15° ¯C(g/ml) [8] | 0.800 | 0.799 | 0.771 | 0.941 | 0.838 |
| Kinematic viscosity at 40 °C ($mm^2/s$) [33] | 2.373 | N/A | 2.115 | 1.461 | 2.0–5.5 |
| Boiling point range (°C) [33] | 119–364 | N/A | 148–355 | 128–179 | 172–350 |
| Cetane index [8] | 65 | 68 | 60 | 10 | 52 |

PE and PP plastics would produce mostly alkane and alkene compounds, whereas PS would mostly produce aromatic compounds. However, there is a difference in the PE and PP alkane to alkene ratio. PE oil has a much higher alkane content at 70% compared to PP oil that has an alkane content of 57% where the remainder is found to be alkenes and a small amount of aromatics [37]. This is attributed to the presence of the methyl group in PP that favors the stabilization of double bonds and not hydrogen addition when compared to PE pyrolysis. It has been observed that the pyrolysis of PE contains mostly 1-alkenes and n-alkanes, whereas the pyrolysis of PP mainly consists of 2,4-dimethyl-1-heptene (19.79%) and other methylated olefins (alkenes) [38].

WPO derived from PS usually contains mostly aromatics due to the presence of the benzene ring in the PS polymer; however, varying amounts of aromatics have been detected in the literature. Pyrolysis experiments for PS at 500 °C have shown that the major compound produced was styrene with a yield of 39.71% [38]. Another study at a temperature of 430 °C has obtained an oil yield that was mainly toluene and ethylbenzene which shows that the secondary reaction was favored as these two compounds are derivatives of styrene [37]. All three aromatics were detected in the pyrolysis of PS at a temperature of 450 °C with styrene being the highest constituent followed by toluene and ethylbenzene [20]. Higher temperature and residence time would favor the production of lighter aromatics when it comes to PS pyrolysis hence the production of toluene and ethylbenzene correlates with an increase in these operating parameters [39].

A study on thermal pyrolysis of HDPE, LDPE, PP, and PS has found that the polyolefins have similar Fourier transform infrared spectroscopy (FT-IR) spectra to that of diesel ranging from 86.18% to 96.89%, whereas PS pyrolysis had the highest similarity to that of gasohol 91 with 63.20% [32]. The FT-IR method identifies the functional groups and covalent bonding information by producing an infrared absorption spectrum [16]. Therefore, the similarities shown depicts how the oil produced has similar functional groups to the commercial fuel which points to similarities in their physicochemical characteristics. This means that it is potentially viable for the oil produced to replace commercial fuel. A further study on liquid fuel from PE pyrolysis observed that the hydrocarbon distribution for the PE oil ranges from $C_7$–$C_{35}$ which constitutes the hydrocarbon number contained in light distillate up until heavy oil content. This large distribution coupled with a relatively higher amount of compounds compared to PS shows that the pyrolysis of PE mainly decomposes through random scission reactions [38].

### *4.3 Other Pyrolysis Products*

The other products from the plastic pyrolysis process consists of gas and char. The gaseous products from pyrolysis of HDPE, LDPE, PP, and PS contained non-condensable hydrocarbon gases usually in the form of alkane and alkene gases within the $C_1$–$C_4$ range [6]. Das & Tiwari [11] also observed that the pyrolysis of LDPE, HDPE, and PP produced short chain hydrocarbon gases with high energy content. Depending on the operating conditions, the gaseous products can have a HHV ranging from 24 to 45 MJ/kg which can be utilized to power up the plastic pyrolysis process [4]. This can help reduce the energy requirements of the pyrolysis process and increase its favorability as a green alternative to incineration and landfilling. Trace amount of CO and $CO_2$ have also been detected in the non-condensable gas fraction of the pyrolysis process. Usually this is attributed to the presence of oxygen gases in the reaction chamber or organic compounds attached to the plastic in the feed [30].

The char product is the residue of the plastic pyrolysis process in the form of carbonaceous residue. The main components of char consist of volatile matter and fixed carbon closely related to its raw plastics counterpart [6]. The char product from HDPE and PS has a HHV of 18.84 MJ/kg and 36.29 MJ/kg, respectively [16]. This shows that the char product can also be utilized in relation with the gaseous products to power up the pyrolysis process. However, it is the most undesired out of the plastic pyrolysis products because it can lead to coking and the deactivation of the catalyst [19].

## 5 Pyrolysis Oil Performance

To test the performance of the WPO in comparison with diesel, a simulated direct injection (DI) engine was used. The comparison of the fuels' performance is usually done using the following three categories: combustion characteristics, engine performance, and exhaust emissions. When it comes to performance testing, most researchers would blend their WPO with diesel for testing as it was found in most cases the WPO would not function in a diesel engine if a blend was not used. Typically, a lower blend ratio of WPO in diesel would perform best compared to higher blends [34].

With regards to combustion characteristics, comparisons were made for the cylinder pressure at different loads and the heat release rates of combustion for the WPO. As the load was lowered below 75%, there were clear signs of delay shown especially for the higher blends of WPO [7]. This may have been caused by the lower cetane value and higher aromatics within the WPO when compared to the diesel. The WPO blends were found to have a higher cylinder pressure and heat release rates in comparison with diesel. As the blend increased in WPO, the cylinder pressure and heat release rate also increased which was attributed to the double bond structure of alkenes and the presence of $C_{21}$–$C_{30}$ found in large quantities in the WPO [8]. The higher WPO blends were found to have an increased combustion delay which in turn led to a higher cylinder pressure and heat release rate. The subsequent delay affects the performance of the diesel engine because it leads to an increase in exhaust temperature and thermal loss.

With regards to engine performance, the parameters compared were the brake specific fuel consumption (BSFC) and the brake thermal efficiency (BTE). Comparing different blends of WPO with diesel, the performance for higher WPO blends led to increased fuel consumption and BSFC [7]. The lower calorific content of the WPO demands a larger amount of fuel. Hence, more fuel is consumed which increased the BSFC. It was found that increasing WPO in the fuel blend would reduce the BTE and increase the BSFC due to WPO having a lower calorific content [40].

Emission comparisons for nitrogen oxide ($NO_x$), unburned hydrocarbon (UHC), carbon monoxide (CO), and carbon dioxide ($CO_2$) emissions were performed on WPO blends. Mangesh et al. [8] showed that the CO, $CO_2$, $NO_x$, and UHC emissions increase with increasing WPO in the blend. This finding is supported by Kumar & Singh [41] who determined all emissions were greater in the WPO blend except for $CO_2$ emissions when compared to diesel which can be explained by late burning of fuel due to incomplete oxidation of CO. The results showed that to some extent the usage of WPO diesel blends is viable to be used as a diesel fuel replacement. However, further improvements need to be made to the blend to create consistent oil quality from the production of plastic wastes.

## 6 Mixed Waste Plastics

The main goal of the plastic pyrolysis process is to create a system that can withstand various mixtures of plastic but still can create a viable fuel replacement for commercial diesel. Multiple studies have been performed throughout the literature to create such a system. Various studies in the literature have shown that the addition of PS in the waste plastic feedstock is advantageous to the production of liquid yield. Wax like material was produced during the pyrolysis of PE but with the addition of PS the amount of liquid oil produced increased because the radicals produced from PS promotes the cyclization process producing cyclic aliphatic hydrocarbons and aromatics [27]. With the addition of PS into the plastic mixture, the amount of gaseous products were reduced as the aromatic structure for PS does not decompose easily into gaseous compounds [37]. This was not the case with a feedstock of only PE since alkanes and alkenes would readily degrade into lighter intermediate radicals resulting in more gaseous products. A similar study has shown that with the addition of PS, the plastic mixture containing mostly polyolefin which had a liquid production of 29.00 wt.% was increased to 72.08 wt.% when the ratio of PS and other plastic mixtures was 1:1 by weight.

Research has also shown blends with PP and PE increase liquid yield. A PP and LDPE mixture showed an increase in liquid product compared to PP alone [42]. There was a synergistic affect in the PP/HDPE mixture where the activation energy of the mixture was reduced with the addition of PP [43]. The reduction of activation energy was also seen in Yuan et al. [44] where a TGA experiment showed that the degradation temperature of the LLDPE/PP mixture was lower when compared to the calculated value because PP has a higher degree of branching resulting in weaker molecular bonds which interacts with the LDPE and reduced the overall degradation temperature of the mixture.

Zhang et al. [35] studied the pyrolysis of simulated mixed plastics (SMP) and observed that despite having a high amount of polyolefin, the produced oil contained a large amount of mono-aromatics instead of aliphatic hydrocarbons which was attributed to the presence of the benzylic ring of PS combining with the alkane and alkene radical to produce aromatic products including styrene and toluene. Adrados et al. [45] observed that despite their plastic mixture containing >70% polyolefin there was a higher percentage of aromatics in the liquid product. The proposed mechanism was due to free radicals in the process in which the $H^+$ ion was either captured or cycled leading to the stabilization of produced aromatics. Hence in this case, the addition of PS into the plastic mixture, even in small amounts, showed that it affects the product distribution greatly.

## 7 Appropriate Technology Reactors

Appropriate technology (AT) is defined as "technological choices and applications that are small scale, labor-intensive, energy efficient, environmentally sound, and locally controlled" [46]. This concept was later used to build a simple technology plastic pyrolysis reactor that focuses on producing ready to use liquid fuel [1]. This comes from the fact that plastic waste management, especially in developing countries, has been ineffective leading to plastic accumulation within the community which in turn adversely affects the environment and human health. Currently those at the forefront of recycling in developing countries are poor inhabitants who live near the landfills such as waste pickers. Waste pickers scourge through the trash in the landfills with little to no protection which detrimentally affects the community in regards to the damaging socio-economic and health impacts of this practice [17].

It has been suggested that using a decentralized approach with regards to plastic fuel production will benefit the local communities especially in developing countries [5]. Using AT pyrolysis has been shown to generate less $CO_2$ emissions when compared to conventional commercial fuel as using WPO as a diesel substitute eliminates the emissions generated during the supply chain transportation of conventional fuel [47].

The concept of the appropriate technology plastic pyrolysis reactor focuses on three main aspects: sourcing local materials, simple construction methodology, and ease of use. A study performed by Joshi & Seay using an AT pyrolysis reactor found that by using clean LDPE plastic bags, an average oil yield of 81.22 wt.% was obtained which can be used as a replacement for local kerosene and petrol diesel [1]. A characteristic comparison with available fuel showed that the liquid fuel produced shares a similar GC–MS peak as well as a similar density to kerosene and diesel fuel; however, the calorific value of the liquid oil was higher than that of kerosene and diesel [48]. It was observed that when it came to their simple pyrolysis reactor, a constant amount of heat supply is needed as it was observed that the fluctuation of the heat supply reduced the liquid fuel produced which may be caused by the fact that the reactor could not maintain the reaction at the constant reaction temperature. The addition of a cover to the reactor decreased the amount of time it took to heat the reactor to the target reaction temperature in which the addition of the cover reduced the time taken to heat up the reactor to 401 °C to 48 min, while at the same time without the cover the reactor would have only reached a reaction temperature of 284 °C as shown by Armadi et al. [49].

A key concept of the appropriate technology reactors is constructing them with accessible and easy to find materials. This is demonstrated by the appropriate technology reactors constructed by Joshi & Seay [1] which consists of a steel drum, whereas Kurniawan et al. [48] used a refrigerant steel tank. The main vessel of the reactor would usually consist of a cylindrical tank either powered by biomass or LPG since these are more readily available in developing areas as opposed to a steady electric supply. The condenser would usually consist of a piping system as seen by Jayswal et al. [50] who built a counter current condenser in order to cool down the vapors produced to room temperature. Ref. [1], however, used a different approach in which the vapors are directed straight into the condensing medium then are separated through gravity separation. Typically using this process, the char products are collected within the reactor, and the gaseous products are released to the surroundings. Figure 3 shows the depiction of the pyrolysis process using an AT method where the condenser depicted follows the description of Joshi & Seay [1].

Since the publication of Joshi & Seay [1], the researchers have modified their processor to be electrically powered; additionally, the condenser has been modified to a copper coil with cooling water as the condensing fluid, instead of a water condensation bath. Figure 4, provided by Jeffrey Seay of Joshi & Seay [1], depicts the upgraded and modified process used in their current research.

Creating the condenser aspect of the pyrolysis reactor from accessible and locally sourced materials in a simple, easy to construct design is fundamental for the appropriate technology. It is important to note that all the appropriate technology reactors

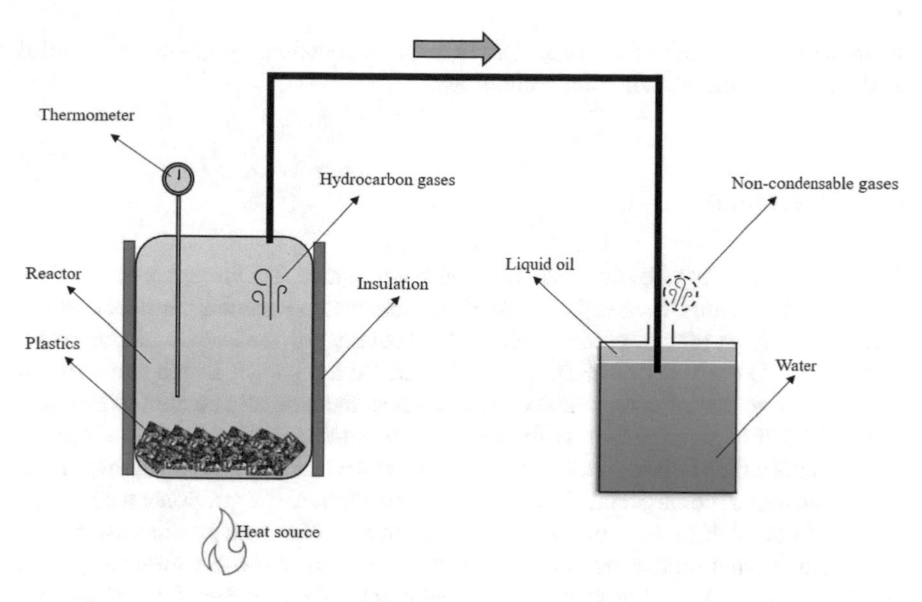


**Fig. 3** General diagram for AT pyrolysis process using Joshi & Seay [1] condensing method to collect the pyrolysis oil

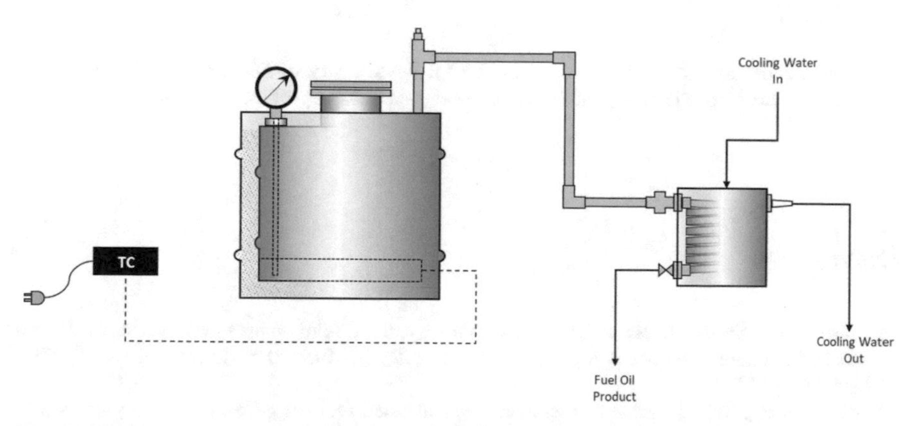


**Fig. 4** Open source electric pyrolysis processor developed at University of Kentucky

studied in the literature used a condensing medium at room temperature and did not use nitrogen gas as a carrier gas in the pyrolysis process.

With regards to fuel quality, Owusu et al. [33] performed a thorough study on the physicochemical characteristics of their oil from single use plastics. An important finding was that the physicochemical properties of the fuel obtained was similar to diesel fuel especially for WPO derived from HDPE and PP. Further studies can be done with regards to the fumes generated from the appropriate technology reactor as the lack of $N_2$ gas as a carrier gas and condensing the pyrolysis product at room temperature might affect the gases released to the surroundings which can be harmful to the communities utilizing this technology.

## 8 Conclusion

The prospect of using plastic waste to produce alternative fuel through pyrolysis has been very promising. The WPO produced can achieve very similar physicochemical characteristics to that of commercial fuel especially for feedstocks of polyolefin plastics. WPO derived from HDPE, LDPE, and PP has shown a high similarity in composition to that of commercial diesel, whereas the aromatic content of PS oil is more like that of gasoline fuel. With regards to performance, blending is still needed for the WPO which is likely because the WPO produced contains a hydrocarbon range exceeding that of commercial diesel. The implementation of appropriate technology reactors in developing communities has the potential to eliminate plastic waste which persists in the environment as well as add value to the plastic waste, additionally there is opportunity for the oil produced to be used practically. However, further research needs to be done including life cycle analysis (LCA) of the AT process in order to maximize benefit and minimize the harmful effects it can have to developing communities.

**Acknowledgements** This work was partly funded by Ministry of Higher Education Malaysia Translational Research (TR@M) and Universiti Malaysia Sabah Special Grant Scheme (SDK0321-2021).

## References

1. Joshi, C.A., Seay, J.R.: An appropriate technology based solution to convert waste plastic into fuel oil in underdeveloped regions. J. Sustain. Dev. **9**, 133–143 (2016). https://doi.org/10.5539/jsd.v9n4p133
2. Singh, R.K., Ruj, B.: Time and temperature depended fuel gas generation from pyrolysis of real world municipal plastic waste. Fuel **174**, 164–171 (2016). https://doi.org/10.1016/j.fuel.2016.01.049
3. Solis, M., Silveira, S.: Technologies for chemical recycling of household plastics—a technical review and TRL assessment. Waste Manag. **105**, 128–138 (2020). https://doi.org/10.1016/j.wasman.2020.01.038
4. Qureshi, M.S., Oasmaa, A., Pihkola, H., Deviatkin, I., Tenhunen, A., Mannila, J., Minkkinen, H., Pohjakallio, M., Laine-ylijoki, J.: J. Anal. Appl. Pyrol. Pyrol. Plast. Waste Opportunities Challenges. **152**, 1–11 (2020). https://doi.org/10.1016/j.jaap.2020.104804
5. Joshi, C., Seay, J., Banadda, N.: A perspective on locally managed decentralized circular economy for waste plastic in developing countries. Environ. Prog. Sustain. Energy. **38**, 3–11 (2019). https://doi.org/10.1002/ep.13086
6. Anuar Sharuddin, S.D., Abnisa, F., Wan Daud, W.M.A., Aroua, M.K.: A review on pyrolysis of plastic wastes. Energy Convers. Manag. **115**, 308–326 (2016). https://doi.org/10.1016/j.enconman.2016.02.037

7. Kalargaris, I., Tian, G., Gu, S.: Combustion, performance and emission analysis of a DI diesel engine using plastic pyrolysis oil. Fuel Process. Technol. **157**, 108–115 (2017). https://doi.org/10.1016/j.fuproc.2016.11.016
8. Mangesh, V.L., Padmanabhan, S., Tamizhdurai, P., Ramesh, A.: Experimental investigation to identify the type of waste plastic pyrolysis oil suitable for conversion to diesel engine fuel. J. Clean. Prod. **246**, (2020). https://doi.org/10.1016/j.jclepro.2019.119066
9. Wong, S.L., Ngadi, N., Abdullah, T.A.T., Inuwa, I.M.: Current state and future prospects of plastic waste as source of fuel: A review. Renew. Sustain. Energy Rev. **50**, 1167–1180 (2015). https://doi.org/10.1016/j.rser.2015.04.063
10. Kalargaris, I., Tian, G., Gu, S.: Experimental evaluation of a diesel engine fuelled by pyrolysis oils produced from low-density polyethylene and ethylene–vinyl acetate plastics. Fuel Process. Technol. **161**, 125–131 (2017). https://doi.org/10.1016/j.fuproc.2017.03.014
11. Das, P., Tiwari, P.: The effect of slow pyrolysis on the conversion of packaging waste plastics (PE and PP) into fuel. Waste Manag. **79**, 615–624 (2018). https://doi.org/10.1016/j.wasman.2018.08.021
12. Polymer Database: Polystyrene. Polymer Database (2018). https://polymerdatabase.com/polymers/polystyrene.html Accessed 31 July 2021
13. Polymer Database: Poly(propylene). Polymer Database (2016). https://polymerdatabase.com/polymers/polypropylene.html. Accessed July 31 2021.
14. Polymer Database: Polyethylene. Polymer Database (2016). https://polymerdatabase.com/polymers/polyethylene.html. Accessed July 31 2021.
15. Muhammad, C., Onwudili, J.A., Williams, P.T.: Thermal degradation of real-world waste plastics and simulated mixed plastics in a two-stage pyrolysis—catalysis reactor for fuel production. Energy Fuels **29**, 2601–2019 (2015). https://doi.org/10.1021/ef502749h
16. Miandad, R., Barakat, M.A., Aburiazaiza, A.S., Rehan, M., Ismail, I.M.I., Nizami, A.S.: Effect of plastic waste types on pyrolysis liquid oil. Int. Biodeterior. Biodegrad. **119**, 239–252 (2017). https://doi.org/10.1016/j.ibiod.2016.09.017
17. Horodytska, O., Cabanes, A., Fullana, A.: Plastic waste management: current status and weaknesses. In: The Handbook of Environmental Chemistry, pp. 1–18. Springer, Berlin, Heidelberg (2019)
18. Anuar Sharuddin, S.D., Abnisa, F., Wan Daud, W.M.A., Aroua, M.K.: Energy recovery from pyrolysis of plastic waste: study on non-recycled plastics (NRP) data as the real measure of plastic waste. Energy Convers. Manag. **148**, 925–934 (2017). https://doi.org/10.1016/j.enconman.2017.06.046
19. Scheirs, J.: Overview of commercial pyrolysis processes for waste plastics. Presented at the (2006)
20. Miandad, R., Nizami, A.S., Rehan, M., Barakat, M.A., Khan, M.I., Mustafa, A., Ismail, I.M.I., Murphy, J.D.: Influence of temperature and reaction time on the conversion of polystyrene waste to pyrolysis liquid oil. Waste Manag. **58**, 250–259 (2016). https://doi.org/10.1016/j.wasman.2016.09.023
21. DeNeve, D., Joshi, C., Samdani, A., Higgins, J., Seay, J.: Optimization of an appropriate technology based process for converting waste plastic in to liquid fuel via thermal decomposition. J. Sustain. Dev. **10**, 116–124 (2017). https://doi.org/10.5539/jsd.v10n2p116
22. Yan, G., Jing, X., Wen, H., Xiang, S.: Thermal cracking of virgin and waste plastics of PP and LDPE in a semibatch reactor under atmospheric pressure. Energy Fuels **29**, 2289–2298 (2015). https://doi.org/10.1021/ef502919f
23. Ding, K., Liu, S., Huang, Y., Liu, S., Zhou, N., Peng, P., Wang, Y., Chen, P., Ruan, R.: Catalytic microwave-assisted pyrolysis of plastic waste over NiO and HY for gasoline-range hydrocarbons production. Energy Convers. Manag. **196**, 1316–1325 (2019). https://doi.org/10.1016/j.enconman.2019.07.001
24. Aboulkas, A., Makayssi, T., Bilali, L., El Harfi, K., Nadifiyine, M., Benchanaa, M.: Co-pyrolysis of oil shale and plastics: influence of pyrolysis parameters on the product yields. Fuel Process. Technol. **96**, 209–213 (2012). https://doi.org/10.1016/j.fuproc.2011.12.001

25. Singh, R.K., Ruj, B., Sadhukhan, A.K., Gupta, P.: Impact of fast and slow pyrolysis on the degradation of mixed plastic waste: product yield analysis and their characterization. J. Energy Inst. **92**, 1647–1657 (2019). https://doi.org/10.1016/j.joei.2019.01.009
26. Efika, E.C., Onwudili, J.A., Williams, P.T.: Products from the high temperature pyrolysis of RDF at slow and rapid heating rates. J. Anal. Appl. Pyrolysis. **112**, 14–22 (2015). https://doi.org/10.1016/j.jaap.2015.01.004
27. Miandad, R., Barakat, M.A., Rehan, M., Aburiazaiza, A.S., Ismail, I.M.I., Nizami, A.S.: Plastic waste to liquid oil through catalytic pyrolysis using natural and synthetic zeolite catalysts. Waste Manag. **69**, 66–78 (2017). https://doi.org/10.1016/j.wasman.2017.08.032
28. Auxilio, A.R., Choo, W.L., Kohli, I., Chakravartula Srivatsa, S., Bhattacharya, S.: An experimental study on thermo-catalytic pyrolysis of plastic waste using a continuous pyrolyser. Waste Manag. **67**, 143–154 (2017). https://doi.org/10.1016/j.wasman.2017.05.011
29. Praveen Kumar, K., Srinivas, S.: Catalytic co-pyrolysis of biomass and plastics (Polypropylene and Polystyrene) using spent FCC catalyst. Energy Fuels **34**, 460–473 (2020). https://doi.org/10.1021/acs.energyfuels.9b03135
30. Parku, G.K., Collard, F., Görgens, J.F.: Pyrolysis of waste polypropylene plastics for energy recovery: influence of heating rate and vacuum conditions on composition of fuel product. **209**, (2020). https://doi.org/10.1016/j.fuproc.2020.106522
31. Lam, S.S., Wan Mahari, W.A., Ok, Y.S., Peng, W., Chong, C.T., Ma, N.L., Chase, H.A., Liew, Z., Yusup, S., Kwon, E.E., Tsang, D.C.W.: Microwave vacuum pyrolysis of waste plastic and used cooking oil for simultaneous waste reduction and sustainable energy conversion: recovery of cleaner liquid fuel and techno-economic analysis. Renew. Sustain. Energy Rev. **115**, (2019). https://doi.org/10.1016/j.rser.2019.109359
32. Budsaereechai, S., Hunt, A.J., Ngernyen, Y.: Catalytic pyrolysis of plastic waste for the production of liquid fuels for engines. RSC Adv. **9**, 5844–5857 (2019). https://doi.org/10.1039/c8ra10058f
33. Owusu, P.A., Banadda, N., Seay, J., Kiggundu, N., Banadda, N., Zziwa, A., Seay, J., Kiggundu, N.: Reverse engineering of plastic waste into useful fuel products. **130**, 285–293 (2018). https://doi.org/10.1016/j.jaap.2017.12.020
34. Syamsiro, M., Saptoadi, H., Norsujianto, T., Noviasri, P., Cheng, S., Alimuddin, Z., Yoshikawa, K.: Fuel oil production from municipal plastic wastes in sequential pyrolysis and catalytic reforming reactors. Energy Procedia. **47**, 180–188 (2014). https://doi.org/10.1016/j.egypro.2014.01.212
35. Zhang, Y., Ji, G., Ma, D., Chen, C., Wang, Y., Wang, W., Li, A.: Liquid oils produced from pyrolysis of plastic wastes with heat carrier in rotary kiln. Process Saf. Environ. Prot. **142**, 203–211 (2020). https://doi.org/10.1016/j.psep.2020.06.021
36. Ahmad, I., Ismail Khan, M., Khan, H., Ishaq, M., Tariq, R., Gul, K., Ahmad, W.: Pyrolysis study of polypropylene and polyethylene into premium oil products. Int. J. Green Energy. **12**, 663–671 (2015). https://doi.org/10.1080/15435075.2014.880146
37. Pinto, F., Costa, P., Gulyurtlu, I., Cabrita, I.: Pyrolysis of plastic wastes. 1. Effect of plastic waste composition on product yield. J. Anal. Appl. Pyrolysis. **51**, 39–55 (1999). https://doi.org/10.1016/S0165-2370(99)00007-8
38. Quesada, L., Calero, M., Martín-Lara, M.Á., Pérez, A., Blázquez, G.: Production of an alternative fuel by pyrolysis of plastic wastes mixtures. Energy Fuels **34**, 1781–1790 (2020). https://doi.org/10.1021/acs.energyfuels.9b03350
39. Onwudili, J.A., Insura, N., Williams, P.T.: Composition of products from the pyrolysis of polyethylene and polystyrene in a closed batch reactor: effects of temperature and residence time. J. Anal. Appl. Pyrolysis. **86**, 293–303 (2009). https://doi.org/10.1016/j.jaap.2009.07.008
40. Kumar, S., Prakash, R., Murugan, S., Singh, R.K.: Performance and emission analysis of blends of waste plastic oil obtained by catalytic pyrolysis of waste HDPE with diesel in a CI engine. Energy Convers. Manag. **74**, 323–331 (2013). https://doi.org/10.1016/j.enconman.2013.05.028
41. Kumar, S., Singh, R.K.: Thermolysis of high-density polyethylene to petroleum products. J. Pet. Eng. **2013**, 1–7 (2013). https://doi.org/10.1155/2013/987568

42. Jadhao, S.B., Seethamraju, S.: Pyrolysis study of mixed plastics waste. IOP Conf. Ser. Mater. Sci. Eng. **736**, 1–7 (2020). https://doi.org/10.1088/1757-899X/736/4/042036
43. Wee, K.P., Ghosh, U.K.: Effect of binary mixture of waste plastics on the thermla behavior of pyrolysis process. Environ. Prog. Sustain. Energy **34**, 1113–1119 (2015). https://doi.org/10.1002/ep.12087
44. Yuan, Z., Zhang, J., Zhao, P., Wang, Z., Cui, X., Gao, L., Guo, Q.: Synergistic effect and chlorine-release behaviors during co-pyrolysis of LLDPE, PP, and PVC. ACS Omega **5**, 11291–11298 (2020). https://doi.org/10.1021/acsomega.9b04116
45. Adrados, A., de Marco, I., Caballero, B.M., López, A., Laresgoiti, M.F., Torres, A.: Pyrolysis of plastic packaging waste: a comparison of plastic residuals from material recovery facilities with simulated plastic waste. Waste Manag. **32**, 826–832 (2012). https://doi.org/10.1016/j.wasman.2011.06.016
46. Hazeltine, B., Bull, C., Wanhammar, L.: Appropriate technology: tools, choices, and implications. Acad. Press. **9**, 32–33 (1999). https://www.proquest.com/docview/236165500
47. Joshi, C.A., Seay, J.R.: Total generation and combustion emissions of plastic derived fuels: a trash to tank approach. Environ. Prog. Sustain. Energy. **39**, 1–9 (2020). https://doi.org/10.1002/ep.13151
48. Kurniawan, A., Sugiarto, B., Perdana, A.: Design of a simple pyrolysis reactor for plastic waste conversion into liquid fuel using biomass as heating source. Eksergi. **17**, 1 (2020). https://doi.org/10.31315/e.v17i1.3080
49. Armadi, B.H., Rangkuti, C., Fauzi, M.D., Permatasari, R.: The effect of cover use on plastic pyrolysis reactor heating process. AIP Conf. Proc. **1826**, (2017). https://doi.org/10.1063/1.4979227
50. Jayswal, A., Kumar, A., Pradhananga, P., Rohit, S., Bahadur, H.: Design, fabrication and testing of waste plastic pyrolysis plant. Proc. IOE Grad. Conf. **5**, 275–282 (2017). https://doi.org/10.13140/RG.2.2.33682.15044

# Characterization of Waste Fractions Found in Landfills

**Julfequar Hussain, Shruti Chatterjee, and Ekramul Haque**

**Abstract** Characterization of waste fractions is important to ascertain the environmental impacts from garbage disposal, cost accounting, and various legal activities. In addition, they enable us to choose the appropriate landfill design and operation of facilities for their disposal. A brief description of sampling, along with the physical characterization of solid waste components such as their composition, particle size distribution, density, and moisture content as well as the chemical components of characterization such as organic content and pH has been presented. Furthermore, the intricacies of the different types of landfills viz. municipal solid waste landfill, industrial waste and construction and demolition (C and D) landfill, and hazardous waste landfill which includes their schematic design, the type and quantity of waste disposed of, and their associated risk factors have been elucidated. The challenges and difficulties in waste characterization have also been touched upon, for example, in the industrial sector, wastes produced in one manufacturing facility may serve as raw materials in a different manufacturing facility thereby reducing the accuracy of waste characterization. Apart from that, a summary of the most common components of municipal solid waste (MSW) generation trends and their percentage-wise distribution in landfills has been exhibited. Finally, an account of hazardous waste imported to the USA from other countries such as Canada, Taiwan, Brazil, Mexico, Belgium, New Zealand, Holland, and Malaysia and those exported from the USA to other countries such as Mexico (762,774 tons), Canada (383,744 tons), South Korea (143,206 tons), and Spain (14,135 tons) has been elucidated; it was observed that ~10–20% of the waste were destined for the hazardous landfills, whereas the remaining 90–80% were subjected to special treatment facilities such as metal recovery, incineration, and

J. Hussain
Department of Biotechnology, School of Life Sciences, Pondicherry University, Puducherry 605014, India

E. Haque
AESD & CIF, CSIR-CSMCRI, Bhavnagar, Gujarat 364002, India

S. Chatterjee (✉)
APBD, CSIR-CSMCRI, Bhavnagar, Gujarat 364002, India
e-mail: schatterjee@csmcri.res.in



A. Z. Yaser et al. (eds.), *Waste Management, Processing and Valorisation*,
https://doi.org/10.1007/978-981-16-7653-6_3

sludge treatment/stabilization, which gives us an idea about the current remediation approaches for the management of hazardous waste.

**Keywords** Landfills · Waste fraction · Waste remediation · Physical characterization · Chemical characterization

# 1 Introduction

Waste characterization enables us to understand the nature of the waste over time and gain new insights that may be helpful in stabilizing the waste. It also enables us to predict the energy recovery potential from a specified area within the landfill. Apart from that, it is a necessity for administrative reasons as well, for instance, in the selection of a suitable landfill site according to the type of waste generated, and also allows provision for a national database to keep track of the waste generated and the appropriate measures taken for their disposal and management. Municipal solid wastes landfills are composed of residential waste and mostly consists of organic wastes such as food, paper, furniture, and yard trimmings which are easier to handle. On the other hand, industrial and C&D waste landfills are generated in goods manufacturing sectors such as plastics, iron, and metal industries and those that are generated during demolition of old buildings, roads, and bridges. Hazardous waste landfills are composed entirely of toxic substances and require special packaging and dumping protocols to be followed in order to minimize the risk of leakage of toxins into the environment and groundwater; they include waste catalysts and filter cakes from waste treatment plants, hazardous byproducts such as aerosols, acids, alkaline, and other liquids generated in chemical manufacturing factories. A brief overview of the typical wastes found in the different types of landfills is given in Table 1 [1, 2]

## *1.1 Waste Behavior*

The behavior of the waste is influenced both by its intrinsic properties as well as by the external conditions it is subjected to. Waste is usually segregated into separate trash group according to their qualities, and each group is referred to as a waste fraction. The potential of the waste fractions to cause pollution is primarily determined by the rate of mobilization of the pollutants [3].

Factors affecting the remediation process in the landfills includes [1, 3]:

- pH
- Water supply
- Temperature
- Particle size
- Landfill design
- Compaction technique

**Table 1** Overview of typical waste components found in different landfills [1, 2]

| Municipal solid waste landfills | Industrial and C&D waste landfills | Hazardous waste landfills |
|---|---|---|
| Food | Plastic | Acids and Alkaline liquids |
| Paper | Iron and metals | Incinerator clinker |
| Textiles | Glass | Cake sludges |
| Yard trimmings | Wood | Contaminated soil and sediments |
| Furniture and furnishings | Concrete | Inorganic solid wastes |
| Metal | Bricks | Oxygen canister |
| Plastic plates and cups | Asbestos | Waste catalysts |
| Rubber and leather | Asphalt | Plating floor debris |
| Glass jar and bottles | Gypsum | Filter cake |
| Electronic wastes | Plumbing fixtures | Aerosols |

- Leachate treatment.

The water supply mediates the transport of gases that drive the biological and chemical processes occurring within the landfills; it influences the movement of dissolved and suspended particles which in turn impacts the oxygen supply and gas transport. Some factors that affect the water supply include temperature, filling techniques, and contour of the surface layer. The duration of contact between the leachate and the garbage determines whether the reactions reach equilibrium or not, and is dependent on the amount of water, pattern of water flow and landfill technology [1].

## *1.2 Waste Characterization*

Sampling

Wastes from industries tend to be more homogenous and a small amount of sample is sufficient. However, municipal solid wastes are diverse in nature and exhibit spatial variation, consequently a huge amount of sample is required for an accurate representation. This may be achieved by employing stratified random sampling, wherein the waste is stratified into distinct sections, each of which is homogenous and then taking random samples from each section. The degree to which the division should be detailed is determined by the investigation's requirement for accuracy. Often this sample is too large to be managed and may require subsampling. This subsampling is essentially reducing the amount of waste from the landfill site by using coarse homogenization equipment such as a shredder. Representative samples are taken from an area which are smaller in size (about 10–30 kg), followed by the transportation of waste sample to an analytical laboratory, wherein half of the samples are taken and subjected to dry blending, subsequently wet blending (at this stage the sample amount is reduced to about 3 kg). For further homogenization, the samples are operated in a high-speed blender with the addition of water and sample is reduced to a brownish pulp. Then, it is subjected to drying at 80 °C on aluminum foil trays to obtain dry cakes of the sample. These cakes are further screened in a hammer mill, which lets out particles smaller than 2 mm in size. Finally, these sample particles are collected and analyzed for their composition [4].

# 2 Characterization Methods

## *2.1 Physical Characterization*

Often characterization of waste fractions is a costly affair owing to its variations in the waste types and their non-homogeneous spatial distribution. Knowledge and data regarding the physical characteristics of the waste are essential for its remediation and selection of suitable facilities that convert the waste to energy or fuels. The most important aspects of physical characterization are the following:

- Physical composition
- Particle size distribution
- Density
- Moisture content.

## *2.2 Physical Composition*

Waste can be segregated into different fractions according to their appearance. However, this approach requires manual labor and consequently is expensive as well as time consuming. Apart from that, waste segregation poses a threat to human health and those involved are at a greater risk of different infections like hepatitis, tetanus, *tuberculosis, histoplasmosis, giardiasis, candidiasis, salmonellosis,* and *leptospirosis* [5, 6]. Therefore, the involved personnel need to take extensive precautionary measures such as wearing a personal protective equipment (PPE) and being vaccinated for hepatitis A, hepatitis B, and tetanus [7]. The waste may be sorted firstly into broad categories such as plastic, glass, metal, paper, wood, and organic waste. Based on the requirement of the exercise, these major waste fractions may be further sub-categorized into smaller fractions, for example: plastics may be further categorized into polyethylene terephthalate (PET), high-density polyethylene (HDPE),

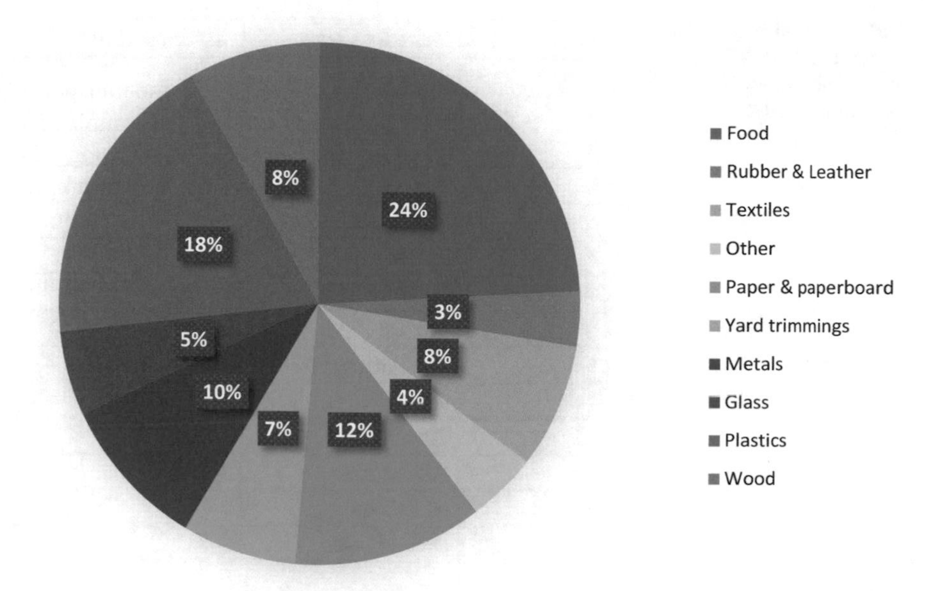


**Fig. 1** Percentage of physically discernable waste fractions in MSW landfills [2]

polystyrene, and polypropylene [8]. A percentage-wise distribution of typical wastes found in municipal solid waste landfills across the USA is shown in Fig. 1 [2].

The pie chart represents a total of about 146.1 million tons of landfilled municipal solid wastes. Among them, organic wastes (food) have the largest share at 24% followed by plastics, paper, and metals at 18.46%, 11.78%, and 9.53%, respectively [2].

## 2.3 *Particle Size Distribution*

Particle size distribution (PSD) gives an outlook for an appropriate mechanical–biological treatment method and their degree of compressibility at the landfill dump-site. In a study, it was found that the most suitable shear shredders for reducing the particle size of municipal solid wastes were those with a 20 mm jaw opening. Any further reduction in jaw size led to clumping of the wet waste particles together to form a ball like structure [9]. PSD is calculated by determining the percentage by weight of particles across various size ranges. The waste is segregated by employing standard sieves having smaller mesh sizes at each level. A smaller PSD (>2 mm) in the organic-rich waste fraction directly influences its biodegradation pattern and helps them reach an advanced degree of compost maturity in a relatively shorter time period [10]. Table 2 shows the most common household or residential waste components and their size range [11].

**Table 2** Typical size range of waste fractions found in household wastes (in descending order) [11]

| Sl. No | Waste fraction | Size range (cm) |
|---|---|---|
| 1 | Cardboard | 20–59 |
| 2 | Paper | 16–46 |
| 3 | Wood | 3–35 |
| 4 | Plastics | 4–34 |
| 5 | Yard waste | 2–29 |
| 6 | Rubber | 3–26 |
| 7 | Textiles | 3–25 |
| 8 | Leather | 3–24 |
| 9 | Other metals | 3–23 |
| 10 | Glass | 3–20 |
| 11 | Tin cans | 3–18 |
| 12 | Food waste | 3–17 |
| 13 | Aluminum | 3–16 |
| 14 | Dirt, ash, etc. | 2–8 |

## 2.4 *Waste Density*

The density ($\rho$) of a sample is defined as mass ($m$) upon volume ($V$). It is useful in designing the layout of waste treatment facilities and landfills [1].

$$\text{Density} = \text{Mass}/\text{Volume}$$

Density is of two types:

- Bulk density ($\rho_b$)
- Material density ($\rho_m$).

The bulk density ($\rho_b$) is the wet weight of the waste packed in a given volume and is also known as apparent density or volumetric density. It is not an intrinsic property and can be altered depending on how the material is handled as it is strongly dependent on the applied pressure [1]. For example, a particular amount of waste, if compressed and packed in a smaller volume, its bulk density increases. Typically, the bulk density of household wastes range between 90 and 150 kg/m$^3$, but when compressed in a landfill, the bulk density may reach upwards of 1000 kg/m$^3$ [12]. The material density ($\rho_m$) is also known as specific density and is an intrinsic property of the material, and it denotes the amount of mass contained in unit volume and is usually expressed in kg/m$^3$ or g/cm$^3$. The material density of a few common materials has been listed in Table 3 [1].

**Table 3** Material densities of some common waste fractions [1]

| Sl. No | Density | Material density (kg/m$^3$) |
|---|---|---|
| 1 | Steel | 7700 |
| 2 | Iron | 5500 |
| 3 | Aluminum | 2700 |
| 4 | Glass | 2500 |
| 5 | Polyvinyl chloride (PVC) | 1250 |
| 6 | Paper | 700–1150 |
| 7 | Polystyrene | 1050 |
| 8 | High-density polyethylene (HDPE) | 960 |
| 9 | Polypropylene | 900 |
| 10 | Cardboard | 700 |

## 2.5 *Moisture Content*

The moisture content of waste is an important characteristic that determines the environmental conditions in the landfill. The wet weight or total weight of a waste sample is composed of the dry weight and the moisture content. In practice, it is determined by weighing the waste sample prior to and after drying at 105 °C (221°F) until constant weight is reached. As moisture is a volatile property and changes quickly, therefore it is tricky to calculate the moisture content as it is often altered upon reaching the landfill from the waste bin, owing to the fact that, a wide variety of wastes are transported together. For instance, a dry paper has a moisture content of about 7%; however, if it remains in close contact with wet food waste, its moisture content may rise up to 20% [12]. Moisture content can be calculated as:

$$\text{Moisture Content} = \text{Total weight} - \text{Dry weight}$$

Depending on the type of landfill, the preferred moisture content varies. For example, conventional landfills require minimum moisture, whereas landfill bioreactors require an increased level of moisture to accentuate the biodegradation potential. A list of common municipal solid waste components and their moisture content is given in Table 4 [13].

## 2.6 *Chemical Characterization*

It gives us an insight to the chemical composition of the waste, which can be helpful in segregating wastes of similar type and dumping them at appropriate places within the landfill. For instance, through chemical characterization, we can predict the chemical and biological reactions that may occur inside a landfill when wastes from different

**Table 4** Moisture content in typical MSW components (in descending order) [13]

| Sl. No | Components | Moisture (%) |
|---|---|---|
| 1 | Food waste | 70 |
| 2 | Grass | 60 |
| 3 | Yard waste | 60 |
| 4 | Leaves | 30 |
| 5 | Wood | 20 |
| 6 | Textiles | 10 |
| 7 | Leather | 10 |
| 8 | Fines (dirt, etc.) | 8 |
| 9 | Paper | 6 |
| 10 | Cardboard | 5 |
| 11 | Aluminum cans | 3 |
| 12 | Steel cans | 3 |
| 13 | Rubber | 2 |
| 14 | Plastic | 2 |
| 15 | Glass | 2 |

sources and varying properties are kept in close contact. Apart, from that, it allows the provision for identification of the inert and non-hazardous wastes, which can be recycled and incorporated in manufacturing processes [14].

## 2.7 *Organic Content*

Organic content or volatile solids provide an outlook of the decomposition processes occurring in a landfill. As the decomposition proceeds, there is a depletion in the organic content. The organic content value of a sample is obtained by determining the loss of dry matter upon ignition, the part that is left after ignition is known as the ash content. In practice, initially the samples are dried at 105 °C (221°F) and then subjected to ignition at 550 °C (1022°F) in a muffle furnace/oven until a constant weight is achieved, which usually takes about 1–2 h [15]. Another method to determine the organic content is by chemical oxidation with the help of potassium dichromate ($K_2Cr_2O_7$). This method is advantageous when organic contents of specific organic compounds are needed to be separately expressed. For example, it can express the organic contents of carbohydrate, acetic acid, protein, and fat [1].

### pH

pH plays an important role in the regulation of leaching in waste and is primarily dependent upon the transfer of acids and bases to and from the landfill site. The pH of a waste sample may be measured by dipping the waste in water, and then testing the resultant water suspension for its Hydrogen-ion activity using a pH meter. In general, the pH of leachate ranges between 4.5 and 9.0 [16]. In a new landfill site, the pH of leachate is around 6.5; however, as time passes, the concentration of volatile fatty acids is reduced which results in a higher pH (>7.5). Typically, an old landfill leachate ranges between 7.5 and 9.0 [17, 18]. Maintaining an optimum pH is essential for the biological degradation of waste in landfill, and care should be taken, so that the pH does not reach either extremes, being too basic (pH > 10) or too acidic (pH < 1) as under such conditions, the biological activity is drastically reduced [3]. Carbonic species, metal oxides, metal hydroxides, organic acids, and aluminum species are all pH buffers found in trash. If acids/bases are introduced to the landfill via rainfall, ground water, or leachate, the pH and buffer capacity may vary over time. Gas interactions and biological degradation may also have an impact on pH. The solubility of chemicals, biological activity, and pH may all be affected by redox relationships. When the redox potential is low, most metals form complexes with poor solubility. The redox potential influences both the breakdown of organic materials and the mobility of metals. Some examples of redox couples that act as buffer systems in waste include iron oxide/iron, nitrate/nitrite, sulfate/sulfide, and $CO_2$/acetic acid [3]. The implications of pH in a landfill setting are shown in Fig. 2 [3].

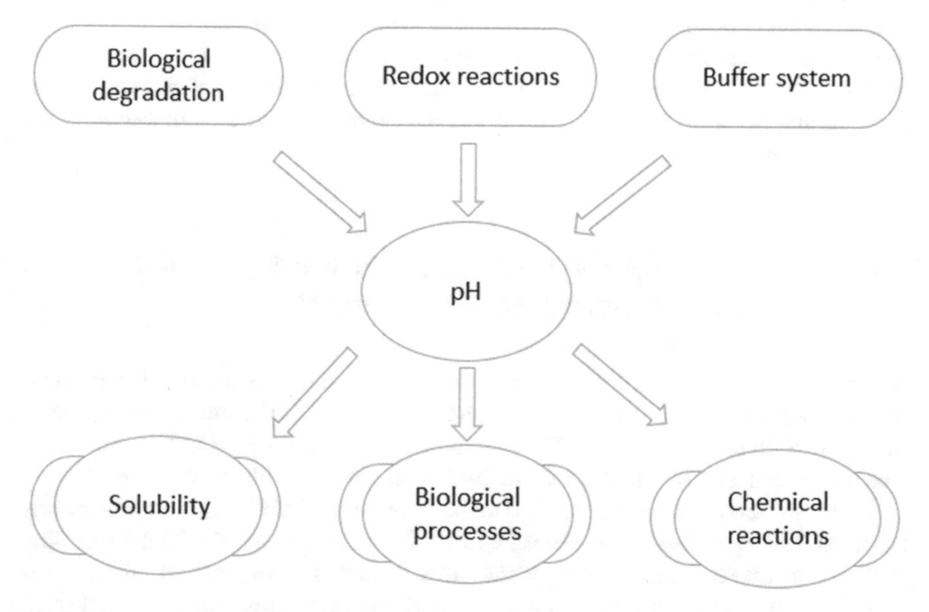


**Fig. 2** Factors affecting pH and their ramifications in a landfill site [3]

**Fig. 3** Schematic diagram of a municipal solid waste landfill

## 2.8 *Municipal Solid Waste (MSW) Landfills*

Municipal solid waste landfills are built in a secluded place away from human settlements, wetlands, and areas prone to flooding. They are the most common types of landfills and collect not only residential or household wastes but also non-hazardous sludge and industrial solid wastes. At present, there are 1,269, 1391, 1121, 604, 468, 221, and 191 landfills in the USA, France, Germany, United Kingdom, Spain, Sweden, and Austria, respectively [2, 19]. A typical layout of a municipal solid waste landfill is shown in Fig. 3.

## 2.9 *Quantity of Different Components in Municipal Solid Waste Landfills from 1960–2018 in the USA*

A sharp increase in the amount of waste dumped in landfills is observed from 1960 to 1990 followed by an overall slow decline from 1990 to 2010 and finally a gradual rise from 2010 to 2018 (Fig. 4). The amount of waste dumped in landfills increased from approximately 90 million tons in 1960 to about 150 million tons in 2018.

In the past few decades, the increase in the generation of MSW has been enormous. Between 1960 and 1990, the generated MSW had more than doubled from 88.1 million tons to ~200 million tons, which further increased up to 292.4 million tons in 2018. The generation rate of municipal solid waste fraction has increased from 2.68 pounds/person/day in the 1960s to 4.9 pounds/person/day in 2018 (Fig. 5).

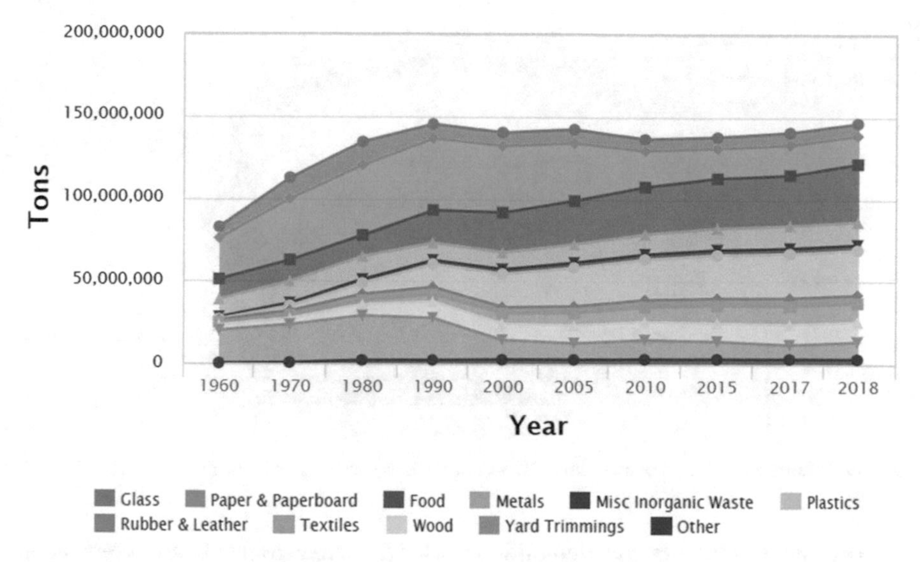


**Fig. 4** Waste fraction trends in MSW landfills from 1960–2018. Adopted from US EPA [2]

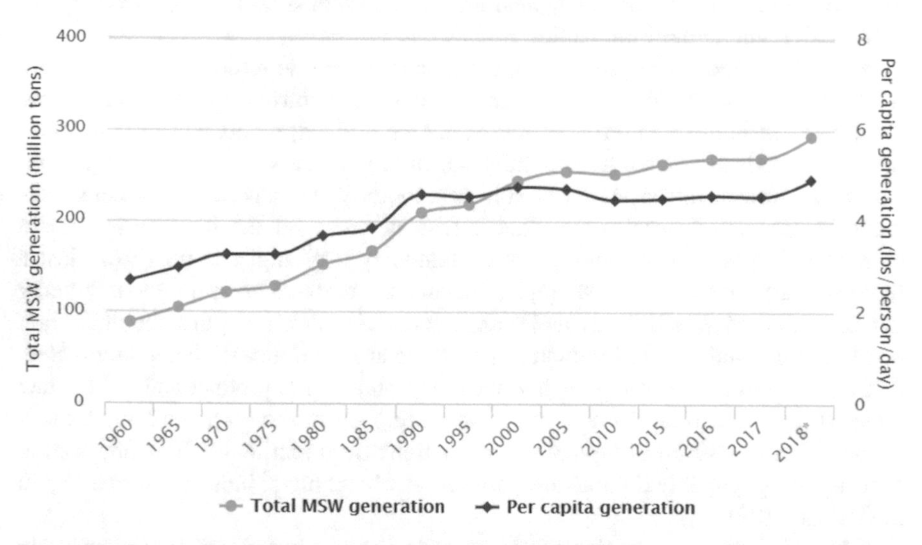


**Fig. 5** Municipal solid waste generation rates from 1960–2018. Adopted from US EPA [2]

## 2.10 *Industrial Waste and Construction and Demolition Landfill*

Industrial waste landfills are aimed to aggregate solid, non-hazardous waste generated from various industries such as plastics; textiles; chemicals; steel and metals; paper; processed food and beverages manufacturing industries. Apart from these, they

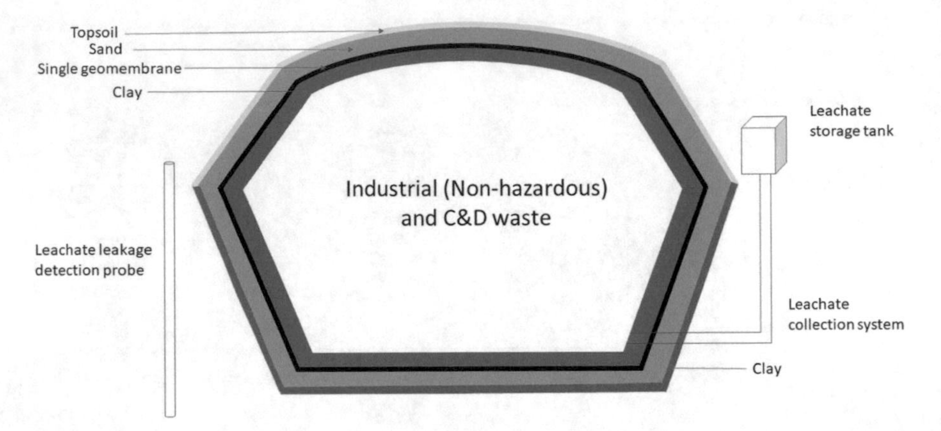


**Fig. 6** Schematic diagram of an industrial (non-hazardous) and construction and demolition landfill

also collect construction and demolition wastes that are generated during repair and remodeling of buildings, roads, and bridges which includes wastes such as concrete, asbestos, wood, and glasses. A typical layout of an industrial (Non-hazardous) and construction and demolition landfill is shown in Fig. 6.

Waste from industrial production and manufacturing is referred to as industrial waste. Industry encompasses a wide range of manufacturing sectors, each having significant differences in terms of raw materials utilized, manufacturing processes used, and goods produced. Additionally, depending on market potential and pricing, a portion of industrial waste may be sold as secondary raw materials. Another significant feature of industrial waste data is that, in order for the industry to remain competitive, industrial technology is continuously evolving, and data from just a few years ago may not always apply. Production methods are continuously being tweaked to improve efficiency and reduce material waste, including the utilization of semi-fabricated goods. Production processes are continuously being automated, and new materials are always being added to enhance the product and reduce raw material costs, resulting in new materials ending up in the waste stream. Table 5 shows the composition of typical waste fractions from textile; clothing and leather industry; pulp; paper and publishing houses; steel and metal industry from a region in Germany [20].

Often it is convenient to classify the construction and demolition waste separately as they can be recycled and incorporated into ongoing civil engineering projects [21]. The most common C&D waste dumped in the landfills are listed in Table 6 [2], and their percentage-wise breakdown is shown in Fig. 7 [2].

Concrete had the maximum fraction at 50% followed by wood, gypsum, asphalt shingles, brick and clay tile, asphalt concrete, and metal at 21%, 9%, 9%, 7%, 3%, and 1%, respectively.

**Table 5** Composition of waste fraction in various industries [20]

| Composition (%) | Textile, clothing, and leather industry | Pulp, paper, and publishing houses | Steel and metal industry |
|---|---|---|---|
| Organic waste | 30 | 19 | 12 |
| Paper and cardboard | 40 | 47 | 20 |
| Plastic | 0 | 9 | 27 |
| Iron and metal | 0 | 6 | 21 |
| Glass | 0 | 2 | 2 |
| Wood | 0 | 1 | 10 |
| Construction and demolition | 0 | 2 | 3 |
| Other | 30 | 14 | 5 |

**Table 6** Composition of construction and demolition debris in landfill [2]

| Sl. No | Construction and demolition debris | Landfill (million tons) |
|---|---|---|
| 1 | Concrete | 71.2 |
| 2 | Wood | 29.6 |
| 3 | Gypsum | 13.2 |
| 4 | Asphalt shingles | 13 |
| 5 | Brick and clay tile | 10.8 |
| 6 | Asphalt concrete | 4.9 |
| 7 | Metal | 1.1 |
| | Total | 143.8 |

## 2.11 *Hazardous Waste Landfills*

Hazardous wastes consist of highly toxic and reactive substances and are usually containerized in 55-gallon drums before being dumped into landfills. They are engineered to properly contain them and prevent any kind of leakage. Some of their unique features include double composite liners (geomembrane), double leachate collection and removal system, and an impermeable cap which prevents precipitation into the landfill as well as restricts waste movement inside the landfill. A typical layout of a hazardous waste landfill is shown in Fig. 8.

USA primarily imports hazardous waste from Canada, Taiwan, Brazil, Mexico, Belgium, New Zealand, Holland, and Malaysia. In 2019, the USA imported a total of 64,864 tons, out of which, 11,902 tons were disposed in landfills (18.34%). Apart from this, the USA also exports hazardous waste to foreign countries. In 2019, it exported a total of 1,307,643 tons of hazardous waste, out of which 142,624 tons were disposed in landfills (10.90%), the rest of the waste (~90%) was subjected to special treatments such as metals recovery, incineration, organics and inorganics

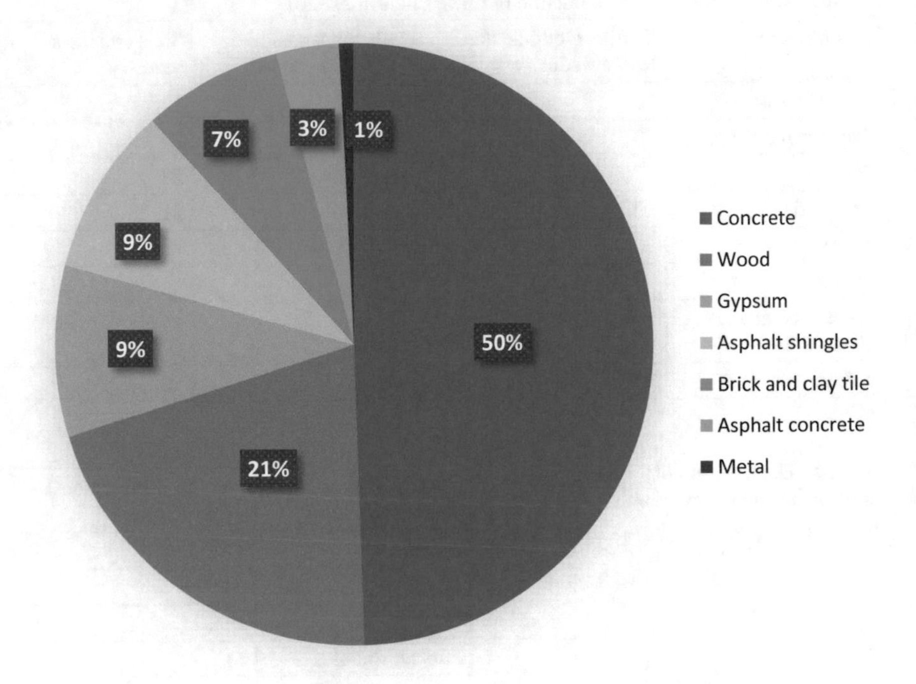


**Fig. 7** Percentage-wise distribution of different components in C&D landfill [2]

recovery, and sludge treatment/stabilization [2]. A percentage-wise breakdown of exported hazardous wastes to major countries in 2019 is depicted in Fig. 9 [2].

**Fig. 8** Schematic diagram of a hazardous waste landfill

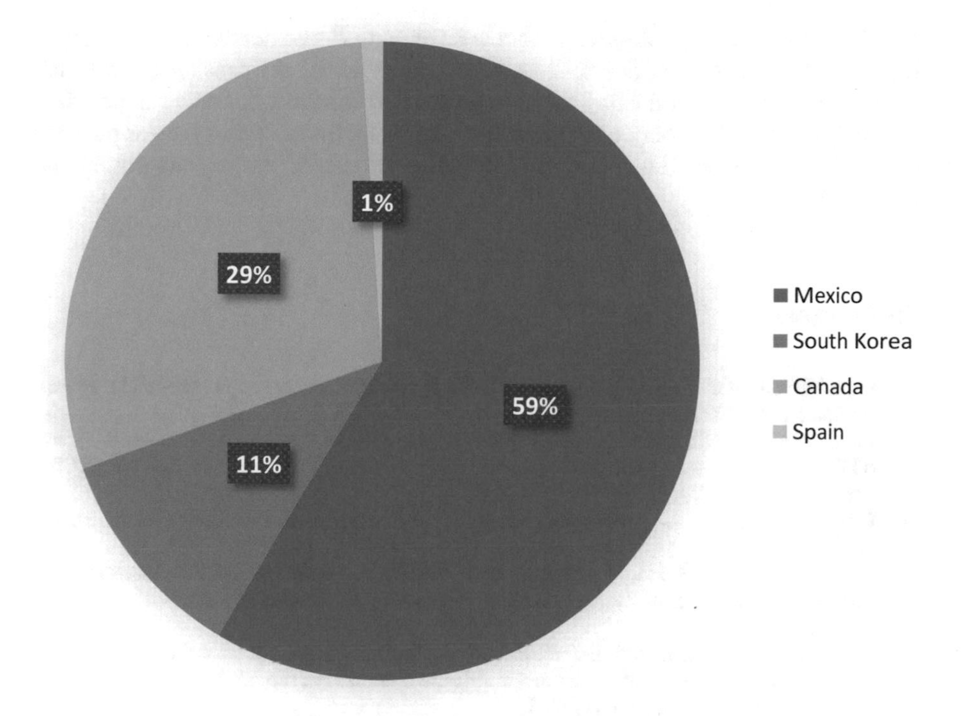


**Fig. 9** Hazardous waste exported to major foreign countries by USA in 2019 [2]

Mexico had the largest fraction at 59% followed by Canada, South Korea, and Spain at 29%, 11%, and 1%, respectively [2].

## 3 Conclusion

This chapter elaborates on the components of physical and chemical characterization of waste such as particle size distribution, waste density, moisture content, organic content, and pH. The challenges in waste characterization have also been dealt with briefly, due to spatial and temporal variations it is difficult to characterize waste with a high degree of accuracy. In addition, a contrast between the typical wastes found in different landfills viz. municipal solid waste (MSW) landfills, industrial, construction, and demolition landfills (IC&D), and hazardous waste landfills (HW) have been presented. Furthermore, the major differences between the three types of landfills have been illustrated using their schematic diagram, for example, the MSW landfills are unique in the sense that they have a landfill gas storage tank owing to the huge percentage of organic waste in them. The IC&D landfills have a single geomembrane layer and a single leachate collection and removal system, whereas

HW landfills have a double geomembrane layer as well as a double leachate collection and removal system due to their increased potential for causing harm to the environment and contaminating the groundwater. In conclusion, waste characterization helps in the designing of appropriate landfill facilities and in the formulation of standard protocols to be observed during the process of waste management and remediation.

## References

1. Lagerkvist, A., Ecke, H., Christensen, T.H.: Waste characterization: approaches and methods. In: Solid Waste Technology and Management, pp. 63–84. Wiley (2011)
2. U.S. Environmental Protection Agency (EPA).: Office of Solid Waste and Emergency Response (OSWER); National Capacity Assessment Report: Capacity Planning Pursuant to CERCLA Section 104(c)(9), Dec (2019)
3. Ecke, H., Bergman, A., Lagerkvist, A.: Waste Characterisation. Luleå tekniska universitet (1999)
4. la Cour Jansen, J., Spliid, H., Hansen, T.L., Svärd, Å., Christensen, T.H.: Assessment of sampling and chemical analysis of source-separated organic household waste. Waste Manage. **24**(6), 541–549 (2004)
5. Boadi, K.O., Kuitunen, M.: Environmental and health impacts of household solid waste handling and disposal practices in third world cities: the case of the Accra Metropolitan Area, Ghana. J. Environ. Health **68**(4) (2005)
6. Dounias, G., Rachiotis, G.: Prevalence of hepatitis A virus infection among municipal solid-waste workers. Int. J. Clin. Pract. **60**(11), 1432–1436 (2006)
7. International Labour Office.: Addressing the Exploitation of Children in Scavenging (Waste Picking): A Thematic Evaluation of Action on Child Labour (2004)
8. Gutberlet, J., Uddin, S.M.N.: Household waste and health risks affecting waste pickers and the environment in low-and middle-income countries. Int. J. Occup. Environ. Health **23**(4), 299–310 (2017)
9. Zhang, Y., Kusch-Brandt, S., Gu, S., Heaven, S.: Particle size distribution in municipal solid waste pre-treated for bioprocessing. Resources **8**(4), 166 (2019)
10. Lopez, R., Hurtado, M.D, Cabrera, F.: Compost properties related to particle size. WIT Trans. Ecol. Environ. **56** (2002)
11. Tchobanoglous, G., Theisen, H., Vigil, S.: Integrated Solid Waste Management: Engineering Principles and Management Issues. McGraw-Hill (1993)
12. Vesilind, P.A., Worrel, W.A., Reinhart, D.R.: Solid Waste Engineering. Brooks/Cole, Pacific Grove, USA (2002)
13. Reinhart, D.R., Townsend, T.G.: Landfill Bioreactor Design & Operation. CRC Press (1997)
14. Borghi, G., Pantini, S., Rigamonti, L.: Life cycle assessment of non-hazardous Construction and Demolition Waste (CDW) management in Lombardy Region (Italy). J. Clean. Prod. **184**, 815–825 (2018)
15. Eaton, A.D., Franson, M.A.H.: Standard Methods for the Examination of Water and Wastewater. American Public Health Association, American Water Works Association, Water Environment Federation, Washington, Denver, Alexandri (2005)
16. Christensen, T.H., Kjeldsen, P., Bjerg, P.L., Jensen, D.L., Christensen, J.B., Baun, A., Albrechtsen, H.J., Heron, G.: Biogeochemistry of landfill leachate plumes. Appl. Geochem. **16**(7–8), 659–718 (2001)
17. Bohdziewicz, J., Kwarciak, A.: The application of hybrid system UASB reactor-RO in landfill leachate treatment. Desalination **222**(1–3), 128–134 (2008)

18. Kulikowska, D., Klimiuk, E.: The effect of landfill age on municipal leachate composition. Biores. Technol. **99**(13), 5981–5985 (2008)
19. Eurostat, portfolio of the European Commissioner for the Economy, pursuant to Regulation (EC) No 223/2009; Number and capacity of recovery and disposal facilities by NUTS 2 regions. Online data code: ENV_WASFAC, Feb (2021)
20. Kranert, M.: Waste from Service Sector and Industry—Waste Type, Amount and Composition Related to the Sector Type; in German. Rhombos-Verlag, Berlin, Germany
21. Bianchini, G., Marrocchino, E., Tassinari, R., Vaccaro, C.: Recycling of construction and demolition waste materials: a chemical–mineralogical appraisal. Waste Manage. **25**(2), 149–159 (2005)

# Assessment and a Case Study of Small-Scale Incinerators for Municipal and Agriculture Waste Disposal in Rural Regions

**Rosalam Sarbatly and Jamilah Sariau**

**Abstract** The absence of solid waste collection and disposal systems in rural areas poses a threat to the environment. Centralised landfills are less practical than the decentralised system for rural areas due to the high cost of garbage collection, transportation, and landfill management. To date, no comprehensive solution exists to the problem of solid waste disposal in most rural areas. Meanwhile, municipal and agricultural waste piles become breeding grounds for rats, cockroaches, flies, mosquitoes, pests, and microorganisms that spread diseases, causing the surrounding areas to become dirty and smelly. Based on these problems, small-scale incinerators using local materials with a focus on municipal and agricultural waste have been evaluated and built to be cost-effective, sustainable, and environmentally-friendly in managing waste disposal. The study found that small-scale incinerators could potentially be used as a method of municipal and agricultural solid waste management in rural areas. It is hoped that the use of small-scale incinerators can help rural communities to manage household and agricultural waste more effectively. However, the concept of reuse, reduce, and recycle (3R) will continue to be on the agenda to foster awareness among the community.

**Keywords** Small-scale incinerator · Agriculture waste · Household waste · Rural solid waste management

## 1 Introduction

In rural and small township areas, the local government authority is managing and collecting solid waste. The main issue is for larger states such as Sabah, Sarawak, Pahang, and Perak in Malaysia where districts are so vast with low population densities. Villages are also far apart from each other. If the municipal solid waste (MSW) collection system is fully implemented, solid waste management requires a high cost. The issue of solid waste disposal in small towns and villages poses a significant

---

R. Sarbatly (✉) · J. Sariau
Faculty of Engineering, Universiti Malaysia Sabah, Kota Kinabalu, Malaysia
e-mail: rslam@ums.edu.my



A. Z. Yaser et al. (eds.), *Waste Management, Processing and Valorisation*,
https://doi.org/10.1007/978-981-16-7653-6_4

challenge to the local government to improve the agriculture and municipal solid waste management system.

Due to the challenges faced, the existing municipal solid waste management practices in small towns are not going well. Municipal solid waste collected from commercial, residential, and industrial premises has been frequently discarded openly in areas without standard landfill disposal practices. There are also garbage disposal activities in hidden areas that are difficult to be controlled by the authorities.

For rural areas, local council authorities do not collect garbage due to scattered and large settlements. Villagers themselves only manage garbage. Given low awareness of waste disposal, garbage containing plastics and hazardous materials would be burned openly or dumped to areas near the house. For those who live by the river, garbage is discarded into it. For those who live in shoreline areas, garbage is dumped under their house. For those who live in hilly areas, garbage is thrown in the gorge areas or burned openly. Agricultural waste is left piled up on the road shoulders becomes breeding grounds for rats, cockroaches, and flies and causes the area to become dirty and smelly. In addition, diseases related to crops and livestock are also easily spread through unsystematic waste disposal.

Based on the challenges and problems faced by the authority in managing solid waste disposal in small and rural areas are different compared to the management problems in large cities; methods and practices of solid waste management in urban and rural areas must be differentiated so that successful solid waste management in rural areas can be achieved.

## 2 Waste Categories and Open Burning

Open burning is the burning of materials in surrounding areas such as on agricultural farms and backyards. Open burning is characterised by combustion at low temperatures between 250 °C and 700 °C and in an oxygen-deprived environment, resulting in incomplete waste combustion. In the Malaysian Act A1030, Environmental Quality (Amendment) Act 1998, "open burning" means any fire, combustion, or smouldering that occurs in the open air and which is not directed there through a chimney or stack but does not include any fire, combustion, or smouldering that occurs for such activities as may be prescribed by the Minister by order published in the Gazette [1].

The types of open burning discussed in this evaluation are fires: (1) caused by anthropogenic activity, which means environmental pollution caused by human activities; and (2) deliberately arranged to dispose non-hazardous waste by open burning. This category does not include combustion in specialised combustion equipment and accidental fires, such as forest fires or fires caused by natural disasters. The subcategory of open burning discussed here involves residential, municipal solid waste, and agricultural waste in rural areas.

Municipal solid waste is non-hazardous waste produced by households. Municipal solid waste covers paper, plastics, glass, textiles, rubber, wood, metals, leather, and food waste. Agricultural waste is unwanted or unsellable materials produced entirely from agricultural operations associated with crop production or livestock breeding for the primary purpose of earning profit or as a source of income.

In Malaysia, the Environmental Quality Act (EQA) provides the most comprehensive regulation of pollution and abatement. Under this act, anyone who allows or causes open burning on properties, including land, commits an offence [2, 3]. Open burning of residential, municipal solid waste is primarily alarming in rural areas, where burning is perceived as an easier or cheaper option to landfilling [4].

## 3 Small-Scale Incinerator

The World Bank (1999) reports on the decision makers' guide to municipal solid waste incineration mentioned that "the municipal solid waste incineration plants tend to be among the most expensive solid waste management options. They require highly skilled personnel careful maintenance. For these reasons, incineration tends to be a good choice only when other, simpler, and less expensive choices are not available" [5].

This statement refers to the large-capacity municipal waste incineration system equipped with an energy recovery system as the last option for municipal waste management due to high construction and operating costs and operating systems that require high technical skills to ensure smooth operation. Since municipal waste incineration systems are high capacity with energy harvesting systems, it is uneconomical if operations are carried out in batch mode; instead, it has to run continuously to operate at optimal cost.

However, in rural areas surrounded by forests and natural beauty with a scattered population distribution without any waste management system where properly managed landfills are not available, open waste and unplanned and not well-managed landfills have caused land and water pollution through leachate. The practice of open dumping and burning of garbage especially for waste categories such as plastics must be eliminated. If open burning of plastics is acted upon, toxins such as dioxins (including PCBs), hydrogen cyanide, furan, hydrochloric acid, and sulfuric acid can be produced. The possible types of toxins are many, and it depends on the type of plastic being burned. The golden rule for households is not to burn plastic at home. Therefore, systematic burning through the usage of small-scale incinerators is expected to best choice.

An incinerator is an apparatus for burning waste materials at high temperatures until it is reduced to ash. Incineration involves burning waste at high temperatures in the range of 750–1100 °C in the presence of oxygen to decrease the weight and volume of the waste and generate heat and energy [6]. This method can reduce waste mass by almost 70% and volume by up to 90% [7].

There are many types and uses of small-scale incinerators. Small scale is generally defined as limited or restricted combustion quantity. Therefore, each small-scale incinerator has a manageable waste capacity limit for incineration. For example, a modular incinerator with a prefabricated unit having a small capacity of between 5 and 120 tons of solid waste per day is equivalent to 12–100 kg per hour. In addition, medical waste incinerators are set at a maximum of 15 kg per hour which is much lower than modular incinerators. Small-scale incinerators have a small capacity compared to large-scale incinerators, with a capacity of up to 3000 tons per day.

Small-scale incinerators can also decrease the volume of non-recyclable waste generated from households, medical operations, slaughterhouses, etc. Also, it generates heat and energy that may be recovered and helps evade the open burning of municipal waste, which contributes to toxic emissions that threaten human health and environment. A systematically constructed waste incinerator system must use inexpensive local technology and can be built by local contractors, is easy to operate, low operating and maintenance costs and incineration can occur periodically or continuously with no significant differences in handling.

A locally made small-scale incinerator has a great potential to solve the issue of solid waste dumping in islands and rural regions apart from recycling and reuse practices and composting of organic waste. The success or failure of the incineration scheme depends on the attitudes of various stakeholders and the current legal and institutional framework [5]. Stakeholders' stakes in solid waste incineration plant projects often have different interests. Therefore, this project can be an environmental and economic issue for many groups. The small-scale incineration system built must also meet the standards set by the local authority. Based on that need, guidelines for the construction of a small-scale waste incineration system specifically for garbage must be developed by the local authority, which will involve the environmental department and the local government.

## 4 Small-Scale Municipal Waste Incinerator

A more versatile small-scale incinerators are a valuable technology to combust municipal waste rather than discharging them into landfills. Moreover, it allows for heat and energy recovery and avoids the open burning of municipal waste, releasing toxic releases that jeopardise human health and pollute the environment. It can be built as permanent in one position or moved by placing them on a truck.

A small-scale municipal waste incinerator being built in Beaufort, Sabah, aims to develop local incinerator technology using local expertise and materials. The research was funded by the Research Center of Universiti Malaysia Sabah in response to the disposal and incineration of waste that does not follow solid waste disposal management. Figure 1 shows the fixed small-scale incinerator built in Beaufort, Sabah.

Small-scale incinerators for household and clinical waste should be designed with recommended operating parameters that follow standards. It is typically equipped with waste feed gates, combustion chambers (primary and secondary), cyclones, air blowers, and chimneys, and the general operating principle consists of filling solid waste in the combustion chamber in a batch process.

**Fig. 1** Fixed small-scale incinerator built in Beaufort, Sabah

The combustion capacity must meet all waste's combustion requirements, especially with high water content and lower heating value (LHV). The lower heating value (also known as the net calorific value) of a fuel is defined as the amount of heat released by burning a specified quantity (initially at 25 °C) and returning the temperature of the combustion product to 150 °C, which assumes latent heat of water vaporisation in the reaction product does not recover [8].

Specifically, the combustion temperature must be high, and the design of the combustion chamber is correct to reach the desired absolute combustion temperature. The main chamber of a small-scale incinerator should have a temperature range of 540–980 °C, while for secondary chambers with a range of 980–1200 °C to follow the recommendations of the Environmental Protection Agency (EPA) [9]. The operating temperature of the incinerator must be in the range of 800–1200 °C when medical waste is incinerated.

Furthermore, effective thermal isolation measures and adequate retention times are required to reduce non-combustible materials. Additionally, gas temperature, oxygen content, and retention time must fulfil the requirements of environmental protection regulations [10]. The operating temperature for small-scale incinerators must exceed 750 °C, and minimal retention time for produced flue gases must be less than 2 s. It is because organic base materials such as household and farm waste burn at that temperature. A complete combustion reaction will be faster if the temperature is set higher and the retention time is longer. At temperatures below 600 °C, toxic fumes such as furans and dioxins are emitted if polyvinyl chloride or other materials are burned [11].

## 5 Fabrication of Heat-Resistant Combustion Chamber Wall

One of the challenges in constructing waste incineration chamber walls is expensive building materials especially mortar and heat-resistant bricks that can withstand long periods and high-temperature operation and long lasting.

A simple method is to make a layer of heat insulation at the internal wall of the combustion chamber. Clay is the material most ideal to use. A layer of clay is pasted on the inner wall about 6 inches thick. An arrangement technique of regular red bricks is required to allow the clay to be easily pasted on the inside to avoid falling. The process of clay pasting in the inner wall must be repeated at least three times, burning and pasting, until crack's formation is no longer observed. Figure 2 below shows the crack that needs to be resurfaced and the burning process.

## 6 Operation of a Small-Scale Incinerator

The small-scale incineration system contains the main parts, namely the first combustion chamber, second combustion chamber, suction fan, carbon filter, cyclone, and chimney, as shown in Fig. 3.

The first combustion chamber was built using red brick and pasted with clay, as explained earlier. It is equipped with a feeding door, forced air supply, and metal tray to segregate the waste for better burning efficiency, and a manhole for maintenance purposes. Figure 4 shows the detail of the design of the first combustion chamber.

The second combustion chamber is metal, covered with fibreglass thermal insulation, and layered with aluminium sheets. The second combustion chamber should be at a high temperature above 850 °C. Temperature can be increased with the help of LPG gas sprayed into the chamber. Figure 5 shows the second combustion chamber with ceramic monolith fitted inside and can be cleaned when necessary.

Ceramic monoliths are placed in the aisle of the second combustion chamber, which aims to retain heat and eliminate incomplete combustion and toxic gas emissions. The use of ceramic monoliths in the second chamber can reduce the use of LPG gas as it can maintain a high temperature and act as a catalyst to enhance the combustion process. The presence of ceramic monoliths acts as a catalyst similar to those used in high-powered vehicle exhaust systems.

Next is the suction fan. This fan is essential to allow the first and second combustion chambers to always be in a state of negative pressure. The negative pressure

**Fig. 2** Brick arrangement, pasting the clay, and hardening process for combustion chamber wall. **a** the brick arrangement, **b** cracking of clay pasted on the internal surface of the combustion chamber, and **c** firing to harden the clay

will allow outside air to enter the first combustion chamber into the second chamber without smoke leakage, and the suction fan pushes the smoke into the cyclone, which will produce a positive pressure there. By using the cyclone effect, larger particles will descend to the bottom of the cyclone, and the lighter flue gas will move up through the nanocarbon fibre filter and out of the chimney. In the inner wall of a cyclone, a sheet of carbon fibre is used as a liner to prevent corrosion when particles

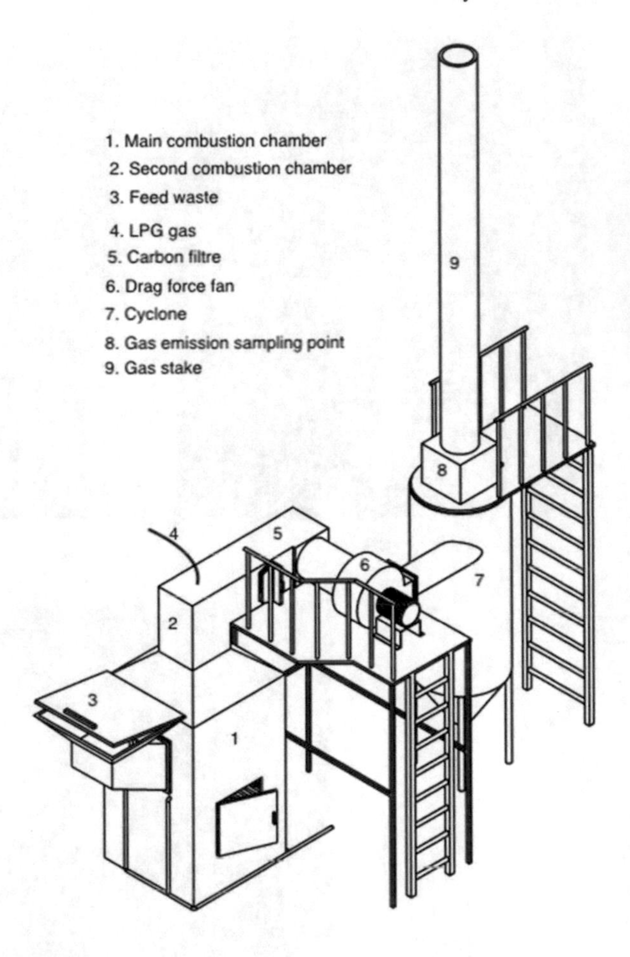


**Fig. 3** Diagram of the small-scale incinerator showing the components for an incinerator

continuously hit the wall to create a cyclone effect. Figure 6 shows the cyclone and chimney.

As a safety measure, combustion operation and testing are performed by simply burning wood and grass to determine the temperature reached the point of the carbon fibre sheet in the chimney. Preliminary observations found no black smoke on the chimney throughout the combustion process. A future test is needed to measure the air suction rate, retention time, pressure drop, temperature changes, and the composition of the smoke in the chimney before the incinerator can be tested to transform the household waste with lower heating value waste.

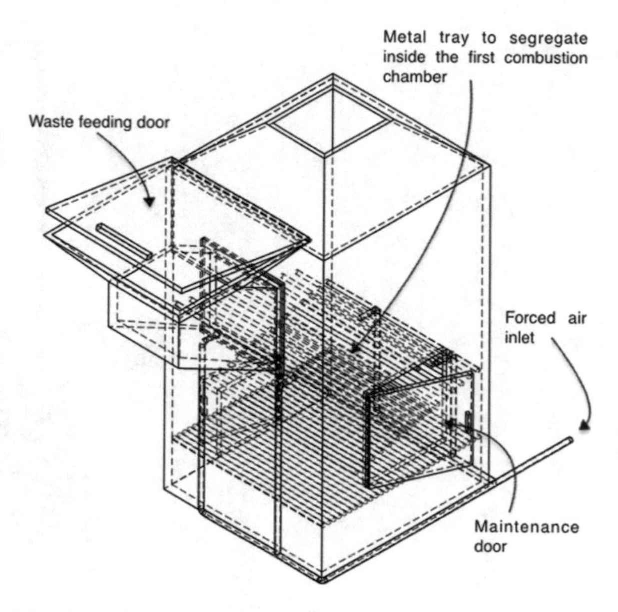


**Fig. 4** Detail of the design of the first combustion chamber

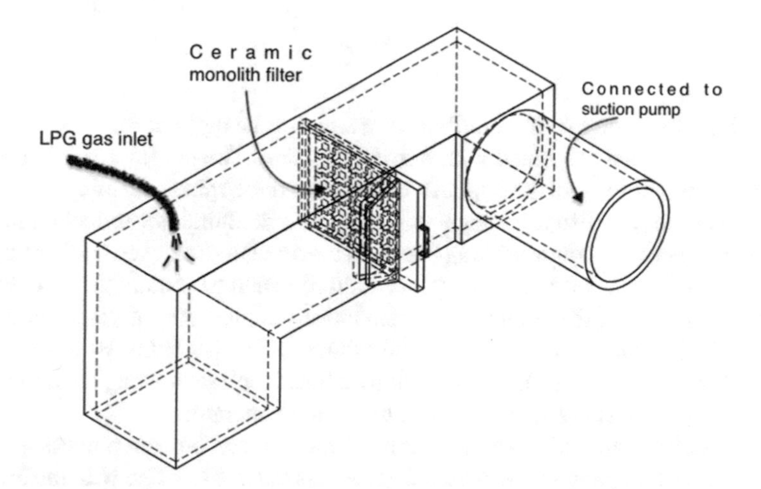


**Fig. 5** Second combustion chamber with ceramic monolith filter

# 7 Small-Scale Agriculture Waste Incinerator

Good agricultural waste management is critical because it aims to prevent the accumulation of agricultural waste into breeding grounds for animals and pests such as rats, flies, mosquitoes, snails, and cockroaches as well as to prevent the production of foul odour due to the process of decomposition of organic matter by bacteria.

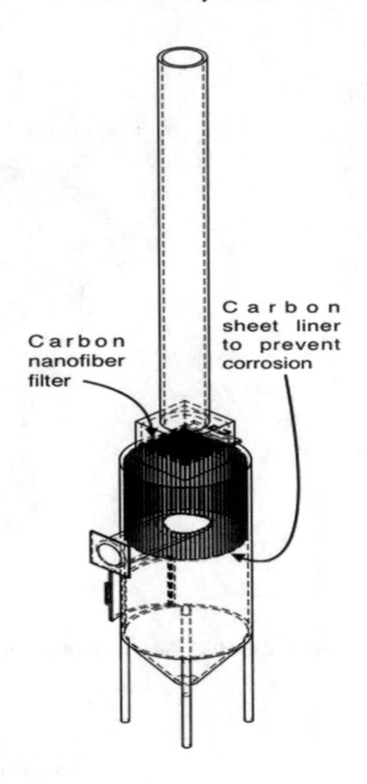


**Fig. 6** Cyclone and chimney

The first-generation incinerator built is a small-scale agricultural waste incineration chamber device combined with a drying system. The development is intended as a complementary tool to support innovative rural mixed farming enterprises involving poultry, catfish, and vegetable crops. Agricultural waste incinerators are needed to dispose of compostable agricultural waste such as chicken feathers, spoiled vegetables, and grass to reduce waste accumulation and to contain the disease.

A trial version of the small-scale incinerator coupled with drying storage was designed, built, and commissioned in the village of Pilajau Baru, Beaufort, Sabah, in 2019. It was designed locally and built by a local contractor using local materials. Figure 7 shows the small-scale agriculture waste incinerator.

The capacity is around 25–50 kg per hour, and the combustion process is carried out in batches, for example, every two days or less than 48 h in line with the breeding period of insects especially flies that usually breed on piles of agricultural waste. The design is based on the concept of easy operation, low construction and operating costs, environmentally-friendly by avoiding open burning, using green technology such as solar energy sources to generate electricity to pump air, using natural gas from sewage waste to help the combustion process, and make full use of agricultural waste to operate.

The drying chamber is also used as a temporary storage place for agricultural waste produced before the burning process, which is done periodically. Storage in this drying chamber also allows the water content of agricultural waste to be reduced

**Fig. 7** Small-scale agriculture waste incinerator

to facilitate the combustion process. The resulting combustion ash residue is to be stabilised before being used as a base material to produce organic fertiliser.

The locally made agricultural waste incinerator is not suitable for household waste containing plastic waste and lower heating value. A single combustion temperature is expected to be low, around 550–800 °C; therefore, there is no guarantee that the plastic waste and the waste with a meagre heating value will be burned entirely into inert gases. The flow of air blown into the chamber produces a positive pressure. Therefore, some smoke will come through the waste entry door. The smoke coming out from the chimney is only based on the principle that heavy air will move up and out through the chimney.

Furthermore, the ash residue should be reused as the primary material to produce organic fertiliser. However, hazardous material like heavy metals in household solid waste makes combustion ash unsuitable to be used as fertiliser because it contains hazardous waste that can be harmful if used as a fertiliser base material. However, the advantage of this incineration system is that it is easy to build and operate. The agricultural waste incineration chamber is also coupled with an agricultural waste drying chamber to reduce the water content before the incineration process.

Figure 8 shows the detailed design of an agricultural waste incinerator equipped with a drying chamber. The description of the figure is accompanied in Table 1.

## 8 Waste Storage and Drying System

Figure 9 shows the heating element unit responsible for delivering heat to the dryer system. The air blower is installed in the space between the heating coil and the metal plate. Blowing hot air is directed to the bottom of the trash can. Hot air that

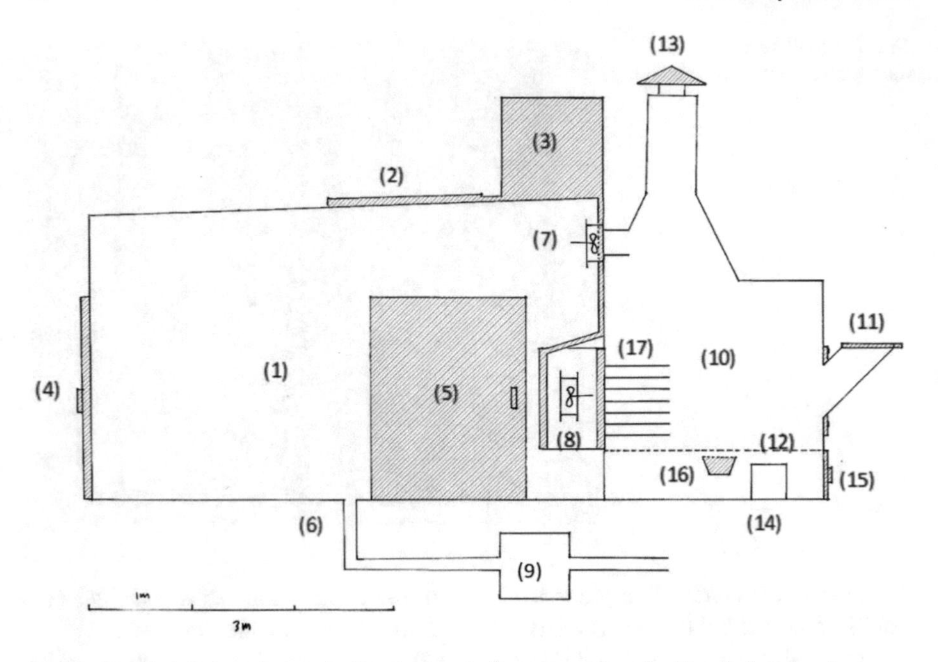


**Fig. 8** Detailed design of an agricultural waste incinerator equipped with a drying chamber

carries moisture is removed immediately through a convection process to increase the drying rate.

The heating element for a storage dryer consists of two parts. The first part is a copper pipe, while the second is a plate and a metal rod. The first part of the heating element is connected to a hot water tank that stores the hot water generated from the solar water heater. Therefore, even if incineration does not operate, the drying process of agricultural waste continues. The heating element is copper pipe due to its high thermal conductivity compared to other standard metal pipes. The high thermal conductivity ensures rapid heat transfer from the hot water through conduction and convection from the copper pipe to the drying chamber assisted by a blower.

Part of the second heating element can be a metal alloy such as carbon steel or stainless steel. Thermal conductivity is not a problem because the heat supplied from the main chamber of the burner is very high at a temperature of 750–850 °C. Copper pipes or copper rods are recommended for better drying due to their melting point and higher thermal conductivity. Using the latest technology known as heat superconductors such as heat pipe heat exchangers will further increase the heat transfer rate. A heat pipe heat exchanger is a tool used to transfer heat from one location to another, using a condensation–evaporation cycle. Heat pipes are called heat "superconductors" because of their fast transferability and low heat loss.

**Table 1** Description as from Fig. 8

| Points | Description |
|---|---|
| 1 | *The Storage Dryer Area*<br>– Space for the large waste bin containment and drying |
| 2 | *Solar Water Heater*<br>– Solar water heater unit is used to heat water using the energy from the sun |
| 3 | *Hot Water Tank*<br>– Heated water from the solar water heater is stored in this tank |
| 4 | *Storage Dryer Service Door 1*<br>– Serves as the inlet door for the large waste bin into the storage dryer |
| 5 | *Storage Dryer Service Door 2*<br>– Serves as the outlet door for the large waste bin |
| 6 | *Leachate Drain Hole*<br>– Drain holes for the leachates sourced from the waste or washed water resulted from cleaning the storage floor. It will be drained into sewage treatment |
| 7 | *Air Ventilation Blower*<br>– Vent moisture and air into the incinerator system, which is then released to the atmosphere from the chimney (13) |
| 8 | *Heating Elements*<br>– Hot heating element consists of:<br>• Hot pipes regulated from the hot water tank (3);<br>• Hot metal rods in which heat is sourced from incineration process within the incinerator chamber (10); and<br>• A hot air blower is used to transfer heat to the storage dryer from the heating elements |
| 9 | *Septic Tank*<br>– Leachate treatment tank. Can opt to be released into the drain |
| 10 | *Solid Waste Incinerator Area*<br>– The area is dedicated to the solid waste incineration process |
| 11 | *Incinerator's Chute Chimney*<br>– An inlet chimney connects to the incineration chamber for the solid waste<br>– Fixed-size is an effort to regulate the sizes of the materials that enter the incinerator |
| 12 | *Incinerator's Mesh Grill*<br>– The incinerator's mesh grill is where the solid waste resides with the system |
| 13 | *Incinerator's Chimney*<br>– The chimney acts as a ventilation pathway from the storage dryer system (1)<br>– Vent flue gas out of the incinerator system (10) to the environment |
| 14 | *Service and Air Hole for the Incinerator*<br>– The service hole can be used to initiate fire within the incineration chamber<br>– Fire can be started manually using matches and some fuels<br>– Fire starting system can also be installed within the incinerator system from this hole depicted as no. (16) (The fire-starting system can be optionally installed with the incinerator in respect to manual fire-starting method. The system uses a gas stove concept to initiate combustion, which can be turned off during the incineration process)<br>– This hole is also used as an air hole that supplies air within the incineration chamber. An air blower can be optionally installed to improve the incineration process |

(continued)

**Table 1** (continued)

| Points | Description |
|---|---|
| 15 | *Incinerator Ash Service Door*<br>– A service door is used for ash and non-combustible material collection purposes |
| 16 | *Fire-Starting System*<br>– A fire-starting system can be optionally installed at this point for ease of operation. Refer to point (14) for more information |
| 17 | *Heating Element Rods*<br>– The rods are essential to provide heat to the storage dryer area. During incineration, the flame within the system (10) will heat the rods. The heat will then be directed to the storage dryer (1). An air blower will be used to transfer the heat from the rods to the surrounding area of the storage dryer |

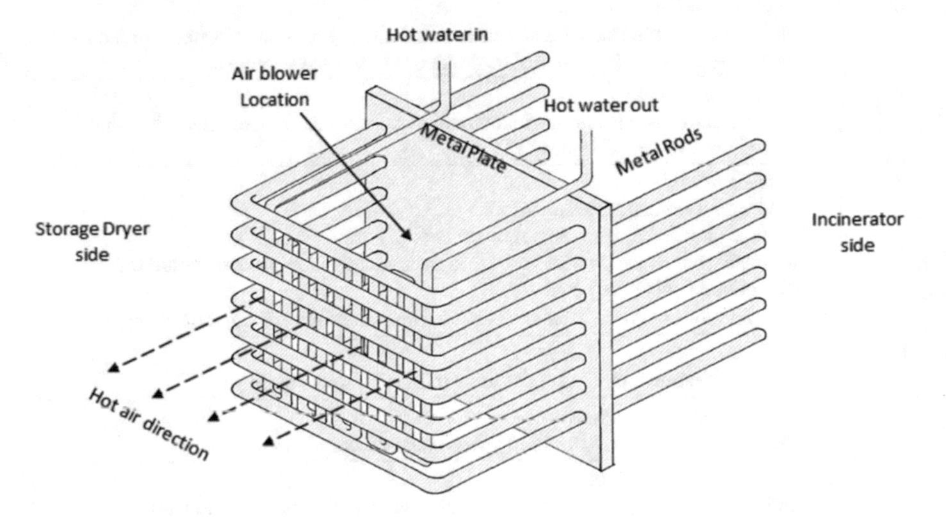


**Fig. 9** The heating element

# 9 Conclusion

The problem of disposal of municipal waste and agricultural waste that occurs in small rural town areas that have scattered residents makes the issue of waste management difficult. Landfill disposal for rural municipal waste is less practical than incineration because the high cost of collection and transportation of solid waste results in no concrete solution to waste disposal in rural areas. Piles of agricultural waste also cause the breeding of animals and pests that are agents of disease spread and cause a bad smell.

Studies have been conducted to design efficient, sustainable, and environmentally friendly combustion technology using local materials with easy operation at a low cost. This study has resulted in a design of incineration technology and experience sharing on practical construction methods of incinerators for agricultural solid waste and household solid waste.

The design of agricultural waste incinerator has introduced the latest technology such as superconductor heat pipe as a heat exchanger to supply heat to the drying compartment of waste storage as a method of heat harvesting. Drying is required to reduce the residual moisture content because the incinerator has only one combustion chamber.

The design of the household waste incinerator, the second combustion chamber, is equipped with a ceramic monolith to ensure that harmful gases due to incomplete combustion of plastic materials that may result during the combustion process in the first combustion chamber can be fully decomposed. On the cyclone side, carbon nanofiber filters are used to ensure that any hazardous materials can be filtered to ensure that smoke emissions on the chimney are safe.

Further studies will be conducted to look in detail at the efficiency of the combustion system in terms of chimney gas emissions and construction and operating costs. Solid waste management systems and policies, starting with the construction of incineration systems, collection to the transformation process using incinerators, and ash waste disposal procedures, must have standard operating methods that need to be developed by the authorities.

**Acknowledgements** This project is partly funded through a UMS grant entitled Showcase of Novel Small–Scale Waste Incinerator and Waste Storage Coupled with Hybrid Dryer (Reference No. SDK0046–2018).

## References

1. Environmental Quality (Amendment) Act 1998 (Act A1030). Laws of Malaysia, Malaysia (1998)
2. Environmental Quality Act 1974 (Act 127). Laws of Malaysia, Malaysia (1974)
3. Ariffin, M., Wan Yacoob, W.N.A.: Assessment of knowledge, attitude and practice of solid waste open burning in Terengganu, Malaysia. EnvironmentAsia. **10**, 25–32 (2017)
4. Eastern Research Group Inc.: Open burning. Emission Inventory Improvement Program, vol. 3 (2001)
5. The World Bank: Decision makers' guide to municipal solid waste incineration. The World Bank (1999)
6. Ghosh, P., Sengupta, S., Singh, L., Sahay, A.: Life cycle assessment of waste-to-bioenergy processes: a review. In: Singh, L., Yousuf, A., and Mahapatra, D.M. (eds.) Bioreactors. pp. 105–122. Elsevier (2020)
7. Jadhao, S.B., Shingade, S.G., Pandit, A.B., Bakshi, B.R.: Bury, burn or gasify: assessing municipal solid waste management options in Indian megacities exergy analysis. Clean Technol. Environ. Policy. **19**, 1403–1412 (2017)
8. Wright, L., Boundy, B., Badger, P.C., Perlack, B., Davis, S.: Biomass Energy Data Book. U.S. Department of Energy, Washington, DC (2009)
9. Batterman, S.: Assessment of small-scale incinerators for health care waste. Water, Sanitation and Health Protection of the Human Environment World Health Organization, Geneva, Switzerland (2004)

10. Youcai, Z.: Municipal solid waste incineration process and generation of bottom ash and fly ash. In: Pollution Control and Resource Recovery: Municipal Solid Wastes Incineration, pp. 1–59. Butterworth-Heinemann (2017)
11. Making Medical Injections Safer: The Incinerator Guidebook: A Practical Guide For Selecting, Purchasing, Installing, Operating And Maintaining Small-Scale Incinerators In Low-Resource Settings. MMIS (2010)

# CFD Assessment of Natural Ventilation Designs for Composting Systems

**Heng-Jin Tham, Mohd Suffian Mohd Misaran Misran, and Christopher Chi-Ming Chu**

**Abstract** Aerobic composting is the decomposition of organic wastes by microorganisms that require oxygen, and therefore, good ventilation during the composting will improve the process. Most systems employ mechanically driven ventilation, and it is envisaged that natural ventilation is more viable for large-scale systems. Four designs of natural ventilation composting vessels have been analysed for typical conditions at Rayleigh number ranging from $5 \times 10^6 - 5 \times 10^9$ and Reynolds number from 7000–40,000. With the CFD analysis, natural ventilation of aerobic composting was found to improve significantly over conventional designs. Out of the four designs analysed, the inclined roof version appeared to be the most promising for large-scale natural ventilating composting systems, with airflow rates and heat discharge that were significantly higher than the rest, and is expected to deliver the fastest composting. Experiments at pilot-scale to validate the results are recommended.

**Keywords** Natural ventilation · Aerobic · Roof · Designs · Vessel · CFD · Compost

## 1 Introduction

The motto "Reduce, Reuse and Recycle" has been widely quoted in the protection of people and the environment. Composting is a recycling process that is simple, natural for processing and adding value to organic wastes to turn them into fertiliser in a matter of weeks. Aerobic composting is the decomposition of organic wastes by microorganisms that require oxygen. The microbes responsible for composting are naturally occurring and live in the moisture surrounding organic matter. Oxygen from the air diffuses into the moisture and is taken up by the microbes. As aerobic digestion takes place the by-products are heat, water and carbon dioxide ($CO_2$). While $CO_2$ can be classified as a greenhouse gas, its evolution from the composting process is not counted in emissions. Additionally, $CO_2$ is only 1/20th as harmful to the

H.-J. Tham · M. S. M. M. Misran · C. C.-M. Chu (✉)
Faculty of Engineering, Universiti Malaysia Sabah, Jalan UMS, 88400 Kota Kinabalu, Sabah, Malaysia
e-mail: chrischu@ums.edu.my



A. Z. Yaser et al. (eds.), *Waste Management, Processing and Valorisation*,
https://doi.org/10.1007/978-981-16-7653-6_5

environment as methane (the main by-product of anaerobic degradation). The heat produced in aerobic composting is sufficient to kill harmful bacteria and pathogens as these organisms are not adapted to these environmental conditions. It also helps support the growth of beneficial bacteria species including psychrophilic, mesophilic and thermophilic bacteria which thrive at the higher temperature levels.

From start to finish, an in-vessel aerobic composting process can take only 8–10 days [1]. No leachate is produced as any surplus moisture is extracted as water vapour which can be condensed and used for watering nearby vegetation. Ahmad et al. [2] were advocates of aerobic systems and noted an optimum range of oxygen level from 10 to 30% had been reported by Willson [3] and Gaur [4]. Most recently an innovative approach of aerobic pre-treatment has been reported to increase the production of methane gas from food waste and sewage sludge Cheng et al. [5]. Forced ventilation of aerobic systems is typically achieved mechanically by using compressors or fans (Lalander et al. [6]; Riley and Forster [7]). Cegarra et al. [8] found that the addition of mechanically forced ventilation to mechanical turning had the benefit of shortening the composting time. Since mechanical ventilation incurs significant capital as well as operating costs, in this study, the natural ventilation of several designs of an aerated composting vessel is assessed by analysing the flow patterns with computational fluid dynamics, to find the configuration with the highest ventilation that is comparable to mechanical ventilation. The most promising design will then be identified by the criteria of flow rate under identical conditions. The scope of the study is to provide an indication of what these typical designs can achieve and not into a detailed study of flow patterns. It lays the groundwork for the actual validation and field tests to reach a final design.

## 2 Methodology

This study will investigate four roof designs of naturally ventilated aerobic composting vessel to assess the ventilation rates. It is expected that with regular turning, good ventilation will accelerate aerobic composting. The gas is assumed to be ideal, consisting of air without moisture or other gases, since most of the gas component is air. Aerobic composting temperature ideally ranges between 40 °C and 60 °C for the bacteria to stay active and oxygen supply must be sufficient to keep the pile hot [9]. A heat source of 50 °C is set assuming steady state of the biochemical reaction is reached. Forced or mechanically driven ventilation is not used.

The model comprises of a 1.524 $m^2$ (5 ft × 5 ft) area of vessel, height of 2.438 m (8 ft) (Fig. 1), with a heat source 1 cm in height spanning the sectional area. The vessel is raised 0.1524 m (6 inch) above floor which acts as an inlet of fresh air, and the air flows through the compost and leaves at the top of the vessel via four different designs: inclined roof, hanging roof, screen on walls and a chimney pipe. The CFD models in the computational domain are shown in Fig. 2. The particulars of the designs are as shown in Table 1.

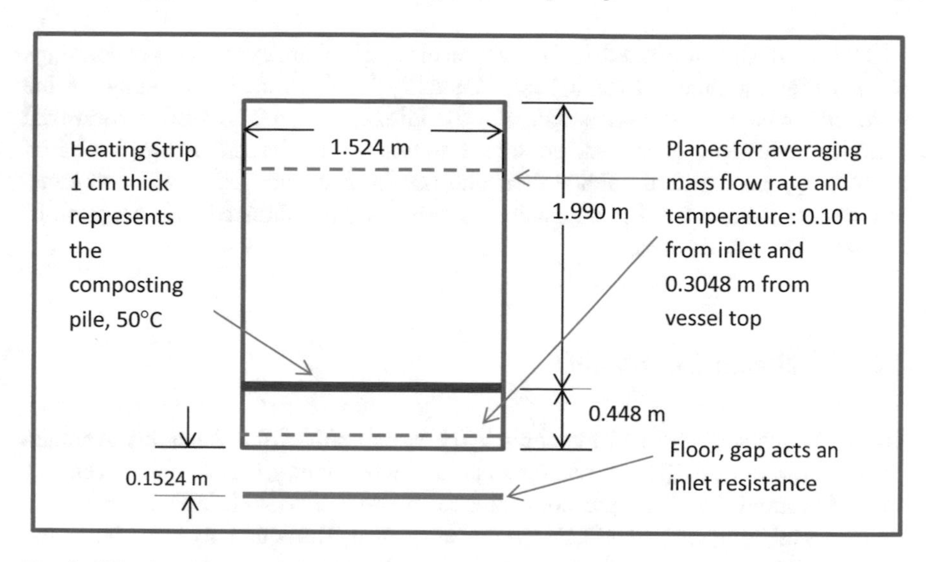


**Fig. 1** Side view of composting vessel common to all designs without the chimney fitted

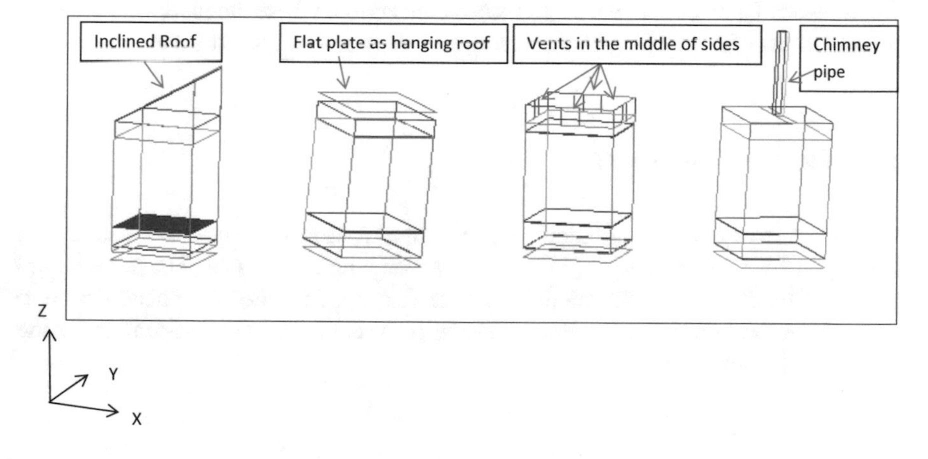


**Fig. 2** CFD geometrical models of the four designs from left to right: Inclined roof, hanging roof, wall vents and a central chimney pipe

**Table 1** Four designs of venting the compost gases

| Design | Description | Remark |
|---|---|---|
| 1 | Inclined roof | 30° from horizontal |
| 2 | Hanging roof | Flat roof lifted at 0.1524 m above vessel |
| 3 | Wall vents | $0.3048 \times 0.\ 3048\ m^2$ in the middle on four sides |
| 4 | Chimney pipe[a] | Pipe diameter = 0.1524 m; Chimney height = 1.5 m |

[a] For design no. 4, the diameter is the equivalent as the CFD pipe is square

The four designs are based on observation of typical chimneys or vents of buildings or industrial equipment. Design 1 is a “Butterfly” roof inclined at 30° angle to the horizontal which is installed in many new buildings; Design 2 is based on industrial chimney stacks with a cap installed to prevent rain from falling in; Design 3 is based on building vents such as public toilets and warehouses, and Design 4 is a chimney of a typical incinerator. The dimensions are reasonably estimated with proportions similar to the real designs.

## 2.1 Mathematical Models

The CFD software is used Phoenics 2019 which solves the Reynolds-averaged Navier–Stokes (RANS) conservation equations of continuity, momentum and energy on a structured grid, and the convergence method of SIMPLEST derived from Patankar and Spalding’s SIMPLE [10] was employed. The equations as implemented by the software are given in Chu et al. [11]. The simulation was solved by the elliptic mode; i.e. upstream values can be affected by downstream values as in the situation of natural convection, unlike in forced convection, where the marching-forward computation suffices. The global convergence criterion was set at 0.1%.

## 2.2 Turbulence Model

The flows are by nature turbulent, and the model used is of type $k$-ε which was developed by Chen and Kim [12] at NASA for plumes and jets. The settings are given in Chu et al. [11]. This model has been chosen since the flow phenomenon to be investigated here involved plumes rising from chimneys and happens to be the default of the software.

## 2.3 Validation

The software has been validated for natural convection simulation employing the Chen-Kim turbulence model using a straight chimney pipe which yielded a mass flow rate error of less than 1 per cent between the simulation value and the value obtained through a widely established friction factor correlation [13].

## 2.4 Boundary Conditions

The ambient pressure is set at 101,325 Pa, temperature at 30 °C and velocities at 0 $ms^{-1}$. The only set value of the process was the heating pile temperature represented by a heating strip at 50 °C. All flows are driven by buoyancy, and the simulation calculates the velocities based on pressure gradient.

## 2.5 Heat Load Calculation

The heat source is modelled in the simulation by a volume of the domain fluid, i.e. air that is constantly at 50 °C and transfer heat instantly to the inlet air which is at ambient temperature. The heat load $\dot{Q}$ is calculated by the single-phase temperature change equation $\dot{Q} =$ mass flow rate × specific heat capacity × temperature change, where the mass flow rate was computed by summing all the cell values across the flow area. The temperature change is averaged based on the area-average outlet temperature over the cross section of the upper composting chamber.

## 2.6 Mesh Grid Optimisation

The structure mesh grid was automatically generated at default intervals for the simulation domain with the dimensions of $X = 12$ m, $Y = 10$ m and $Z = 15$ m. The mesh grid independence test was run using the inclined roof design since it was the most likely to encounter turbulence at the exhaust port. It was observed that there would be minimal fluid flow along the $Y$-axis of the exhaust port (tangent to the port face), but much would be happening along the $X$- (normal to the exhaust port area) and $Z$ (vertically)-axes. The grid densities were doubled and then tripled at the relevant regions where turbulence was anticipated (Table 2). Region 3 of $X$-axis and Region 8 of $Z$-axis were considered relevant for the mesh grid independence test, and hence, region 3 of $X$-axis was set at 11, 22 and 33 cells while the corresponding $Z$-axis region 8 was set at 6 for default, then 12 (2 × ) and 18 (3 × ), respectively. The comparison in Table 3 based on the air mass flow rate shows that for this purpose the default mesh grid generated by the software can be applied in comparing the simulations between the four designs.

Table 3 compares the air mass flow rates of three mesh grid densities at the relevant regions of Design 1. The displayed percentage errors of "Global Convergence" are the highest of all the monitored parameters at the probe position among pressure, temperature, turbulence and velocities and were usually the vertical velocity. The throughput flow behaviour was not very stable since the best percentage error was still 8 times higher than the set criterion of 0.1 per cent, perhaps because of the inlet resistance put up by the 6-inch air gap causing flow instability at the discharge port. It has been found that vertical velocity is an inherently unstable parameter with the possible presence of flow reversals and turbulent fluctuation [11]. The comparison shows that at 3 times the grid density of the default, the mass flow rate did not change from the default value significantly, by 3%, and therefore can apply default mesh grid structure in the regions.

**Table 2** Mesh grid structure according to axes and regions in the grid independence test (Design 1)

| Axis/region | Default | | | | | | | | | |
|---|---|---|---|---|---|---|---|---|---|---|
| | 1 | 2 | 3 | 4 | 5 | 6 | 7 | 8 | 9 | 10 |
| X | 16 | 1 | **11** | 17 | | | | | | |
| Y | 15 | 1 | 12 | 1 | 20 | | | | | |
| Z | 11 | 1 | 1 | 3 | 1 | 10 | 2 | **6** | 1 | 31 |
| | **2× XZ** | | | | | | | | | |
| X | 16 | 1 | **22** | 17 | 17 | | | | | |
| Y | 15 | 1 | 12 | 1 | 20 | | | | | |
| Z | 11 | 1 | 1 | 3 | 1 | 10 | 3 | **12** | 2 | 31 |
| | **3× XZ** | | | | | | | | | |
| X | 16 | 1 | **33** | 17 | 17 | 17 | | | | |
| Y | 15 | 1 | 18 | 1 | 20 | | | | | |
| Z | 11 | 1 | 1 | 3 | 1 | 10 | 4 | 18 | 3 | 31 |

**Table 3** Test of mesh grid dependence using inclined roof (Design 1)

| Mesh grid | Mass flowrate (kg/s) | Heat load (kW) | Ratio | Global convergence % (Highest) |
|---|---|---|---|---|
| Default | 1.19 | 23,764 | 1.00 | 0.81 |
| 2× XZ | 1.26 | 25,042 | 1.05 | 14.8 |
| 3× XZ | 1.22 | 24,388 | 1.03 | 3.74 |

# 3 Results and Discussion

The simulations were completed with reasonable convergence and satisfied the simulation mass and energy balances, where the maximum error was ±2% in Design 2's energy balance. Some of the iterations failed to reach the set convergence %. Design 1 showed the highest flow throughput than the others while encountering the most flow instability. The results are summarised in Table 4. Not surprisingly, the outlet temperatures of all were found to be the same as the heater temperature of 50 °C since the heat transfer was assumed complete and instantaneous. With a more or less constant compost pile temperature, the amount of heat discharged had depended on the flow rate and theoretically had no cap limit.

Design 4 has a small exhaust port area, while it is a simple and commonly seen design for natural ventilation, which seems to discharge only 750 W, while that of Design 1, the heat discharge was 23,764 W, or 31 times more. The estimated velocity 1.4 $ms^{-1}$ at discharge of Design 4 using the basic chimney formula of Height × Density Diff. × Gravitation Acceleration constant = ½ Density × Square of velocity was close to the simulated value of 1.47 $ms^{-1}$ to within ± 5%. This adds confidence to the entire simulation set up in this study. Design 3 installed wall vents, which

**Table 4** Natural ventilation air mass flow rates through each of the four designs under identical ambient conditions

Compost set at 50 °C Ambient temp 30 °C Ambient pressure 101,325 Pa

| | Design 1 | Design 2 | Design 3 | Design 4 |
|---|---|---|---|---|
| | 30° inclined roof | 6-inch hanging roof | 4-units of Wall Vents 12-inch sq | 1.5 m × 6-inch dia. Chimney pipe |
| Mass flow (kg/s) | 1.1926 | 0.8763 | 0.3986 | 0.0373 |
| Heat discharged (W) | 23,764 | 17,594 | 8004 | 750 |
| Characteristic length | Vessel width | Roof gap | Vent size | Chimney diameter |
| Rayleigh number | $5.097 \times 10^9$ | $5.097 \times 10^6$ | $4.078 \times 10^7$ | $5.097 \times 10^6$ |
| Reynolds number | $4.12 \times 10^4$ | $7.57 \times 10^3$ | $1.72 \times 10^4$ | $1.29 \times 10^4$ |
| Global convergence error % (set at 0.1%) | 0.81 | 0.09 | 0.73 | 0.08 |

are popularly installed for ventilation in buildings and merely performed at 1/3rd of Design 1's heat discharge. The nearest to the performance of Design 1 was Design 2, at 35 per cent less than Design 1, perhaps due to the flow area being much closer to Design 1. Design 1's superior performance can be attributed to the larger exhaust port area and aerodynamic shape of the exhaust port. Contour plots of these designs may give further insight into the different performances of ventilation (Figs. 3, 4, 5, and 6). The velocity profiles provide the most insight as they reveal where the turbulence and resistances are, while the temperature profiles show whether any cold air reversal flow occurs. None of the designs appears to suffer from adverse cold inflow as the temperature contour plot showed no reduction of the temperature by cold air at the outlet, or the velocity vector plots exhibited flow reversals (Figs. 3d,

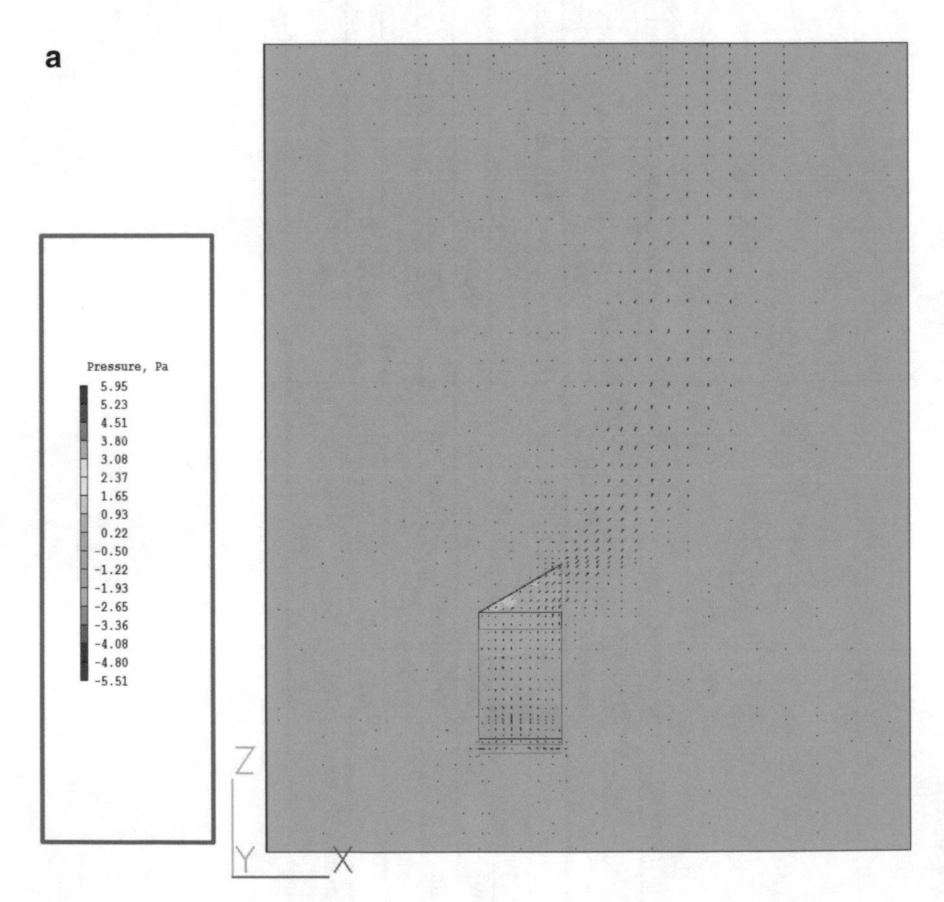


**Fig. 3** **a** Pressure contour plots of Design 1 at mid-plane of Y-dimension of the composting system. **b** Temperature contour plots of Design 1 at mid-plane of Y-dimension of the composting system. **c** Velocity contour plots of Design 1 at mid-plane of Y-dimension of the composting system, **d** Velocity vector plot of Design 1 at mid-plane of Y-dimension of the composting system showing the general one-way direction of the vectors

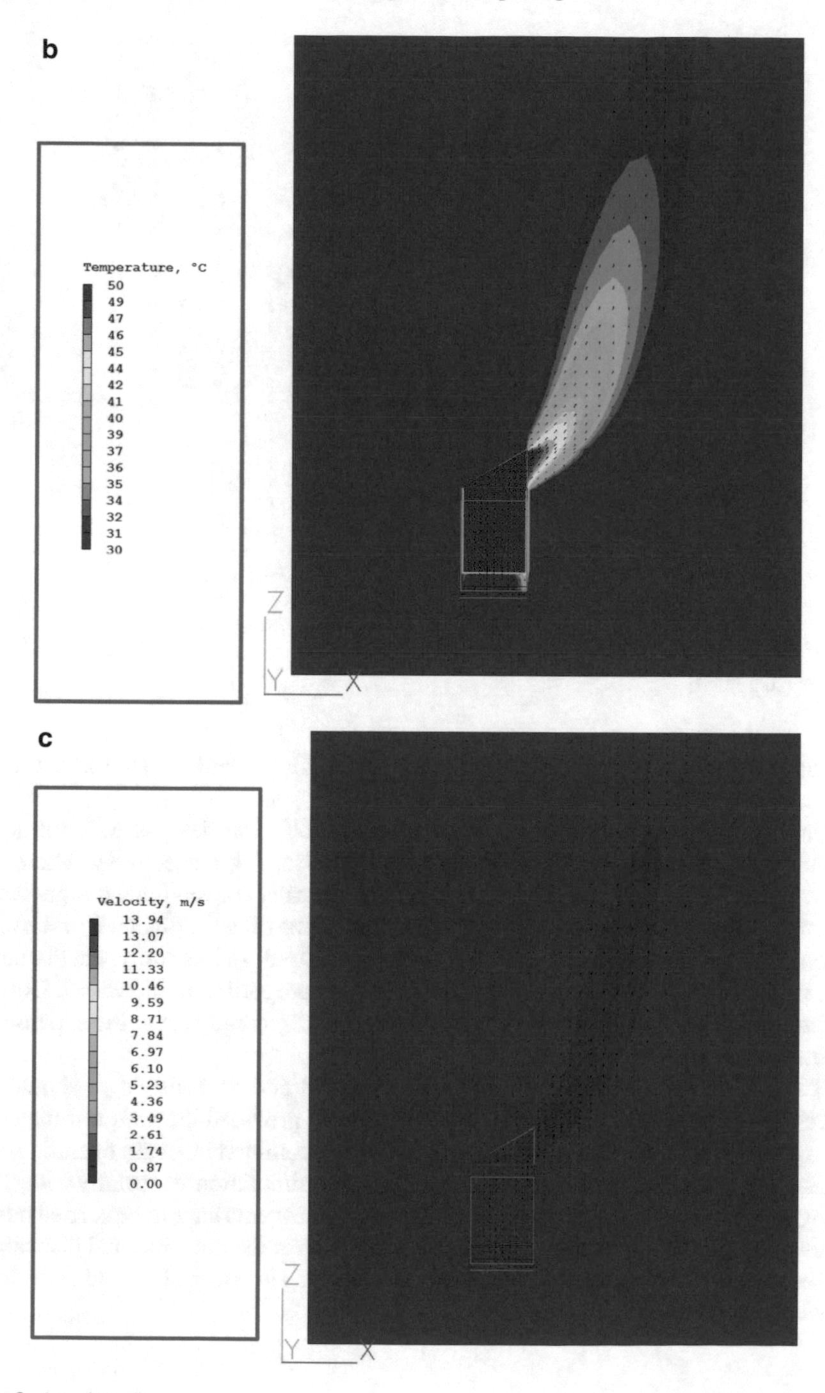


**Fig. 3** (continued)

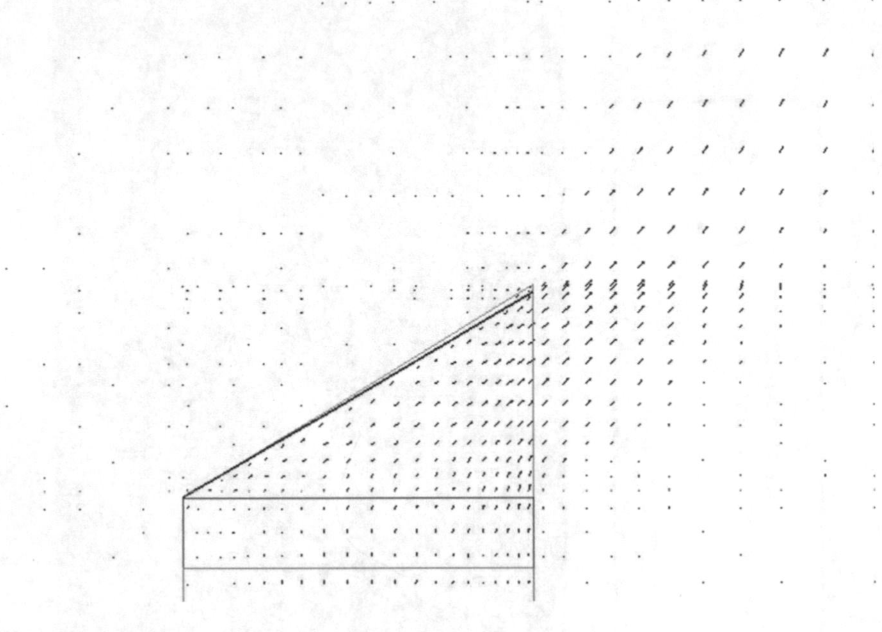


**Fig. 3** (continued)

4d, 5d and 6d) such as found in the case studied by Khanal and Lei [14] where there was a clear reversal in the velocity vectors.

From the velocity plots (Figs. 3b, 4b, 5b and 6b), the designs with the least turbulent disruption at the discharge were Designs 1 and 4, respectively, where the flow direction only changed gently. Design 4 is a typically conventional design that is guaranteed to achieve a steady discharge but most of the time the diameter of the pipe is too small for the full discharge potential to be realised. In this study, the diameter was set at 0.1524 m (6-inch) which is reasonable in proportion to the vessel. Design 1 removes some of the restriction to the flow at the discharge port at the expense of an insignificant loss of stability.

The chimney design with the highest throughput and heat discharge should be selected to enhance the aerobic composting process provided the compost material is turned regularly. Design 1 appears promising for commercial-scale operation.

As there were several assumptions to simplify the simulation, especially with $CO_2$ being mixed with the hot air will make it denser, experiments have to be carried out to validate the simulation results that Design 1 can deliver the most thermal discharge, and hence the composting is expected to be the fastest. However, all the other designs should be workable.

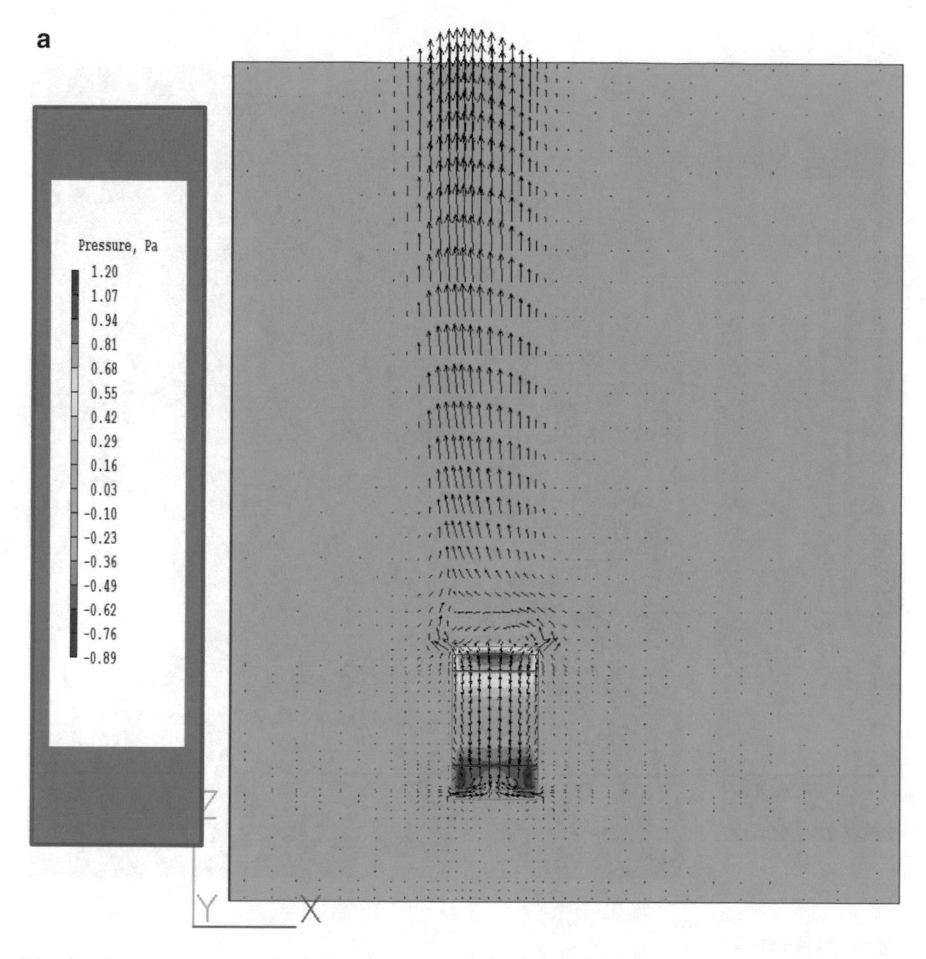


**Fig. 4** **a** Pressure contour plots of Design 2 at mid-plane of Y-dimension of the composting system. **b** Temperature contour plots of Design 2 at mid-plane of Y-dimension of the composting system. **c** Velocity contour plots of Design 2 at mid-plane of Y-dimension of the composting system, **d** Velocity vector plot of Design 2 at mid-plane of Y-dimension of the composting system showing the general one-way direction of the vectors as the flow discharges

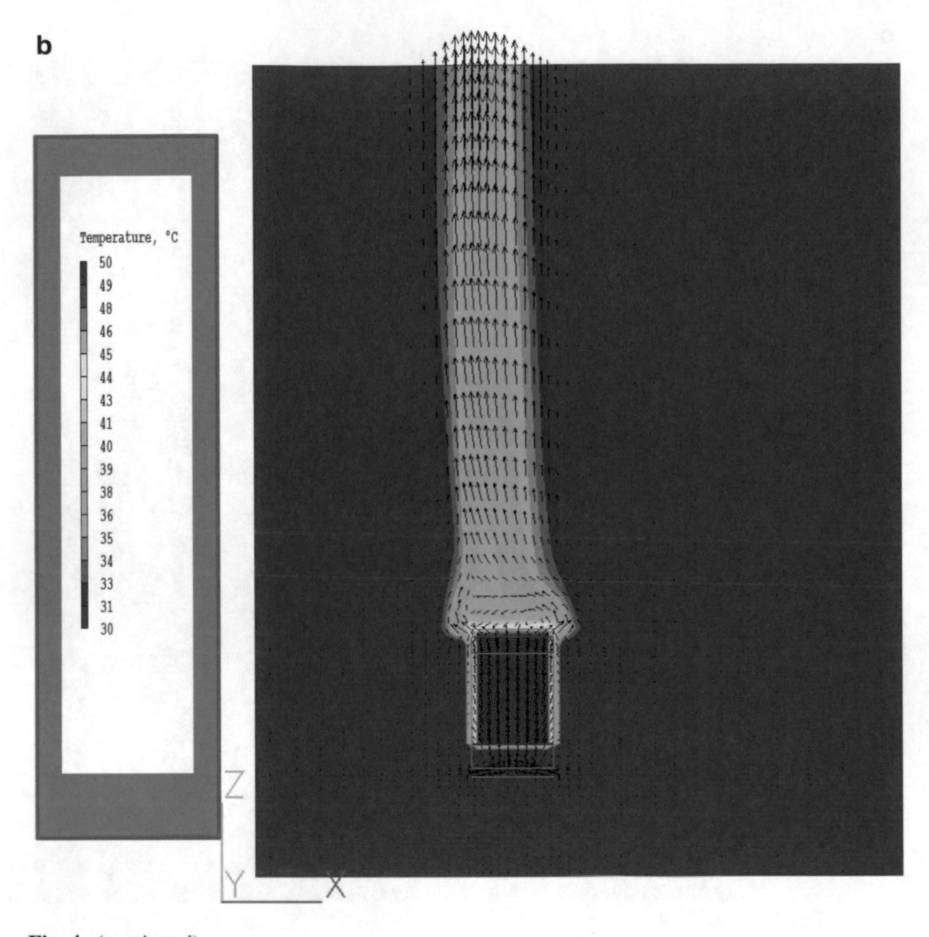


**Fig. 4** (continued)

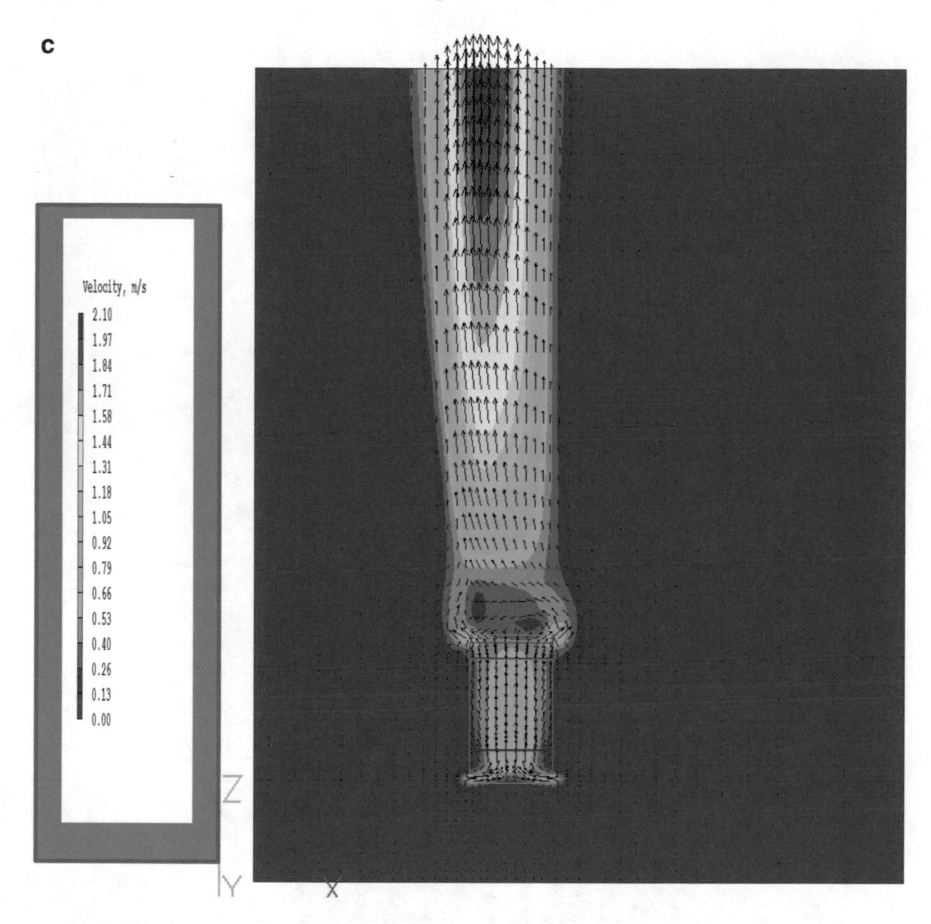


**Fig. 4** (continued)

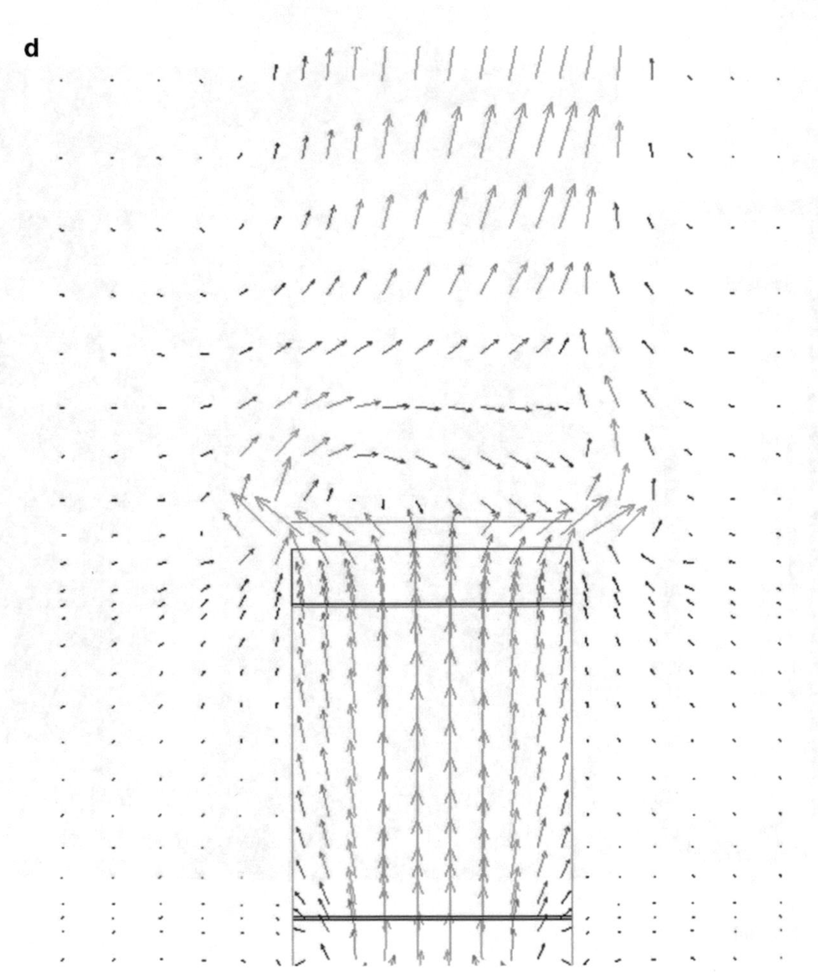


**Fig. 4** (continued)

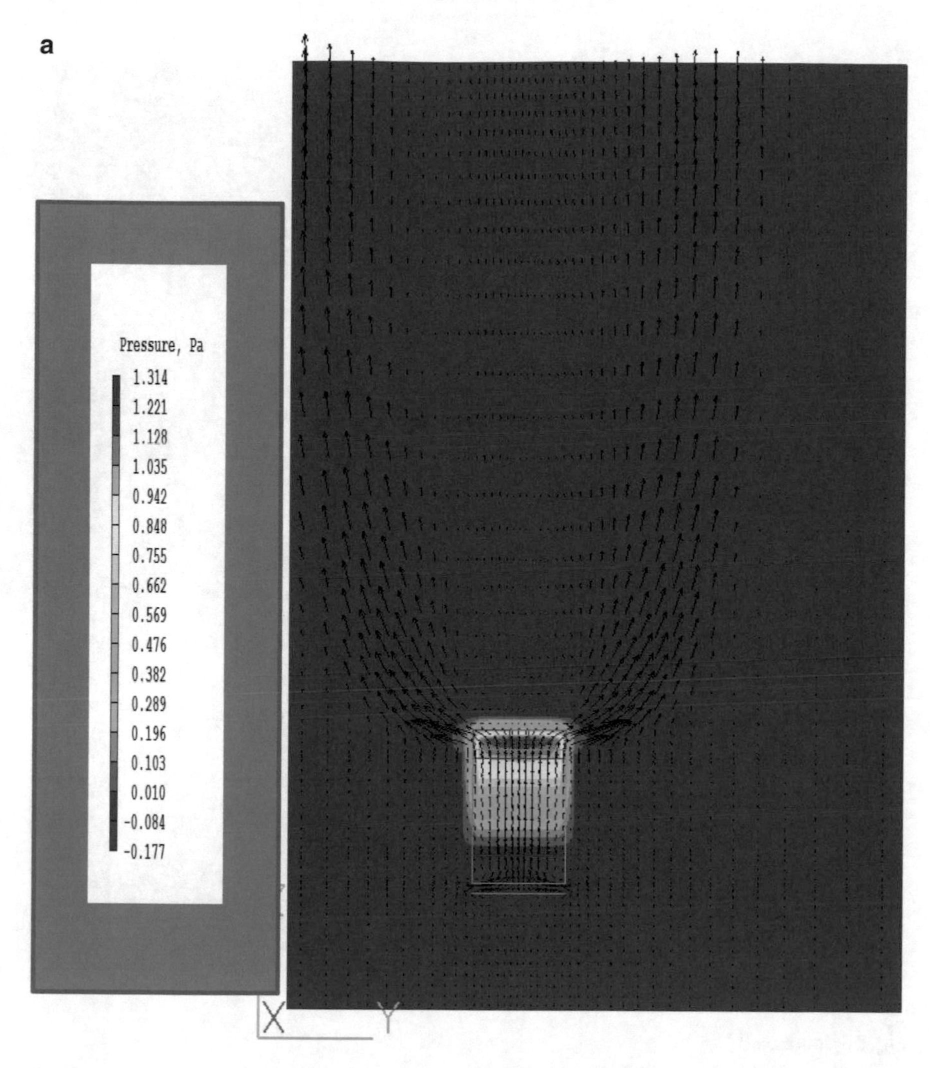


**Fig. 5** **a** Pressure contour plots of Design 3 at mid-plane of Y-dimension of the composting system, **b** Temperature contour plots of Design 3 at mid-plane of Y-dimension of the composting system, **c** Velocity contour plots of Design 3 at mid-plane of Y-dimension of the composting system, **d** Velocity vector plot of Design 3 at mid-plane of Y-dimension of the composting system showing the general one-way direction of the vectors as the flow discharges

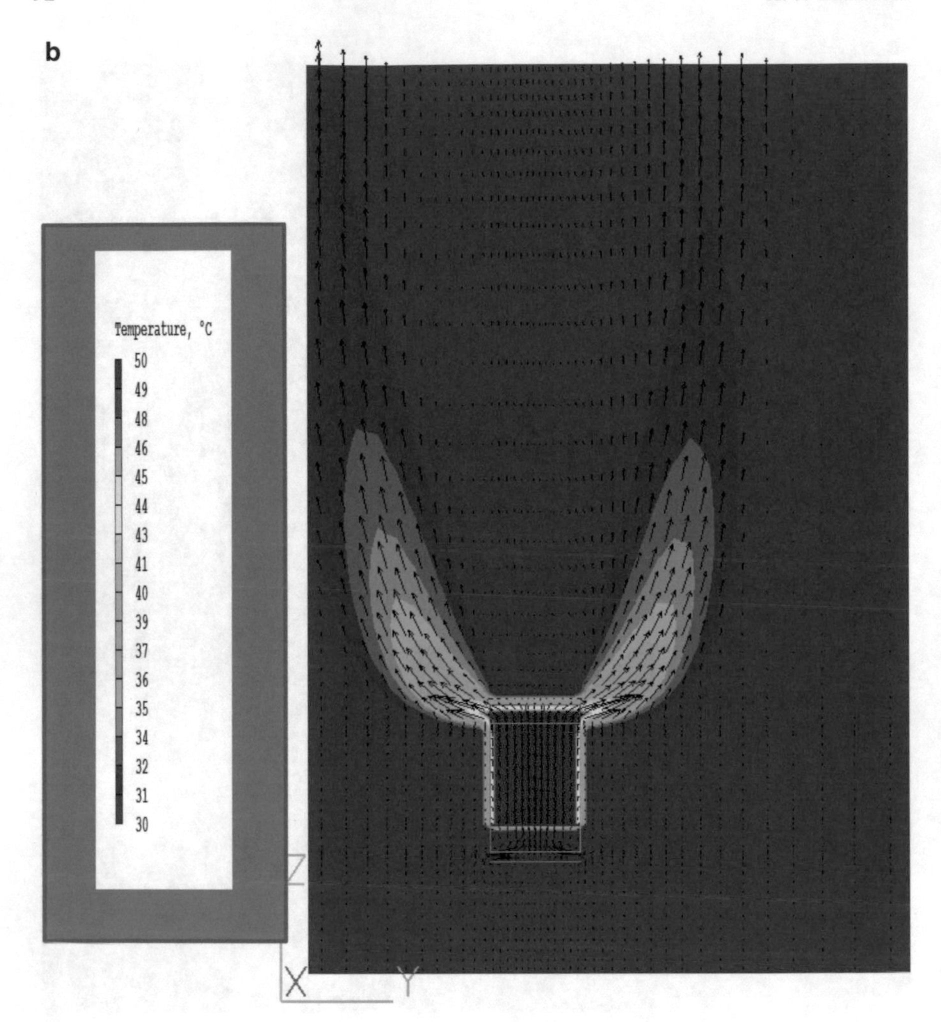


**Fig. 5** (continued)

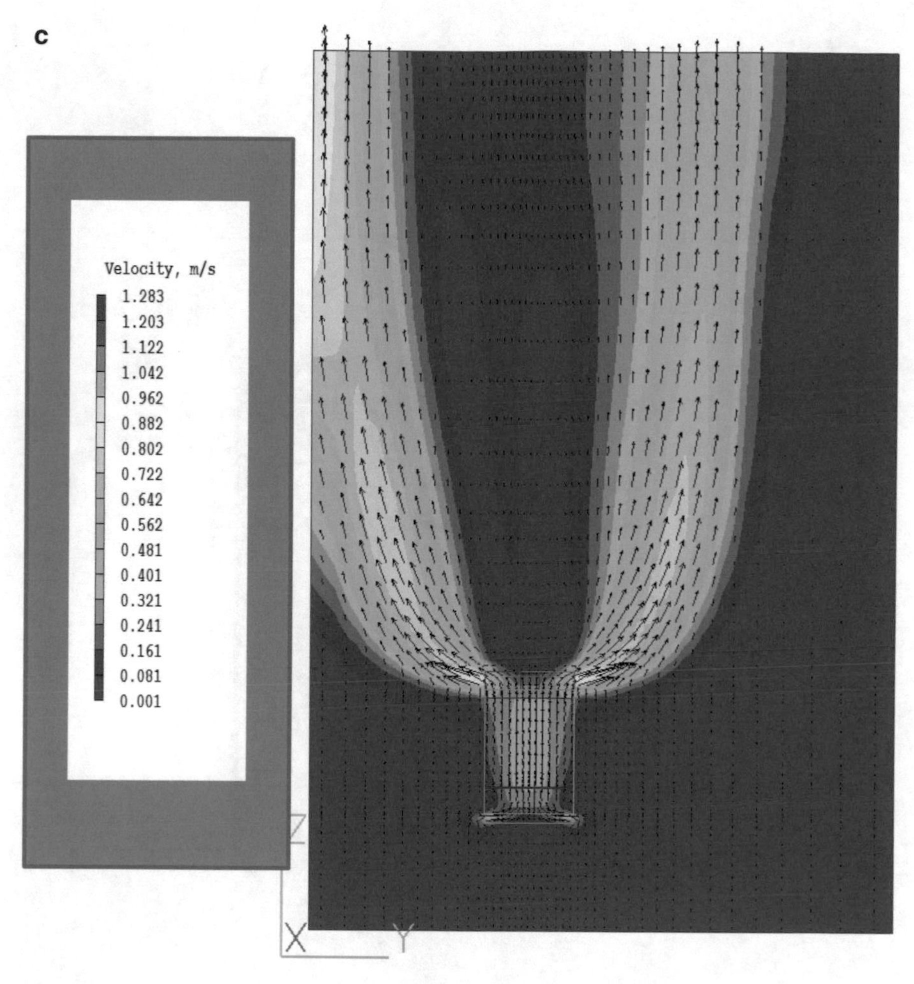


**Fig. 5** (continued)

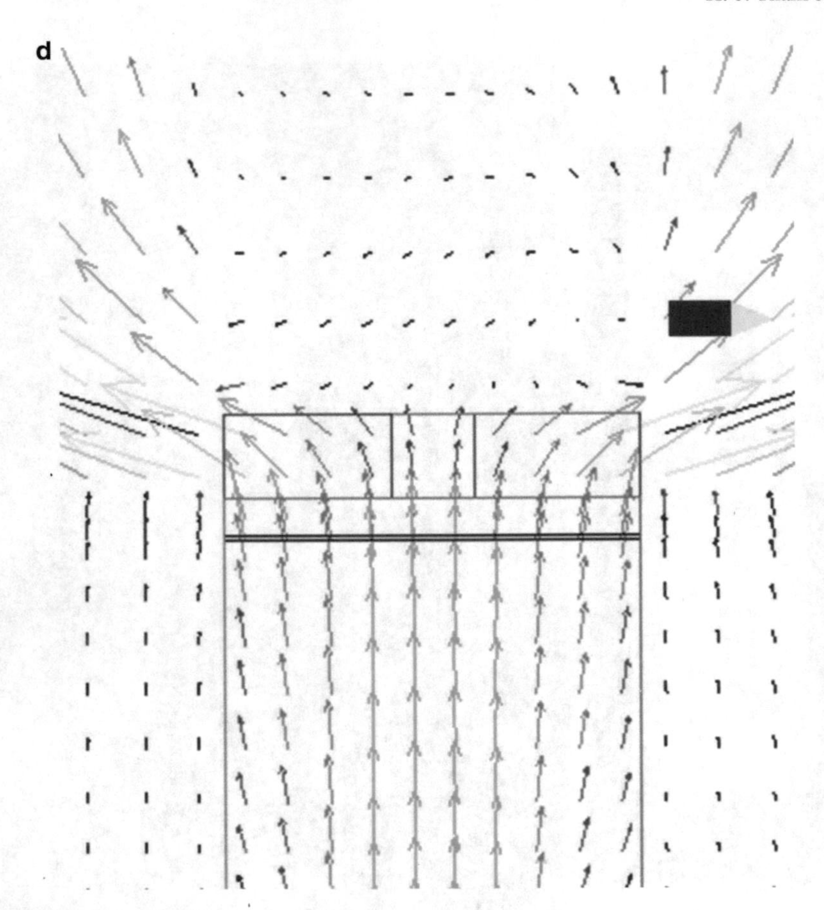


**Fig. 5** (continued)

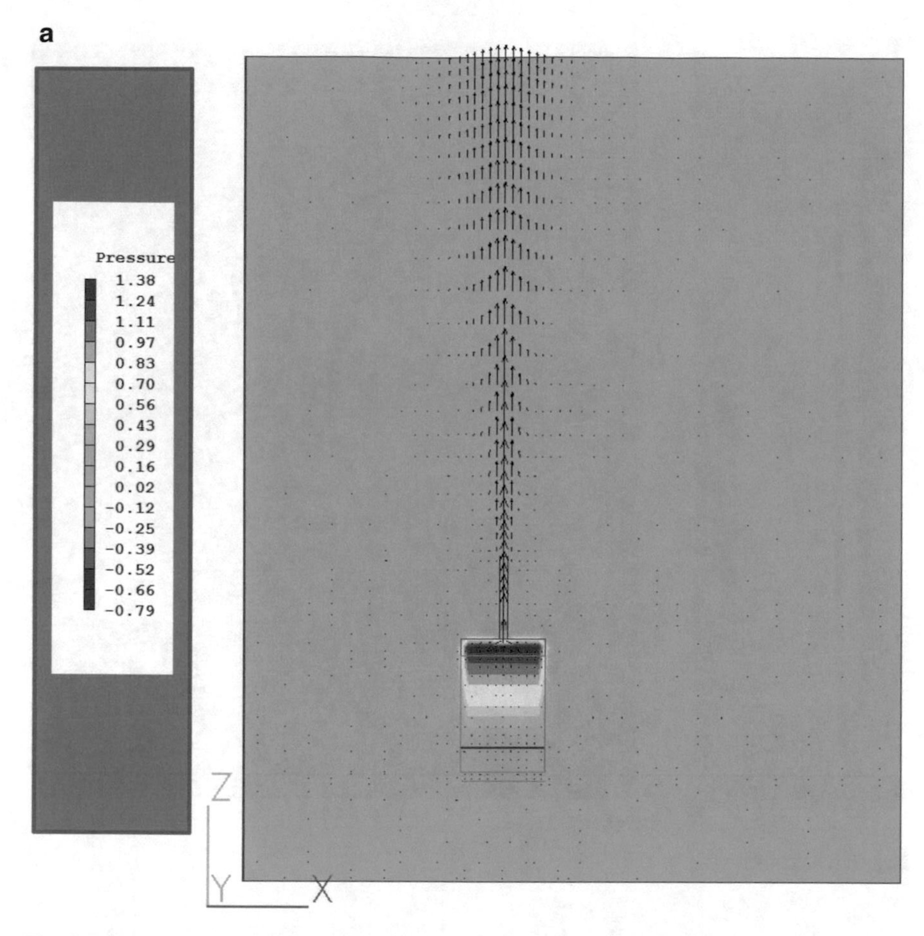


**Fig. 6** **a** Pressure contour plots of Design 4 at mid-plane of Y-dimension of the composting system. **b** Temperature contour plots of Design 4 at mid-plane of Y-dimension of the composting system. **c** Velocity contour plots of Design 4 at mid-plane of Y-dimension of the composting system, **d** Velocity vector plot of Design 4 at mid-plane of Y-dimension of the composting system showing the general one-way direction of the vectors as the flow discharges

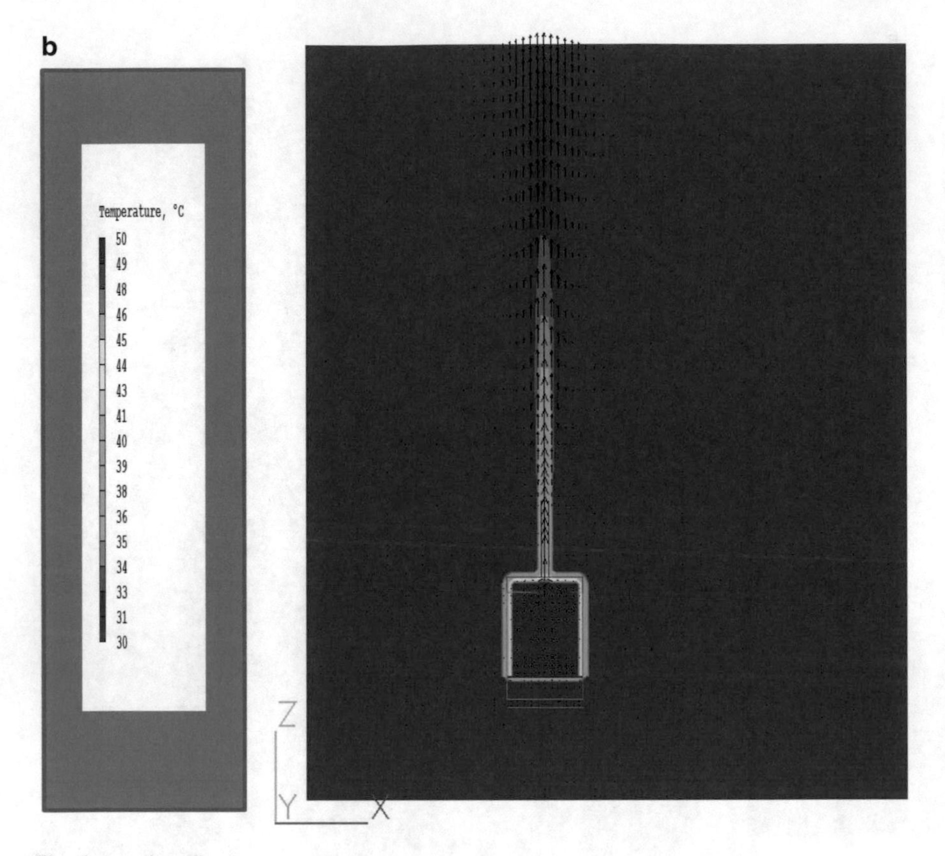


**Fig. 6** (continued)

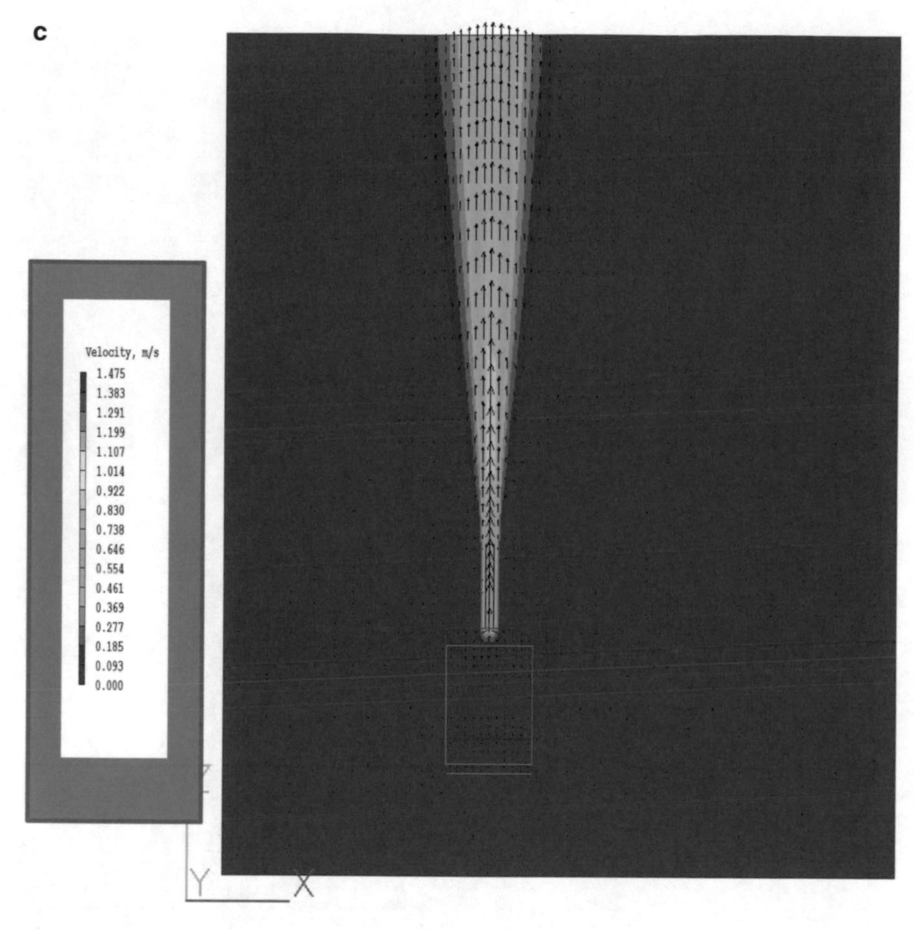


**Fig. 6** (continued)

**Fig. 6** (continued)

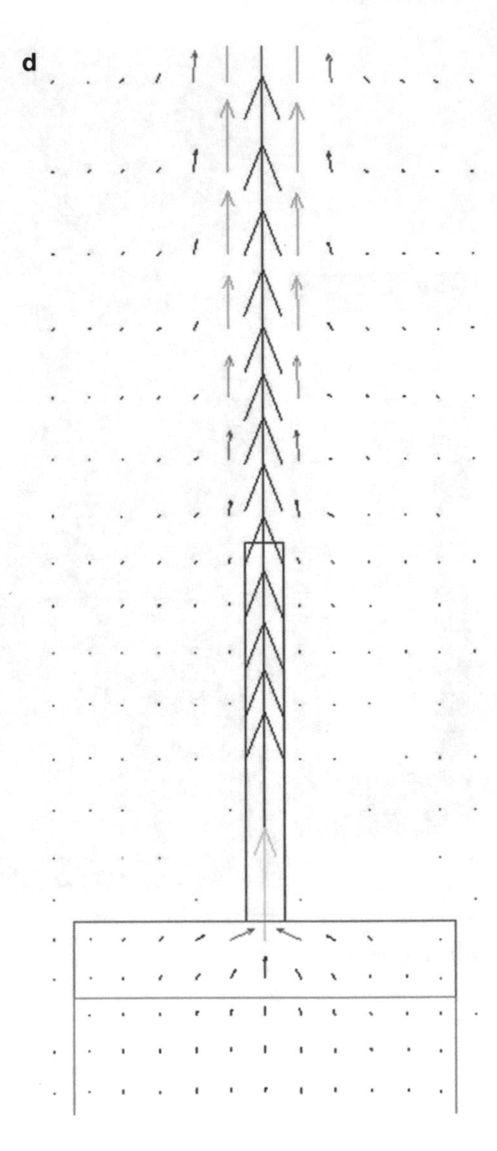


## 4 Conclusion and Recommendation

With CFD analysis, natural ventilation of aerobic composting can be improved significantly over conventional designs and should be able to replace mechanical ventilation.

Out of the four designs analysed, the inclined roof version appears to be the most promising for large-scale natural ventilating composting systems, with airflow rates and heat discharge that is significantly higher than the rest and is expected to deliver the fastest composting time.

It is recommended to validate the simulation results with experiments at a pilot scale.

**Acknowledgements** The authors wish to thank Universiti Malaysia Sabah Centre of Research and Innovation for providing a grant No. SGA0034 towards this study.

## References

1. Global Composting Solutions: https://www.globalcomposting.solutions/aerobic-vs-anearobic-composting. Accessed 15 June 2021
2. Ahmad, R., Jilani, G., Arshad, M., Zahir, Z.A., Khalid, A.: Bio-conversion of organic wastes for their recycling in agriculture: an overview of perspectives and prospects. Ann. Microbiol. **57**(4), 471–479 (2007). Springer
3. Willson, G.B.: Combining raw materials for composting. Biocycle Aug 82–85 (1989)
4. Gaur, A.C.: Bulky organic manures and crop residues. In: Tandon, H.L.S. (ed.) Fertilizers, Organic Manures, Recyclable Wastes and Biofertilizers, pp. 37–51. FDCO, New Delhi, India (1997)
5. Cheng, L.J., Gao, N.B., Quan, C., Chu, H., Wang, G.J.: Promoting the production of methane on the co-digestion of food waste and sewage sludge by aerobic pre-treatment. **292**, 120197, 8 (2021). 15 May 2021, Elsevier
6. Lalander, C., Ermolaev, E., Wiklicky, V., Vinnerås, B.: Process efficiency and ventilation requirement in black soldier fly larvae composting of substrates with high water content. Sci. Total Environ. **729**, 13896, 8 (2020). Elsevier
7. Riley, D.W., Forster, C.F.: An evaluation of an autothermal Aerobic digestion system. Trans IChemE, Part B Inst Chem Eng **80**, 100–104 (2002)
8. Cegarra, J., Alburquerque, J.A., Gonza´lvez, J., Tortosa, G., Chaw, D.: Effects of the forced ventilation on composting of a solid olive-mill by-product ("alperujo") managed by mechanical turning. Waste Manag. **26**, 1377–1383, Elsevier
9. Chester County Solid Waste Authority, Pennsylvania, USA. https://www.chestercountyswa.org/156/Aerobic-Composting-Temperature. Accessed 15 June 2021
10. Patankar, S.: Numerical heat transfer and fluid flow. Taylor & Francis (1980)
11. Chu, C.M., Rahman, M.M., Kumaresan, S.: Improved thermal energy discharge rate from a temperature-controlled heating source in a natural draft chimney. Appl. Therm. Eng. **98**, 991–1002, (2016), Elsevier
12. Chen, Y.S., Kim, S.W.: Computation of turbulent flows using an extended k-e turbulence closure model. NASA CR-179204 (1987)
13. Chu, C.M.: Lazy plume stack effect above chimneys. In: Rahman, M.M., Chu, C.M. (eds.) Cold Inflow-Free Solar Chimney. Springer, Singapore (2021). https://doi.org/10.1007/978-981-33-6831-6_5
14. Khanal, R., Lei, C.W. (2012) Flow reversal effects on buoyancy induced air flow in a solar chimney. Sol. Energy **86**, 2783–2794. Elsevier

# Preparation, Characterisation and Application of Palm Oil Mill Solid Waste as Sustainable Natural Adsorbent for the Removal of Heavy Metal

**H. L. H. Chong, R. Idris, and N. Surugau**

**Abstract** The current global annual production of fresh fruit bunch was 410.70 million ton/year. This translated to the by-production of 92.41, 55.44, 22.59 and 1.64 million ton/year of empty fruit bunch, oil palm mesocarp fibre, oil palm shell and palm oil mill fly ash, respectively, indicating sustainable supply. Unfortunately, these by-products were of low commercial value due to the limited application. Thus, the idea of valorising the empty fruit bunch, oil palm mesocarp fibre, oil palm shell and palm oil mill fly ash, but with a minimum environmental footprint, which preparation was via water washing and minor size reduction into unmodified adsorbents codenamed OPF, OPMF, OPS and OPMFA, respectively, was born. The objectives of this chapter were to show some physicochemical characterisation and heavy metal adsorption properties of the palm oil mill solid wastes-based adsorbent. While the physicochemical properties of these adsorbents were as expected from an agricultural waste-based unmodified adsorbent, the OPF, OPMF, OPS and OPMFA have the adsorption capacities of 14.39, 5.93, 1.76 and 153.85 mg/g for Cu(II) ions and 11.12, 39.53, 3.39 and 181.82 mg/g for Pb(II) ions, respectively. All the absorbents were well-fitted into the pseudo-second-order kinetic model indicating chemisorption was the rate-limiting factor. Future research should be focused on the application of these adsorbents for slow- or batch-flow wastewater treatment system such as constructed wetland and adsorption bed.

**Keywords** Preparation · Characterisation · Palm oil mill solid waste · Capacities · Cu(II) · Pb(II) · Chemisorption

H. L. H. Chong (✉) · R. Idris · N. Surugau
Faculty of Science and Natural Resources, Universiti Malaysia Sabah, Jalan UMS, 88400 Kota Kinabalu, Sabah, Malaysia
e-mail: hlhchong@ums.edu.my

H. L. H. Chong
Water Research Unit, Universiti Malaysia Sabah, Jalan UMS, 88400 Kota Kinabalu, Sabah, Malaysia

Sustainable Palm Oil Research Unit, Faculty of Sustainable Agriculture, Universiti Malaysia Sabah, Jalan UMS, 88400 Kota Kinabalu, Sabah, Malaysia

 
A. Z. Yaser et al. (eds.), *Waste Management, Processing and Valorisation*,
https://doi.org/10.1007/978-981-16-7653-6_6

# 1 Introduction

Globally palm oil is the most widely consumed edible oil. No doubt according to the Food and Agricultural Organisation of United Nations [1], the palm oil crop is planted in 44 countries worldwide (Table 1). The total matured planted area which produced harvestable fresh fruit bunch (FFB) was 28.31 million ha, and this figure is on the rise. Of these harvested planted areas, the top four significant planters were Indonesia, Malaysia, Nigeria and Thailand which have a matured planted area of at least 14.68, 5.22, 3.93 and 0.90 million ha, respectively.

The worldwide FFB production recorded from the harvested planted area was 410.70 million ton. Indonesia is the lead producer which produced 59.81% of the

**Table 1** Global harvested planted area and fresh fruit bunch production [1]

| Country | Harvested planted area | | | FFB production | | |
|---|---|---|---|---|---|---|
| | Million ha | Percentage | Rank | Million ton | Percentage | Rank |
| Indonesia | 14.6776 | 51.8411 | 1 | 245.6331 | 59.8089 | 1 |
| Malaysia | 5.2168 | 18.4258 | 2 | 99.0654 | 24.1213 | 2 |
| Nigeria | 3.9349 | 13.8982 | 3 | 10.0252 | 2.4410 | 4 |
| Thailand | 0.8963 | 3.1658 | 4 | 16.7724 | 4.0839 | 3 |
| Columbia | 0.5041 | 1.7805 | 5 | 8.3903 | 2.0429 | 5 |
| Ghana | 0.3743 | 1.3219 | 6 | 2.6554 | 0.6466 | 8 |
| Guinea | 0.3179 | 1.1227 | 7 | 0.8475 | 0.2064 | 18 |
| Côte d'Ivoire | 0,3151 | 1.1128 | 8 | 2.0552 | 0.5004 | 13 |
| Democratic Republic of The Congo | 0.2805 | 0.9907 | 9 | 1.8354 | 0.4469 | 14 |
| Ecuador | 0.2009 | 0.7096 | 10 | 2.2759 | 0.5542 | 11 |
| Honduras | 0.2000 | 0.7064 | 11 | 2.3372 | 0.5691 | 10 |
| Papua New Guinea | 0.1985 | 0.7010 | 12 | 2.6819 | 0.6530 | 7 |
| Guatemala | 0.1880 | 0.6640 | 13 | 3.2696 | 0.7961 | 6 |
| Brazil | 0.1775 | 0.6268 | 14 | 2.5833 | 0.6290 | 9 |
| Cameroon | 0.1550 | 0.5475 | 15 | 2.1655 | 0.5273 | 12 |
| Mexico | 0.0855 | 0.3021 | 16 | 1.1942 | 0.2908 | 15 |
| Cota Rica | 0.0767 | 0.2710 | 17 | 1.0818 | 0.2634 | 16 |
| Peru | 0.0729 | 0.2573 | 18 | 0.9129 | 0.2223 | 17 |
| Philippines | 0.0628 | 0.2219 | 19 | 0.4993 | 0.1216 | 21 |
| China | 0.0503 | 0.1778 | 20 | 0.6659 | 0.1621 | 20 |
| Venezuela | 0.0416 | 0.1469 | 21 | 0.4355 | 0.1060 | 22 |
| Benin | 0.0401 | 0.1417 | 22 | 0.7329 | 0.1784 | 19 |
| Sierra Leone | 0.0338 | 0.1195 | 23 | 0.2573 | 0.0626 | 26 |

(continued)

**Table 1** (continued)

| Country | Harvested planted area | | | FFB production | | |
|---|---|---|---|---|---|---|
| | Million ha | Percentage | Rank | Million ton | Percentage | Rank |
| Angola | 0.0233 | 0.0823 | 24 | 0.2807 | 0.0684 | 25 |
| Solomon Island | 0.0220 | 0.0777 | 25 | 0.3089 | 0.0752 | 23 |
| Dominican Republic | 0.0207 | 0.0731 | 26 | 0.2884 | 0.0702 | 24 |
| Togo | 0.0186 | 0.0655 | 27 | 0.1586 | 0.0386 | 29 |
| Liberia | 0.0177 | 0.0626 | 28 | 0.1761 | 0.0429 | 27 |
| Paraguay | 0.0167 | 0.0590 | 29 | 0.1579 | 0.0384 | 30 |
| Cambodia | 0.0149 | 0.0525 | 30 | 0.1600 | 0.0390 | 28 |
| Senegal | 0.0126 | 0.0445 | 31 | 0.1450 | 0.0353 | 32 |
| Congo | 0.0125 | 0.0443 | 32 | 0.1558 | 0.0379 | 31 |
| Guinea-Bissau | 0.0096 | 0.0338 | 33 | 0.0807 | 0.0196 | 35 |
| Burundi | 0.0089 | 0.0316 | 34 | 0.0886 | 0.0216 | 33 |
| Nicaragua | 0.0082 | 0.0291 | 35 | 0.0862 | 0.0210 | 34 |
| Gabon | 0.0061 | 0.0214 | 36 | 0.0203 | 0.0049 | 41 |
| United Republic of Tanzania | 0.0055 | 0.0195 | 37 | 0.0756 | 0.0184 | 36 |
| Equatorial Guinea | 0.0035 | 0.0124 | 38 | 0.0359 | 0.0088 | 37 |
| Gambia | 0.0035 | 0.0124 | 39 | 0.0351 | 0.0085 | 38 |
| Panama | 0.0027 | 0.0097 | 40 | 0.0278 | 0.0068 | 39 |
| Sao Tome and Principe | 0.0019 | 0.0067 | 41 | 0.0180 | 0.0044 | 42 |
| Madagascar | 0.0018 | 0.0064 | 42 | 0.0212 | 0.0052 | 40 |
| Suriname | 0.0006 | 0.0020 | 43 | 0.0011 | 0.0003 | 44 |
| Central African Republic | 0.0002 | 0.0007 | 44 | 0.0017 | 0.0004 | 43 |
| Total | 28.3126 | 100 | – | 410.6968 | 100 | – |

global FFB while Malaysia, Thailand and Nigeria produced 24.12, 4.08 and 2.44%, accordingly. In this regard, Indonesia and Malaysia collectively produced more than 83% of the global FFB.

## 2 Palm Oil Mill Solid Wastes

It is the industry practice that the harvested FFB be processed on the same day in order to obtain the best quality crude palm oil (CPO). Once the FFB arrived at the palm oil mill, it will be graded and then sterilised prior to entering the series of palm oil extraction processes. The extraction of the high-value CPO by-produced wastes such as palm oil mill effluent (POME) which is in liquid form and other palm oil mill solid wastes such as empty fruit bunch (EFB), oil palm mesocarp fibre, oil palm shell and palm oil mill fly ash.

The POME and its related handlings and treatments are beyond the scope of our discussion here; however, one can read on the treatment of POME as discussed in the book by Chong [2]. For every FFB that entered the palm oil mill CPO extraction processes, the by-productions of EFB, oil palm mesocarp fibre, oil palm shell and palm oil mill fly ash were estimated at the range of 20–25%, 12–15%, 4–7% and 0.3–0.5% weight-to-weight ratio, accordingly. These EFB, oil palm mesocarp fibre and oil palm shell to FFB ratios of 22, 13.5 and 5.5%, respectively, were in agreement with the previous report by Yusoff [3].

To date, applications on the alternative use of these palm oil mill solid wastes are limited. For example, the EFB was processed into long fibre for utilisation as mattress fibre, but with limited success, while the mixture of oil palm mesocarp fibre and oil palm shell was used as boiler fuel to generate self-sustaining electricity supply for the palm oil mill. The palm oil mill fly ash, however, has no industrial application at the current moment. In the light of these observations of limited uses, there is an opportunity and motivation to valorise and convert these palm oil mill solid wastes into sustainable true low-cost adsorbents which is the focus of this chapter.

## 3 Towards True Low-Cost Adsorbent

The market demand for adsorbent is on the rise for the commercialised zeolite and activated carbon types of adsorbent. However, the zeolite and activated carbon types of adsorbent are not cheap due to the cost and time spent in their productions.

The search for cheaper adsorbent has begun, and it is not uncommon to find research articles on alternative adsorbent which often claimed to be low cost. One of the ways to remain low cost is to use a no- or low-value solid waste, and thus, agricultural waste is often being the target of research. Nonetheless, often these researches involved some modification which range from simple single physical or chemical modification step to more complicated combination of physical and chemical modification steps. The reason for such modifications was to increase the adsorption performance; however, it is difficult to be on par with the zeolite and activated carbon types of adsorbent if one insists on the mg/g (mass of adsorbate adsorbed /mass of adsorbent) performance ratio. But, if one were to look at performance from the cost versus benefit point of view where mg/cent (mass of adsorbate adsorbed/cost of adsorbent) performance ratio is put into perspective, there is surely a potential for true low-cost adsorbent to be as good as the zeolite and activated carbon types of adsorbent.

If one were to search for a true low-cost alternative, the modification should be kept to minimum as physical and chemical modifications will utilise electricity and chemical which not only will increase the production cost but could also harm the environment via carbon and chemical footprints. Unfortunately, it is rare to find published research article on unmodified adsorbent. In this chapter, the authors intend to address this data gap by presenting some data on adsorbent derived from palm oil mill solid wastes.

# 4 Preparation of Adsorbent

The palm oil mill solid waste which consists of EFB fibre, oil palm mesocarp fibre, oil palm shell and palm oil mill fly ash was collected from local palm oil mill. Six cycles of washing and overnight soaking cleaning process utilising clean domestic tap water were performed on the EFB fibre, oil palm mesocarp fibre and oil palm shell to eliminate impurities which included the residual palm oil. The residual palm oil was removed during the overnight soaking where the density gradient forced the less dense oil to leave the denser EFB fibre, oil palm mesocarp fibre and oil palm shell, and to float on the water. By the sixth cycle which was the sixth day, there was no longer any trace of floating oil in the overnight soaked water surface. The collected palm oil mill fly ash was, however, not subjected to the washing and soaking process.

The impurities-free EFB fibre and oil palm mesocarp fibre were then oven dried at 65 °C in batches until constant weight was achieved. The temperature setting of 65 °C was selected instead of 103 °C to eliminate over-drying and to reduce electricity cost. The cooled-down oven-dried impurities-free EFB fibre and oil palm mesocarp fibre were subsequently kept into respective airtight containers and codenamed as OPF (Fig. 1a) and OPMF (Fig. 1b) hereafter.

Unlike the fibrous palm oil mill solid wastes-based adsorbent, the impurities-free oil palm shell and palm oil mill fly ash were sieved into various sizes prior to oven

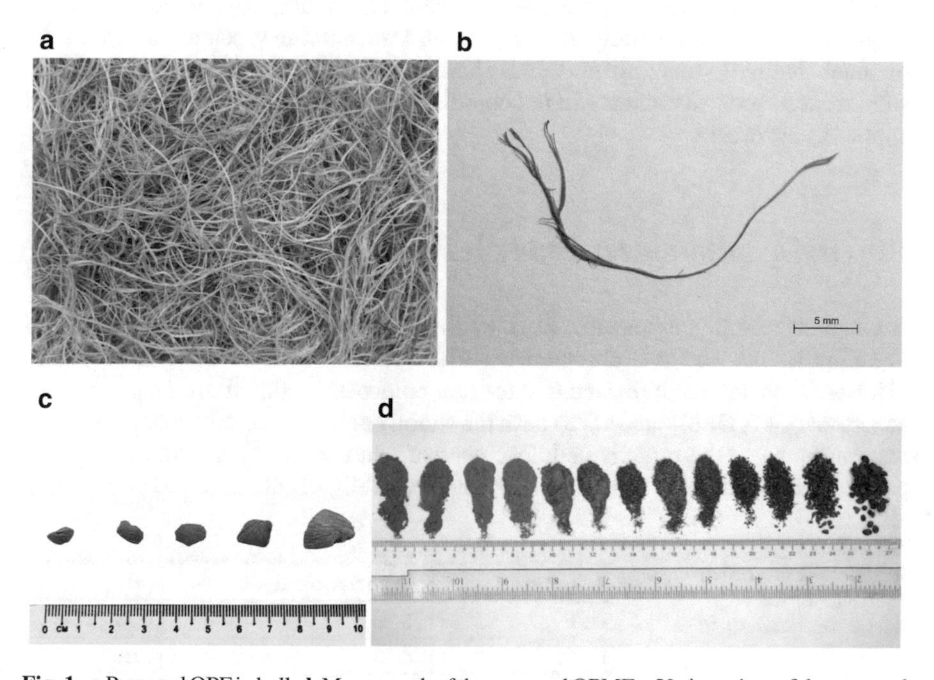


**Fig. 1** **a** Prepared OPF in bulk, **b** Macrograph of the prepared OPMF, **c** Various sizes of the prepared OPS, **d** Various sizes of the prepared OPMFA

drying at 65 °C and storage in the respective airtight containers. The oven-dried sieved impurities-free oil palm shell and palm oil mill fly ash were codenamed as OPS (Fig. 1c) and OPMFA (Fig. 1d), accordingly.

For practical reasons, all the adsorption-related data reported in this chapter originated from experiments that utilised the OPF, OPMF, OPS and OPMFA of 2 cm length, 2 cm length, $6.5 < Ø \leq 8.0$ mm and $0.150 < Ø \leq 0.200$ mm (where Ø refers to geometrical mean particle size), respectively.

## 5 Some Physicochemical Properties of the Palm Oil Mill Solid Wastes-Based Adsorbent

### 5.1 *Leaching of Heavy Metal*

The palm oil mill solid wastes-based adsorbent, namely OPF, OPMF, OPS and OPMFA, were tested for the potential leaching of heavy metal by reacting 1.0 g of adsorbent with five types of eluent which consist of 100 mL 0.01 N $HNO_3$, HCl, $H_2SO_4$, NaOH and ultrapure water, respectively, at 150 rpm for 24 h [4].

The heavy metal analysis revealed that there was no leaching or presence of arsenic, beryllium, calcium, cadmium, cobalt, chromium, copper, iron, lithium, magnesium, manganese, molybdenum, nickel, lead, antimony, selenium, strontium, titanium, thallium, vanadium and zinc. These results suggested that the palm oil mill solid wastes-based adsorbent will not contribute heavy metal to the waterbody when applied as adsorbent.

### 5.2 *Water-Adsorbent pH and pH Point of Zero Charge*

An amount of 1.0 g palm oil mill solid wastes-based adsorbent was equilibrated at 150 rpm with 100 mL ultrapure water for 12 h to obtain the water-adsorbent pH. Results in Table 2 indicated that the non-combusted palm oil mill solid wastes-adsorbent (OPF, OPMF and OPS) have the slightly acidic water-adsorbent pH which was attributed to the organic-cellulose content. On the other hand, the combusted palm oil mill solid waste adsorbent (OPMFA) has a slightly alkaline water-adsorbent

**Table 2** pH properties of palm oil mill solid wastes-based adsorbent

| Adsorbent | Water-adsorbent pH | $pH_{pzc}$ |
|---|---|---|
| OPF | 5.76 | 4.55 |
| OPMF | 5.53 | 5.60 |
| OPS | 5.07 | 4.10 |
| OPMFA | 10.10 | 9.10 |

pH which was attributed to its higher silica content. These observations suggested that palm oil mill solid wastes-based adsorbent, when applied into a waterbody, will not significantly affect the pH of the waterbody into any extreme pH.

The batch equilibrium technique or solid addition method [5] was employed to determine the pH point of zero charge ($pH_{pzc}$) of each produced adsorbent. The non-combusted palm oil mill solid wastes-based adsorbent has acidic $pH_{pzc}$ while the combusted oil palm solid waste-based adsorbent has alkaline $pH_{pzc}$.

## *5.3 Surface Morphology*

The OPF, OPMF, OPS and OPMFA samples were individually sputter coated with Au, and the surface morphologies were observed utilising scanning electron microscope. The captured OPF, OPMF and OPMFA micrographs exhibited the expected repetitive squarish segmentation of fibre cells which were common characteristic surface morphology among cellulosic plant tissues.

In the palm oil mill production line, the EFB has gone through rough physical handling which included the thresher that separated the fruitlet from the FFB and the screw press that further optimised CPO extraction. These rough physical processes disintegrated the original physical form of the EFB into loose separated individual fibre. It was this separated individual fibre that was eventually processed into the OPF. Due to the exposed fibrous structure, the micrograph exhibited a rough OPF surface morphology (Fig. 2a).

Palm oil is mainly extracted from the mesocarp of the fruitlet. The extraction process by-produced spent oil palm mesocarp fibre and oil palm shell which were then separately processed into the OPMF and OPS. Unlike the rough OPF surface morphology, the surface morphologies of the OPMF and OPS appeared undulated smooth (Fig. 2b). These smoother surfaces were attributed to the palm oil that was once present in the oil palm mesocarp fibre and oil palm shell.

The spent oil palm mesocarp fibre and oil palm shell were used as the boiler fuel at the weight-to-weight ratio of 70% and 30%, respectively. The high temperature, pressure and updraft in the boiler combustion chamber well mixed the boiler fuel which was inserted from the top of the combustion chamber. While the heavier combusted residue settled down as palm oil mill bottom ash, the lighter combusted residue was updraft to exit the combustion chamber as palm oil mill fly ash which was processed into the OPMFA. In the combustion chamber, the silica that presents in the spent oil palm mesocarp fibre and oil palm shell were not combusted. Under high temperature and pressure, the silica melted and mixed up with the fibrous residue. This explained the observed surface morphology of the OPMFA (Fig. 2c).

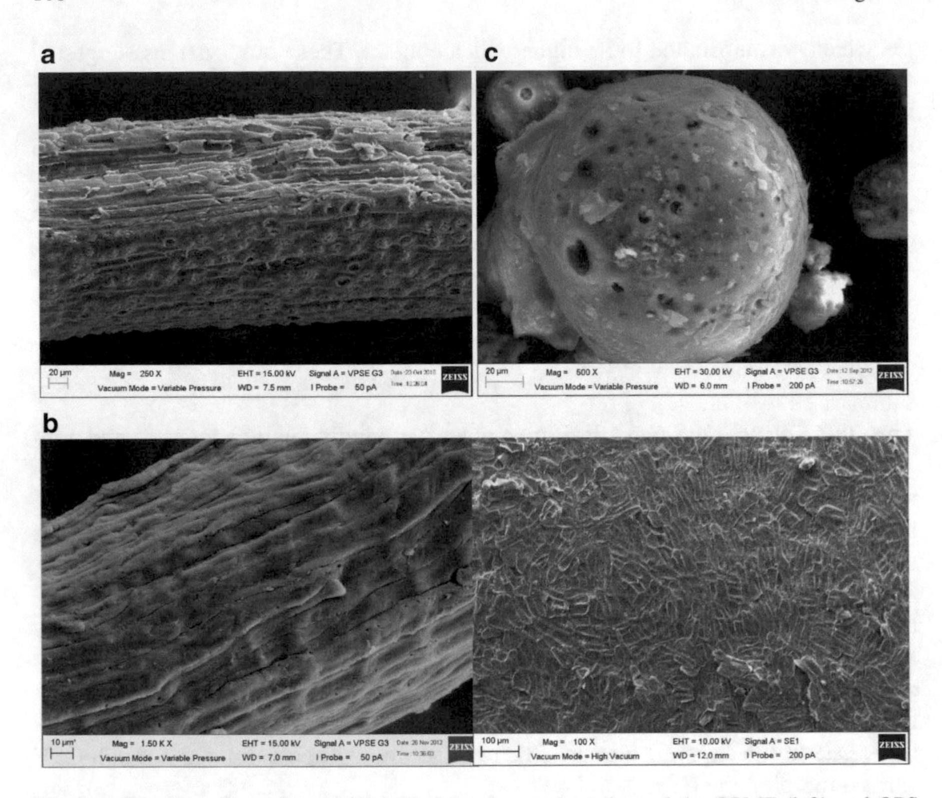


**Fig. 2** **a** Rough surface of the OPF, **b** Undulated smooth surface of the OPMF (left) and OPS (right), **c** Surface morphology of the OPMFA

## *5.4 Surface Area and Average Pore Diameter*

The palm oil mill solid wastes-based adsorbent was analysed using micropore sensitive porosimeter, and the results obtained were given in Table 3. The non-fibrous adsorbent (OPS and OPMFA) appeared to have a higher Brunauer–Emmett–Teller (BET) surface area than that of the fibrous adsorbent (OPF and OPMF).

According to the recommendation of the International Union of Pure and Applied Chemistry (IUPAC) [6], based on the obtained average pore diameter size, the OPMF,

**Table 3** Pore properties of palm oil mill solid wastes-based adsorbent

| Adsorbent | BET surface area ($m^2$/g) | Average pore diameter (nm) |
|---|---|---|
| OPF | 2.33 | 52.40 |
| OPMF | 3.63 | 4.89 |
| OPS | 104.41 | 16.96 |
| OPMFA | 24.04 | 2.37 |

OPS and OPMFA were categorised as mesoporous while the OPF was macroporous. Nevertheless, it is worthy to note that the OPF and OPMFA were both slightly overshot the borderline of 50 nm and 2 nm which otherwise will be categorised as mesoporous and microporous, respectively.

### 5.5 *Carbon, Hydrogen and Nitrogen Contents*

The CHNS/O analyser was used to analyse the carbon (C), hydrogen (H) and nitrogen (N) contents of the palm oil mill solid wastes-based adsorbent, and the results were summarised in Table 4. The non-combusted palm oil mill solid wastes-based adsorbent (OPF, OPMF and OPS) have higher contents of C and H, respectively. These were attributed to their cellulosic organic parent material.

In contrast, the combusted palm oil mill solid wastes-based adsorbent (OPMFA) has lower C and H contents. These observations were due to the heat pyrolysis process that occurred in the boiler combustion chamber where the combustible carbon was removed as carbon dioxide and the hydrogen was removed as water vapour. The non-combustible silica although changed physically, the silica was retained (see Fig. 2c).

In all cases, the OPF, OPMF, OPS and OPMFA have low nitrogen content which is good news for adsorbent intended for water-based application. In other words, these palm oil mill solid wastes-based adsorbent will not contribute nitrogen into the waterbody which it came to be in contact with, even if it is applied in long hydraulic retention time application such as constructed wetland for wastewater treatment [7].

## 6 Adsorption of Heavy Metal

### 6.1 *Optimum Adsorption pH*

The Cu(II) and Pb(II) ions were, respectively, prepared from copper nitrate hemipentahydrate and lead nitrate salts, adjusted to various pH (3–9) by adding negligible amount of nitric acid or sodium hydroxide of appropriate concentration and reacted

**Table 4** Elemental content of palm oil mill solid wastes-based adsorbent

| Adsorbent | Percentage content (w/w) | | |
|---|---|---|---|
| | C | H | N |
| OPF | 45.45 | 6.71 | 0.34 |
| OPMF | 45.46 | 5.51 | 0.43 |
| OPS | 48.48 | 6.00 | 0.37 |
| OPMFA | 13.86 | 0.23 | 0.16 |

**Table 5** Optimum pH for adsorption of Cu(II) and Pb(II)

| Adsorbent | Optimum adsorption pH | |
|---|---|---|
| | Cu(II) | Pb(II) |
| OPF | 5.0 | 6.0 |
| OPMF | 5.0 | 5.0 |
| OPS | 5.0 | 5.7 |
| OPMFA | 5.0 | 5.0 |

with the palm oil mill solid wastes-based adsorbent. Optimum adsorptions were achieved at the initial pH of the prepared solutions as given in Table 5.

At lower pH, particularly at pH below 4, the $H^+$ ions from the acid were competing against the Cu(II) and Pb(II) ions for adsorption sites; thus, the adsorption performance was not at optimum [8]. At pH slightly higher than the optimum adsorption pH, the removal of the Cu(II) and Pb(II) ions in the form of metal hydroxides precipitations took place and thus reduced the role of adsorption [9]. It is worthy to note that the optimum adsorption pH for OPMF and OPMFA occurred below their $pH_{pzc}$ of 5.60 and 9.10 while for OPF and OPS, they occurred slightly above their $pH_{pzc}$ of 4.55 and 4.10, respectively. These observations indicated that physisorption was unlikely the main adsorption mechanism.

## *6.2 Optimum Adsorption Contact Time*

The mixture of OPF, OPMF, OPS and OPMFA with the Cu(II) and Pb(II) ions which were prepared from nitrate salts was, respectively, reacted at different contact time ranging from 5 to 1440 min in order to determine the optimum contact time of each respective adsorption system. Unlike the optimum adsorption pH results which were almost similar to one another, the optimum contact time results were more diverse (Table 6). For example, the OPMFA has a relatively shorter optimum contact time as compared to the OPS; this was attributed to the finer verses coarser geometrical mean particle size of the adsorbent. When the optimum contact time data were compared to the data in Table 3, it was found that higher BET surface area did not result in a shorter optimum contact time; hence, physisorption was unlikely the rate-controlling step.

**Table 6** Optimum contact time for adsorption of Cu(II) and Pb(II)

| Adsorbent | Optimum contact time (min) | |
|---|---|---|
| | Cu(II) | Pb(II) |
| OPF | 90 | 120 |
| OPMF | 240 | 120 |
| OPS | 240 | 480 |
| OPMFA | 60 | 60 |

**Table 7** Langmuir and Freundlich adsorption isotherm constants

| Adsorbent | Adsorbate | Langmuir | | | Freundlich | | |
|---|---|---|---|---|---|---|---|
| | | $Q_m$ (mg/g) | $K_L$ (L/mg) | $R^2$ | $n$ | $K_F$ [mg/g(L/mg)$^{1/n}$] | $R^2$ |
| OPF | Cu(II) | 14.39 | 0.013 | 0.910 | 1.817 | 0.592 | 0.994 |
| | Pb(II) | 11.12 | 0.110 | 0.993 | 5.277 | 4.142 | 0.992 |
| OPMF | Cu(II) | 5.93 | 0.087 | 0.999 | 4.143 | 1.531 | 0.956 |
| | Pb(II) | 39.53 | 0.034 | 0.975 | 2.574 | 4.121 | 0.992 |
| OPS | Cu(II) | 1.76 | 0.013 | 0.860 | 2.371 | 0.124 | 0.956 |
| | Pb(II) | 3.39 | 0.016 | 0.891 | 3.866 | 0.561 | 0.893 |
| OPMFA | Cu(II) | 153.85 | 0.041 | 0.994 | 2.185 | 13.549 | 0.948 |
| | Pb(II) | 181.82 | 0.068 | 0.998 | 2.170 | 18.599 | 0.906 |

## 6.3 Adsorption Isotherms

Langmuir [10] and Freundlich [11] were the two most commonly used adsorption isotherm models. The linear equations of the Langmuir and Freundlich adsorption isotherm models were represented in Eqs. (1) and (2), respectively:

$$\frac{C_e}{Q_e} = \frac{1}{K_L Q_m} + \frac{C_e}{Q_m} \tag{1}$$

$$\text{Log}\, Q_e = \text{Log}\, K_F + \frac{1}{n}\text{Log}\, C_e \tag{2}$$

where the $C_e$ (mg/L), $Q_e$ (mg/g), $K_L$ (L/mg), $Q_m$ (mg/g), $K_F$ (mg/g)(L/mg)$^{1/n}$ and $n$.

were the concentration of adsorbate ions at equilibrium, amount of adsorbate ions adsorbed at equilibrium, Langmuir energy sorption constant, monolayer adsorption capacity, Freundlich adsorption capacity constant and Freundlich intensity constant, respectively.

The palm oil mill solid wastes-based adsorbent was, respectively, equilibrated with Cu(II) and Pb(II) ions of various concentrations (5–500 mg/L) via batch experiments. The data obtained were processed and fitted into the Langmuir and Freundlich adsorption isotherm models. The adsorption isotherm constants derived from these fittings were given in Table 7.

While the OPF-Pb(II), OPMF-Cu(II), OPMFA-Cu(II) and OPMFA-Pb(II) adsorption systems were better fitted into Langmuir adsorption isotherm model, the OPF-Cu(II), OPS-Pb(II), OPS-Cu(II) and OPS-Pb(II) adsorption systems were better fitted into Freundlich adsorption isotherm model. In other words, the fibrous palm oil mill solid wastes-based adsorbent (OPF and OPMF) showed good fitting to both adsorption isotherm models simultaneously, while the coarse and fine particle palm oil mill solid waste-based adsorbent (OPS and OPMFA) showed consistent better fitting to Freundlich and Langmuir adsorption isotherm models, respectively. The n value of more than one ($n > 1$) denoted a favourable adsorption [12]. The $K_L$ and $K_F$ values indicated adsorption affinity where higher value meaning better adsorption affinity [13, 14].

The $Q_m$ value represented the monolayer adsorption capacity of the respective adsorbent-adsorbate adsorption system. It was noted that higher BET surface area (Table 3) did not result in higher $Q_m$ value. For example, the BET surface area and Pb(II) monolayer adsorption capacity of OPS were 104.41 $m^2/g$ and 3.39 mg/g as compared to OPMFA which was 24.04 $m^2/g$ and 181.82 mg/g. It appeared that the Cu(II) and Pb(II) adsorption capacities of palm oil mill solid waste-based adsorbent were not positively influenced by the magnitude of BET surface area, hence suggesting influence of chemisorption instead of physisorption.

Table 8 compares the monolayer adsorption capacities of various unmodified adsorbents for the adsorption of Cu(II) and Pb(II) ions. The OPS was in $6.5 < Ø \leq 8.0$ mm geometrical mean particle size; thus, it is not surprising that it has relatively low monolayer adsorption capacities. Nonetheless, it is worthy to note that the OPS outperformed coral-, coconut coir-, rice husk- and onion seed-based adsorbents which were in finer sizes. The OPMFA, being the finest size among the experimented palm oil mill solid wastes-based adsorbent, has the maximum adsorption capacities that outperformed most unmodified adsorbent except some coal and bagasse fly ashes.

**Table 8** Maximum adsorption capacities of various unmodified adsorbents

| Adsorbate | Adsorbent | $Q_m$ (mg/g) | Reference |
|---|---|---|---|
| Cu(II) | Coral | 0.43 | [15] |
| | Coconut coir | 1.34 | [16] |
| | Rice husk | 1.56 | [16] |
| | OPS | 1.76 | This work |
| | Maple sawdust | 1.79 | [17] |
| | Bagasse fly ash | 2.36 | [18] |
| | Coal fly ash | 2.80 | [19] |
| | OPMF | 5.93 | This work |
| | Poplar sawdust | 6.59 | [20] |
| | Linden sawdust | 9.90 | [21] |
| | Grape stalk | 10.12 | [22] |
| | Spent barley grain | 10.47 | [23] |
| | OPF | 14.39 | This work |
| | Coal fly ash | 21.00 | [24] |
| | Pine sawdust | 24.30 | [25] |
| | Peanut shell | 25.39 | [9] |

(continued)

**Table 8** (continued)

| Adsorbate | Adsorbent | $Q_m$ (mg/g) | Reference |
|---|---|---|---|
| | Sunflower leaves | 89.37 | [26] |
| | OPMFA | 153.85 | This work |
| | Coal fly ash | 207.30 | [27] |
| | Coal fly ash | 249.10 | [28] |
| Pb(II) | Coral | 1.14 | [15] |
| | Onion seed | 1.68 | [29] |
| | OPS | 3.39 | This work |
| | OPF | 11.12 | This work |
| | Coal fly ash | 18.00 | [30] |
| | Poplar sawdust | 21.05 | [20] |
| | Coal fly ash | 22.00 | [24] |
| | Mustard husk | 30.48 | [31] |
| | OPMF | 39.53 | This work |
| | Orange peel | 89.77 | [32] |
| | Mango peel | 96.32 | [33] |
| | Cucumber peel | 133.60 | [34] |
| | OPMFA | 181.82 | This work |
| | Coal fly ash | 249.10 | [28] |
| | Coal fly ash | 444.70 | [27] |
| | Bagasse fly ash | 566.00 | [35] |

### *6.4 Adsorption Kinetics*

The pseudo-first-order [36] and pseudo-second-order [37] were the two most commonly used adsorption kinetic models. The linear equations of the pseudo-first-order and pseudo-second-order adsorption kinetic models were represented in Eqs. (3) and (4), respectively:

$$\text{Log}\,(Q_e - Q_t) = \text{Log}\,Q_e - \frac{k_1}{2.303}t \quad (3)$$

$$\frac{t}{Q_t} = \frac{1}{k_2 Q_e^2} + \frac{1}{Q_e}t \quad (4)$$

where $k_1$ (1/min) was the pseudo-first-order rate constant and $k_2$ (g/mg min) was pseudo-second-order rate constant.

The OPF, OPMF, OPS and OPMFA were individually reacted with Cu(II) and Pb(II) ions prepared from nitrate salts for a series of varied contact time (5–480 min) via batch adsorption experiments. The experimental results were computed and fitted

**Table 9** Pseudo-first- and pseudo-second-orders adsorption kinetics constants

| Adsorbent | Adsorbate | Pseudo-first-order | | Pseudo-second-order | | |
|---|---|---|---|---|---|---|
| | | $k_1$ (1/min) | $R^2$ | $k_2$ (g/mg min) | $u$ (mg/g min) | $R^2$ |
| OPF | Cu(II) | 0.044 | 0.972 | 0.040 | 0.653 | 0.997 |
| | Pb(II) | 0.044 | 0.984 | 0.022 | 0.738 | 0.999 |
| OPMF | Cu(II) | 0.014 | 0.962 | 0.048 | 0.297 | 0.999 |
| | Pb(II) | 0.178 | 0.991 | 0.343 | 3.829 | 0.999 |
| OPS | Cu(II) | 0.012 | 0.958 | 0.204 | 0.013 | 0.997 |
| | Pb(II) | 0.009 | 0.875 | 0.063 | 0.027 | 0.995 |
| OPMFA | Cu(II) | 0.053 | 0.982 | 0.002 | 0.200 | 0.998 |
| | Pb(II) | 0.058 | 0.966 | 0.004 | 0.307 | 0.986 |

into the two most commonly applied adsorption kinetic models, namely pseudo-first-order and pseudo-second-order. All the adsorption kinetic constants derived from the fittings were given in Table 9 where $k_1$, $k_2$ and u are pseudo-first-order rate constant, pseudo-second-order rate constant and initial adsorption rate, respectively.

Relatively higher $k_1$, $k_2$ and $u$ values were obtained from the OPMF-Pb(II) adsorption system. These were attributed to its combination of higher maximum adsorption capacity and shorter optimum contact time characteristic which translated into higher adsorption rate constants ($k_1$ and $k_2$ values) and higher preference for adsorption ($u$ value). The pseudo-second-order adsorption kinetic model gave high correlation coefficient, which almost unity, for all the experimented adsorption systems, thus indicating that chemisorption was the rate-limiting factor [38].

## 7 Conclusion and the Way Forward

The global statistic indicated that the palm oil industry will continue to expand. In this regard, it means that there will be a continuous expanding sustainable by-production of the palm oil mill solid wastes for the purpose of valorisation as adsorbent material.

The preparation procedure of the palm oil mill solid wastes-based adsorbent, namely OPF, OPMF, OPS and OPMFA, was explored. In fact, it was a natural, without any heat or chemical modification, hence keeping the adsorbents' environmental footprint as small as possible. The washing and soaking time will be reduced in future commercial scale which can be implemented within the palm oil mill compound by utilising the spent boiler warm water. Physicochemical investigations indicated that the palm oil mill solid wastes-based adsorbent will not leach out heavy metal and nitrogen into the waterbody that it came to be in contact with. This suggested that the adsorbent will not be a source for heavy metal and eutrophication-related pollution.

Based on the results of Cu(II) and Pb(II) ions adsorption, it has been proven that the palm oil mill solid wastes-based adsorbent were functional. The adsorbents gave impressive adsorption capacities, and the adsorption is believed to be of chemisorption.

Future research should be looking into the application of the OPF, OPMF, OPS and OPMFA for slow- or batch-flow wastewater treatment system. For example, the potential application of OPF and OPMF as adsorption or filter bed media for the improvement of water quality in wastewater treatment system. The application of OPS as alternative media to enhance the heavy metal removal performance and operational lifespan of the wastewater treatment constructed wetland should be investigated. The OPMFA shall be tested in wastewater treatment system that needs adsorption assistance such as the sequential batch reactor.

**Acknowledgements** The research grants that funded the study on the application of palm oil mill solid wastes as unmodified adsorbent, namely the FRG0270-ST-2/2010, SBK0112-STWN-2013, SBK0194-ST-2015, GUG0070-SG-2/2016 and SBK0415-2018, are highly appreciated. The inputs as well as laboratory and research assistances of Chia, P.S., Wong, K.H., Sohail, R., Thoe, J.M.L. and Lim, J.M. are cordially acknowledged.

## References

1. Food and Agriculture Organization of the United Nation (FAO).: FAOSTAT Statistical Database. http://www.fao.org/faostat/en/#data/QC (2020). Accessed 16 May 2021
2. Chong, H.L.H.: Treatment Processes of Palm Oil Mill Effluent: A Case Study of Medium-sized Palm Oil Mill in Northern Sabah. Universiti Malaysia Sabah Publisher, Kota Kinabalu (2020)
3. Yusoff, S.: Renewable energy from palm oil—innovation on effective utilisation of waste. J. Clean. Prod. **14**, 87–93 (2006)
4. Chong, H.L.H., Chia, P.S.: Basic Characterisation of Natural Bornean Oil Palm Shell and Its Heavy Metal Adsorption. Universiti Malaysia Sabah Publisher, Kota Kinabalu (2014)
5. Balistrieri, L.S., Murray, J.W.: The surface chemistry of geothite (αFeOOH) in major ion seawater. Am. J. Sci. **281**, 788–806 (1981)
6. International Union of Pure and Applied Chemistry (IUPAC).: Reporting physisorption data for gas/solid systems with special reference to the determination of surface area and porosity. Pure Appl. Chem. **54**(11), 2201–2218 (1982)
7. Chong, H.L.H.: Basic characterisation of oil palm shell as constructed wetland media and the effect of biofilm formation on its adsorption of copper (II). Ph.D. thesis, Universiti Sains Malaysia (2008)
8. Chong, H.L.H., Chia, P.S., Ahmad, M.N.: The adsorption of heavy metal by Bornean oil palm shell and its potential application as constructed wetland media. Bioresour. Technol. **130**, 181–186 (2013)
9. Witek-Krowiak, A., Szafran, R.G., Modelski, S.: Biosorption of heavy metals from aqueous solutions onto peanut shell as a low-cost biosorbent. Desalination **265**, 126–134 (2011)
10. Langmuir, I.: The adsorption of gases on plane surfaces of glass, mica and platinum. Am. Chem. Soc. **40**, 1361–1403 (1918)
11. Freundlich, H.M.F.: Uber die adsorption in lösungen (On adsorption in solutions). Z. Phys. Chem. **57**, 385–470 (1906)

12. Arami, M., Limaee, N.Y., Mahmoodi, N.M., Tabrizi, N.S.: Equilibrium and kinetics studies for the adsorption of direct and acid dyes from aqueous solution by soy meal hull. J. Hazard. Mater. **B135**, 171–179 (2006)
13. Ferreira, L.S., Rodrigues, M.S., Carvalho, J.C.M.D., Lodi, A., Finocchio, E., Perego, P., Converti, A.: Adsorption of $Ni^{2+}$, $Zn^{2+}$ and $Pb^{2+}$ onto dry biomass of *Arthrospira* (*Spirulina*) *platensis* and *Chlorella vulgaris*. I. Single metal systems. Chem. Eng. J. **173**, 326–333 (2011). Insert references for $K_L$ and $K_F$ interpretations
14. Kwon, J., Yun, S., Lee, J., Kim, S., Jo, H.Y.: Removal of divalent heavy metals (Cd, Cu, Pb, and Zn) and arsenic(III) from aqueous solutions using scoria: Kinetics and equilibria of sorption. J. Hazard. Mater. **174**, 307–313 (2010)
15. Ahmad, M., Usman, A.R.A., Lee, S.S., Kim, S., Joo, J., Yang, J.E., Ok, Y.S.: Eggshell and coral wastes as low cost sorbents for the removal of $Pb^{2+}$, $Cd^{2+}$ and $Cu^{2+}$ from aqueous solutions. J. Ind. Eng. Chem. **18**, 198–204 (2012)
16. Tokay, B., Akρınar, I.: A comparative study of heavy metals removal using agricultural waste biosorbents. Bioresour. Technol. Rep. **15**, 100719 (2021). https://doi.org/10.1016/j.biteb.2021.100719
17. Yu, B., Zhang, Y., Shukla, A., Shukla, S.S., Dorris, K.L.: The removal of heavy metal from aqueous solutions by sawdust adsorption—removal of copper. J. Hazard. Mater. **B80**, 33–42 (2000)
18. Gupta, V.K., Ali, I.: Utilisation of bagasse fly ash (a sugar industry waste) for the removal of copper and zinc from wastewater. Sep. Purif. Technol. **18**, 131–140 (2000)
19. Lin, C.J., Chang, J.E.: Effect of fly ash characteristics on the removal of Cu(II) from aqueous solution. Chemosphere **44**, 1185–1192 (2001)
20. Li, Q., Zhai, J., Zhang, W., Wang, M., Zhou, J.: Kinetic studies of adsorption of Pb(II), Cr(II) and Cu(II) from aqueous solution by sawdust and modified peanut husk. J. Hazard. Mater. **141**, 163–167 (2007)
21. Bozic, D., Stankovic, V., Gorgievski, M., Bogdanovic, G., Kovacevic, R.: Adsorption of heavy metal ions by sawdust of deciduous trees. J. Hazard. Mater. **171**, 684–692 (2009)
22. Villaescusa, I., Fiol, N., Martinez, M., Miralles, N., Poch, J., Serarols, J.: Removal of copper and nickel ions from aqueous solutions by grape stalks wastes. Water Res. **38**, 992–1002 (2004)
23. Lu, S., Gibb, S.W.: Copper removal from wastewater using spent-grain as biosorbent. Bioresour. Technol. **99**, 1509–1517 (2008)
24. Alinnor, I.J.: Adsorption of heavy metal ions from aqueous solution by fly ash. Fuel **86**, 853–857 (2007)
25. Hansen, H.K., Arancibia, F., Gutierrez, C.: Adsorption of copper onto agriculture waste materials. J. Hazard. Mater. **180**, 442–448 (2010)
26. Benaissa, H., Elouchdi, M.A.: Removal of copper ions from aqueous solutions by dried sunflower leaves. Chem. Eng. Process. **46**, 614–622 (2007)
27. Apak, R., Tutem, E., Hugul, M., Hizal, J.: Heavy metal cation retention by unconventional sorbents (red muds and fly ash). Water Res. **32**, 430–440 (1998)
28. Hsu, T.-C., Yu, C.-C., Yeh, C.-M.: Adsorption of $Cu^{2+}$ from water using raw and modified coal fly ashes. Fuel **87**, 1355–1359 (2008)
29. Sheikh, Z., Amin, M., Khan, N., Khan, M.N., Sami, S.K., Khan, S.B., Hafeez, I., Khan, S.A., Bakhsh, E.M., Cheng, C.K.: Potential application of *Allium cepa* seeds as a novel biosorbent for efficient biosorption of heavy metals ions from aqueous solution. Chemosphere **279**, 130545 (2021). https://doi.org/10.1016/j.chemosphere.2021.130545
30. Wang, S., Terdkiatburana, T., Tadé, M.O.: Single and co-adsorption of heavy metals and humic acid on fly ash. Sep. Purif. Technol. **58**, 353–358 (2008)
31. Meena, A.K., Kadirvelu, K., Mishraa, G.K., Rajagopal, C., Nagar, P.N.: Adsorption of Pb(II) and Cd(II) metal ions from aqueous solutions by mustard husk. J. Hazard. Mater. **150**, 619–625 (2008)
32. Liang, S., Guo, X., Feng, N., Tian, Q.: Application of orange peel xanthate for the adsorption of $Pb^{2+}$ from aqueous solutions. J. Hazard. Mater. **170**, 425–429 (2009)

33. Iqbal, M., Saeed, A., Zafar, S.I.: FTIR spectrophotometry, kinetics and adsorption isotherms modeling, ion exchange, and EDX analysis for understanding the mechanism of $Cd^{2+}$ and $Pb^{2+}$ removal by mango peel waste. J. Hazard. Mater. **164**, 161–171 (2009)
34. Basu, M., Guha, A.K., Ray, L.: Adsorption of lead on cucumber peel. J. Clean. Prod. **151**, 603–615 (2007). https://doi.org/10.1016/j.jclepro.2017.03.028
35. Gupta, V.K., Mohan, D., Sharma, S.: Removal of lead from wastewater using bagasse fly ash—A sugar industry waste material. Sep. Purif. Technol. **33**, 1331–1343 (1998)
36. Lagergren, S.: Kungliga Svenska Vetenskapsakademiens (About the theory of so-called adsorption of soluble substances). Handlingar **24**, 1–39 (1898)
37. Ho, Y., McKay, G.: Pseudo-second order model for sorption processes. Proc. Biochem. **34**, 451–465 (1999)
38. Anirudhan, T.S., Radhakrishnan, P.G.: Kinetics, thermodynamics and surface heterogeneity assessment of uranium(VI) adsorption onto cation exchange resin derived from a lignocellulosic residue. Appl. Surf. Sci. **255**, 4983–4991 (2009)

# Comparison Between Fresh and Degraded Biochar for Ammonium Ion ($NH_4^+$) Removal from Wastewater

**Noor Maizura Ismail, Nurliyana Nasuha Safie, Manjulla Subramaniam, Nur Syafidah Junaidi, and Abu Zahrim Yaser**

**Abstract** The fresh bamboo biochar (FBB) and degraded bamboo biochar (DBB) on $NH_4^+$ removal was investigated. FBB and DBB were produced at an average of 500 °C for 2 h with a slow rate using a self-fabricated setup, Top Lit Up Draft Kiln (TLUD) brick kiln. The physicochemical properties of raw fresh bamboo (FB), raw degraded bamboo (DB), FBB, and DBB such as elemental analysis, morphological structure-specific surface area ($S_{BET}$), pore-volume ($P_V$), and diameter ($P_z$), as well as the surface functional group, were characterized using scanning electronic microscope-energy dispersive X-ray spectroscopy (SEM–EDX), Brunauer-Emmet-Teller (BET) and Fourier-transform infrared radiation (FTIR). N–H stretching at 2665.49 $cm^{-1}$ was indicated after the adsorption for DBB and has shown that the $NH_4^+$ has been adsorbed on the surface of DBB. The adsorption capacity and percentage of $NH_4^+$ removal of FBB and DBB was 7.29 mg/g, 65.5% and 5.98 mg/g, 61.7%, respectively. The performance of FBB and DBB in $NH_4^+$ removal was also studied. This has shown clearly that the FBB has higher adsorption capacity and percentage removal as compared to DBB. Both FBB, as well as DBB, can adsorb $NH_4^+$. However, DBB may require a longer time to achieve equilibrium. The adsorption of $NH_4^+$on the FBB and DBB is a function of oxygen-containing functional groups, and physisorption is not the dominant mechanism.

**Keywords** Biochar · Ammonium ion · Adsorption · Sewage

## 1 Introduction

$NH_4^+$ is actively present in sewage originated from urine, stools, and washing water. The varying concentration due to the population increment has caused a decline in the efficiency of conventional sewage treatment in $NH_4^+$ removal. Particularly, the average concentration of $NH_4^+$ in sewage can range between 20 and 50 mg/L [1]. The conventional biological treatment by the nitrification–denitrification process has an

N. M. Ismail · N. N. Safie · M. Subramaniam · N. S. Junaidi · A. Z. Yaser (✉)
Chemical Engineering Programme, Faculty of Engineering, Universiti Malaysia Sabah, Kota Kinabalu, Sabah, Malaysia
e-mail: zahrim@ums.edu.my

 
A. Z. Yaser et al. (eds.), *Waste Management, Processing and Valorisation*,
https://doi.org/10.1007/978-981-16-7653-6_7

efficiency of less than 60% in the removal of $NH_4^+$ due to the varying concentration of $BOD_5$ daily [2]. Moreover, the increment of the population equivalent (PE) will cause unsteady changes in influent concentration [3]. The $NH_4^+$ concentration by 0.15–1.8 mg/L will subsequently reduce the rate of nitrification in the trickling filter by 60% [3]. Other than temperature, the efficiency of denitrification by conventional activated sludge treatment depending on the C: N ratio [2]. Due to this, tertiary treatment such as adsorption is needed to increase the efficiency in $NH_4^+$ removal before discharge.

Malaysia has approximately 50 bamboo species in Peninsular Malaysia, 30 species in Sabah, and 20 species in Sarawak [4]. The most common bamboo species are *Gigantochloa scortechinii Gamble, Dendrocalamus asper Backer ex K.Heyne and Bambusa blumeana Schult.f.*, [5]. Bamboo is a vernacular or common term for members of a particular taxonomic group of large woody grassed (subfamily *Bambusoideae*, family *Andropogoneae/Poaceae*) [6]. Sabah consists of 5200 ha of bamboo forest that located mostly in the Tambunan, Ranau, Keningau, Sipitang, and Telupid that has huge potential for industrial application [7].

The limitation of bamboo is its susceptibility to the biological and physical deterioration [8]. Biological deterioration on bamboo during storage can affect the utilization, toughness, shelf life, and quality of the finalized product. More than 40% of bamboo is destroyed by xylophagous boring insects (e.g., termites and beetles) as well as microorganisms (e.g., *Postia placenta* fungi, *Polyporus fumosus*) during storing in untreated conditions [9]. The most common species commonly found on bamboo culms during storage and use is *S. commune* and has the highest decaying rate at $T =$ 30 °C–35 °C [10]. Unused or wasted bamboo can reach up to 19.1% [11]. The industrial exploitation of 4.5 million tons of bamboo, resulting in substantial quantity of bamboo residual materials [12]. Hence, recycling deteriorated bamboo and left-over fresh bamboo into biochar is an economical and environment-friendly solution to the large unexploited biomass feedstock and waste.

The preparatory process of biochar conducted through slow pyrolysis of crop residues under an oxygen-limited condition will alter the physicochemical properties and provide the ability to be used in the removal of $NH_4^+$ [13]. TLUD kiln or stove is a simple method to produce biochar, low solid particles emissions, and usage of waste such as dried leaves, twigs, and bark of trees as combustibles to provide heat for carbonization leading to the negative balance to the low emissions of $CO_2$ into the atmosphere during carbonization [14]. As compared to the muffle furnace, carbonization temperature cannot be controlled when using TLUD kiln. The carbonization will occur at $T > 450$ °C, leading to biochar that is more microporous, larger surface area, and higher ash content [15]. There is a few research being conducted to produce biochar from crop wastes such as cottonseed [15], rice husk [16], and coffee waste [17] using the TLUD kiln.

Few biochar adsorbents derived from assorted agricultural residues that have been used in $NH_4^+$ removal are pine sawdust [18], coconut shell [19], rice husk [20], and bamboo [21, 22] with an adsorption capacity of 1.27 mg/g, 2.3 mg/g, 3.76 mg/g, 5 mg/g, and 9.8 mg/g, respectively. Nevertheless, the comparison of bamboo biochar derived from fresh bamboo and degraded bamboo using TLUD kiln to remove $NH_4^+$ is still limited. Hence, the physicochemical as well as performance of FBB and DBB in $NH_4^+$ removal will be investigated.

# 2 Methodology

## 2.1 Materials and Preparation of FBB and DBB

The bamboo was collected at Sepanggar, Kota Kinabalu. The synthetic solution of $NH_4^+$ was prepared using ammonium chloride ($NH_4Cl$) salt, ChemAr. The pH of the synthetic $NH_4^+$ solution was altered accordingly using NaOH and HCl. Nessler reagent (Hach, USA) was used to analyze $NH_4^+$. The bamboo was sun-dried and chipped using a woodchipper into the size of approximately 10–5 cm length and 3–5 cm wide. The DB chips were left at room temperature for 30 days to be naturally degraded. This was done to impose the real situation during the restoration of bamboo in a warehouse that is easily being exposed to biological deterioration. Bamboo is usually stored for less than two years, but it is treated chemically to avoid the destruction of hemicellulose and cellulose structures that maintain the bamboo strength. The degradation period of 30 days was picked as only lignin will be degraded at this period, without altering the structure of bamboo [23]. The FB and DB were washed using tap water and dried at 104 °C and placed in a small metal barrel. This is to ensure the carbonization process proceeds under the limited oxygen and avoid direct combustion of precursors. The carbonization of the precursors was carried out in the TLUD brick kiln (W: 0.53 m, H: 0.72 m, L: 0.53) with a chimney (Ø: 0.20, H: 0.50 m). The fabrication and setup are done by following the method by Romo [24]. TLUD kiln has been used by few authors to produce biochar [15, 25, 26]. The carbonization occurred at average $T = 500$ °C approximately for 2 h. The collected biochar was ground to powder form using mortar and pestle and stored in a desiccator.

## 2.2 Characterization of Physicochemical Properties of DB, FB, DBB, and FBB

The comprehensive surface morphology of DB, FB, DBB, and FBB are attained via morphology, and elemental composition of FB and DB were attained by via a scanning electron microscope (SEM, S-3400 N, Hitachi) that is combined with an energy dispersive X-ray (EDX) spectrometer with 10.0 kV acceleration voltage at magnification range of 100–500 and 100–2.3 k for FBB and DBB, respectively. The samples were coated with gold by a sputter coater Q150RS before analysis to intensify the conductivity of the samples [27]. The moisture content of DBB and

FBB was determined by following the standard method ASTM D1762-84 [28] and calculated using Eq. (1).

$$\text{Ash } (\%) = \left( \frac{M_{\text{initial}} - M_{\text{ash}}}{M_{\text{initial}}} \right) \tag{1}$$

$$\text{Moisture } (\%) = \left( \frac{M_{\text{initial}} - M_{\text{OD-BC}}}{M_{\text{initial}}} \right) \times 100 \tag{2}$$

where

$M_{\text{initial}}$ = Initial sample mass (g)

$M_{ash}$ = Sample mass after combustion at 500 °C for 8 hours (g)

$M_{\text{OD-BC}}$ = Mass of the sample after oven-drying at 105 °C for 2 hours (g)

The ash content of DBB and FBB was tested using the method by Dai et al. [29] and calculated using Eq. (2). The percentage yield was identified and calculated by using Eq. (3) below:

$$\text{Percentage yield } (\%) = \frac{\text{Weight of biochar}}{\text{Weight of raw bamboo}} \times 100\% \tag{3}$$

The pore structure characteristics such as $S_{\text{BET}}$, $P_z$, and pore volume $P_v$ of FB, DB, FBB, and DBB were determined from the nitrogen adsorption isotherms obtained at 77.3 K [30]. The $S_{\text{BET}}$ was estimated by applying the BET method, and $P_v$ was evaluated at a relative pressure ($P/P_o$) of 0.99 [30, 31]. The surface functional group presence in biochar was observed by using Fourier transform infrared spectrometer (FT-IR) analysis. The FT-IR spectra of the biochar sample were assessed within the range 4000–400 $cm^{-1}$ at a resolution of 4 $cm^{-1}$. FTIR analysis was conducted before and after the adsorption experiment to observe the presence of ammonia nitrogen components [32].

## 2.3 *Adsorption*

The consequence of contact time on the adsorption process was examined. 1.0 g of adsorbent was mixed with 200 mL of a solution containing 50 mg/L of the initial concentration of $NH_4Cl$ solution at pH 7 similarly to our previous study [33]. The alteration of pH was done by applying 0.1 M of NaOH. The $NH_4^+$ synthetic solution was shaken using an electronic for 24 h with an agitation speed of 150 rpm at room temperature. 1 mL of samples were taken out every 0.5, 1, 2, 3, 5, 22, and 24 h. The $NH_4^+$ was analyzed and estimated by using UV–Vis Spectrophotometer DR6000 with a maximum wavelength of 425 nm following the standard USEPA Nessler

Method No. 8038 [34]. The concentration of $NH_4^+$ adsorbed at equilibrium was calculated as in Eq. (4) [35].

$$q_e = \frac{(C_o - C_e)V}{W} \quad (4)$$

where

$C_o$: initial concentration (mg/L)

$C_e$: final concentration (mg/L)

$q_e$: Equilibrium adsorption capacity (mg/g)

$V$: volume (L)

$W$: weight of adsorbent used (g)

# 3 Result and Discussion

## *3.1 Determination of Elemental Atomic Composition, Carbon Char Yield, Ash Content, and Moisture Content*

The elemental atomic composition of FB, DB, FBB, and DBB was tabulated in Table 1. FB contains carbon (C), hydrogen (H), oxygen (O), nitrogen (N), potassium (K), magnesium (Mg), and silica (Si). While for DB, H, N, and Si were undetected. There are 81.18% of C, 15.65% of O, 0.36% of K, 0.28% of Mg, and 0.05% of calcium (Ca) in DB. After the carbonization of fresh bamboo (FB), an increment of C was noticed for FBB by 49.41% and a decrement of C by 2.97% for DBB.

Element of H, O, K, and Mg has been decreased after the carbonization of FB. This is because, during the carbonization process, non-carbon elements such as H, O, K, and Mg have been decreased are eliminated and more condensed structures are formed including amorphous C (which presides at lower pyrolysis temperatures), turbostratic C (shaped at higher temperature and graphite C [36, 37]. The aromatization of carbon is a result of the reduction of volatile components that produce more compact and dense structures due to the increased pressure of heat during the pyrolysis that produces graphitic crystalline structures [38]. Throughout the carbonization, inorganic constituents as for instance K, Mg, and Ca will be utilized to facilitate in the development of O-containing functional groups, such as hydroxyl and carboxyl fractions on the biochar surface [39].

While the concentration of N increased after the carbonization, Si was not detected in the FBB. DBB has a lower percentage of C, O, and Mg compared to DB. The element K has been increased by 93.83%. A significant increase in K is noticed that may be due to the higher carbonization temperature as the temperature in TULD was

**Table 1** The elemental composition of FB, FBB, DB (DB), FBB and DBB

| | C (at. %) | H (at. %) | O (at. %) | N (at. %) | K (at. %) | Mg (at. %) | Ca (at. %) | Si (at. %) | Carbon char yield (%) | Ash content (%) | Moisture content (%) |
|---|---|---|---|---|---|---|---|---|---|---|---|
| FB | 61.65 | 16.31 | 12.10 | 1.15 | 0.60 | 0.20 | UD | 0.05 | NA | | |
| DB | 81.18 | UD | 15.65 | UD | 0.36 | 0.28 | 0.05 | UD | | | |
| FBB | 92.11 | 0.09 | 5.03 | 1.51 | 0.30 | 0.13 | UD | UD | 46.95 | 2.34 ± 0.23 | 0.09 ± 0.03 |
| DBB | 78.77 | UD | 8.59 | UD | 5.83 | 0.09 | 0.70 | 0.06 | 26.40 | 4.11 ± 1.00 | 1.63 ± 2.00 |

* Remarks: UD: undetected, NA: not applicable

hard can be controlled [40]. Mostly, the highest element in ash content in biochar is the inorganic K along with other minerals in particular Ca and Mg in the form of oxides that will influence the pH value of biochar [41].

The percentage of carbon char yield, ash content, and moisture content of FBB and DBB is listed in Table 1. FBB has a greater C percentage yield by 44% compared to DBB. The presence of C is highly attributed by the presence of cellulose (~47.5%), lignin (~26.25%), and hemicellulose (~15.35%) originated from the raw bamboo [42]. DBB has a lower carbon char yield as compared to FBB possibly due to the deposition of lignin and some hemicellulose as well as cellulose during the biological degradation process. Weight losses of bamboo were recorded by 35–40% after the degradation process [43].

As listed in Table 1, the moisture content of DBB is higher than DBB which are 0.09% and 1.63%, respectively. The ash content of DBB is higher than FBB which are 4.11% and 2.34%, respectively. This is likely due to the higher composition of inorganic elements such as K and Mg in DBB as compared to FBB. The high ash content in DBB, $4.11 \pm 1.00$ is conceivably impacted by the larger composition of inorganic compounds such as K, Ca, and Mg in DBB as compared to FBB as in Table 1. Ash mostly aggregated after the volatilization of C, O, and H compounds [38]. The maximum ash content in commercial activated carbon is 5% [44]. Therefore, the ash content in DBB is considerably high.

The lower percentage of C char yield and high ash content of DBB as compared to FBB is suspected due to the higher temperature during the carbonization process. In this study, the temperature of carbonization using TLUD brick kiln is affected by the weight of biomass used as a fuel outside of the metal barrel that contains precursors to be carbonized as biochar, and the temperature cannot be controlled manually or set as compared to using the electrical furnace. The higher pyrolysis temperature will lead to higher C and ash content and produced a more condensed carbon structure in the biochar.

### *3.2 Determination of Functional Groups Through Fourier-Transform Infrared Spectroscopy (FTIR) Analysis for Bamboo Biochar Before Adsorption and After Adsorption*

Figure 1 shows the FTIR analysis of DBB and FBB before and after the $NH_4^+$ adsorption. Before the adsorption, DBB has stretching vibration at 3600–3200 $cm^{-1}$ that indicates the presence of O–H [45]. The peak at 1565 $cm^{-1}$ indicates that there C = O attached with aromatic [45]. A peak at 885–714 $cm^{-1}$ indicates C–H bending aromatic out-of-plane deformation which might be caused by carbonates [46]. Before the adsorption, the peak observed on FBB at 1110.74 $cm^{-1}$ gives symmetric C–O stretching for cellulose, hemicellulose, and lignin [47] Moreover, the transmittance at 748 $cm^{-1}$ gives alkynes with C–H bending that was possibly due to the loss of

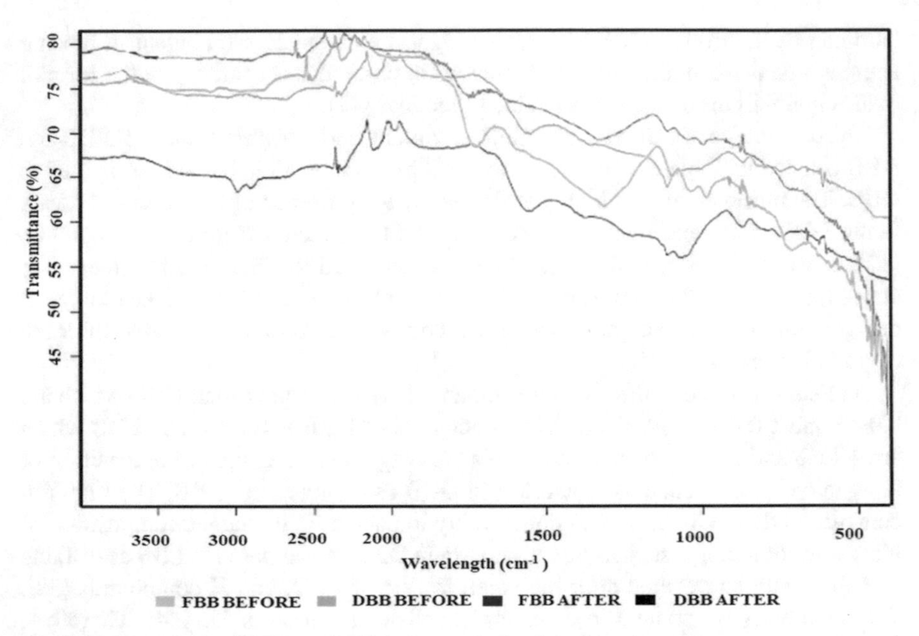


**Fig. 1** FTIR analysis of DBB and FBB before and after the adsorption

aromatic carbon in cellulose, hemicellulose, and some lignin decomposed into the biomass carbon dioxide and methane gas [48].

Comparing with the FTIR spectrum before the adsorption, DBB shows a peak at 2665.49 $cm^{-1}$ that was absent before the adsorption which is related to the N–H stretching [49]. DBB has stretch bending of N–H at 1564 $cm^{-1}$. This is indicated as an overlapped band of bending vibration of N–H and stretching vibration of C–N and signified the attachment of $NH_4^+$ after the adsorption on DBB [50]. A peak of 1585 $cm^{-1}$ appears in the DBB after adsorption which shows the presence of aromatic C = C stretching vibrations and the peak of aromatic C–H groups at 873.24 $cm^{-1}$ [51]. While for FBB, after the adsorption, there is a shift of band at 1110 $cm^{-1}$ from 1113 $cm^{-1}$ that might indicate the OH and –COOH groups [50]. The adsorption of cations on carbon char is relatively related to its –OH and –COOH content [52]. Hence, the adsorption of $NH_4^+$ on DBB and FBB might be contributed by the presence of –OH and –COOH.

## 3.3 *Determination of Surface Morphology DB, FB, FBB, and DBB*

The structure of the samples for DB, DBB, FB, and FBB is depicted in Fig. 2 (a), (b), (c), and (d), respectively. It was observed that an ellipsoidal spore structure formed

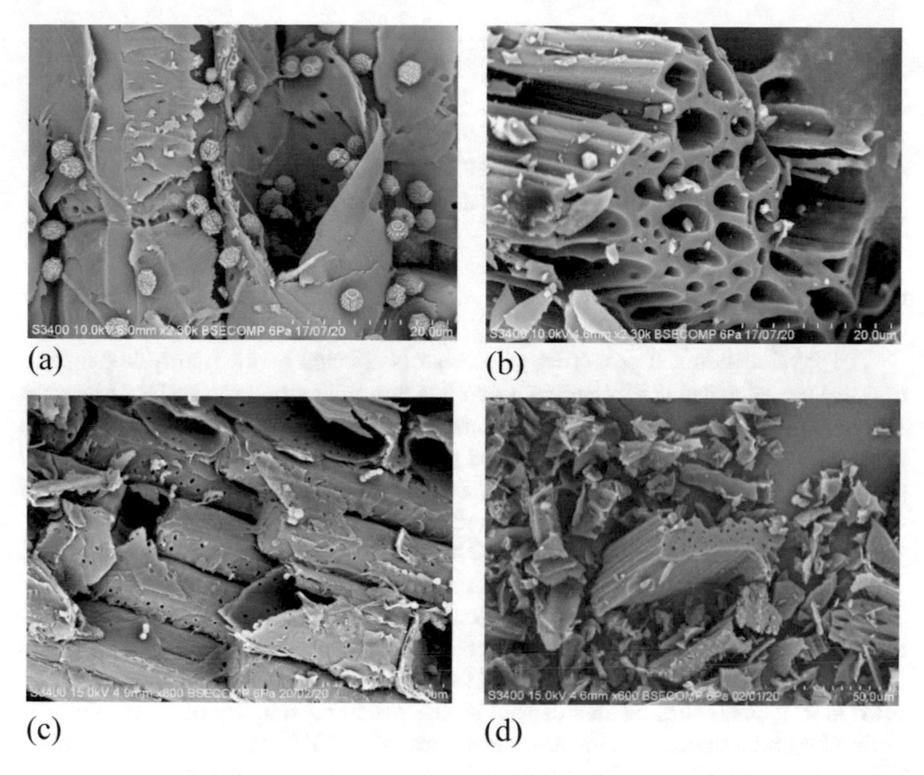


**Fig. 2** SEM image DB (**a**) and DBB (**b**) at magnification 2.3 k, respectively; FB (**c**) and FBB (**d**) at magnification of 800 and 600, respectively

within the surface and hyphae in the wall of bamboo of DB after exposure to the microorganisms.

However, after DB is being pyrolyzed, the morphology of DBB and FBB display as a honeycomb-like structure with various pore sizes formed on the surface of DBB. This is because, during the pyrolysis process, rapid volatile compounds were released; as a result, an internal overpressure was made which contributed to a combination of small pores and thus lead to inner cavities and a more open structure [53].

### 3.4 Determination of $S_{BET}$, $P_z$, and $P_v$

The $S_{BET}$, $P_z$, and $P_v$ are listed in Table 2. The average $P_z$ of FB has decreased after the carbonization from 13.67 to 10.1 nm. The decrement was noticed. While the $P_v$ was slightly increased from 0.001 to 0.002 $cm^3/g$, $S_{BET}$ has increased after the carbonization for FB from 0.33 to 0.62 $m^2/g$. The surface area increases with residence time and pyrolytic temperature [54]. The increment of $S_{BET}$ and $P_V$ with pyrolysis temperature and heating period is associated by the passive degradation of

**Table 2** Textural properties of FB, DB, FBB, and DBB

| Sample | $S_{BET}$ (m²/g) | $P_z$ (nm) | $P_v$ (cm³/g) |
|---|---|---|---|
| FB | 0.33 | 13.67 | 0.001 |
| DB | 6.00 | 5.40 | 0.008 |
| FBB | 0.62 | 10.1 | 0.002 |
| DBB | 4.33 | 6.07 | 0.007 |

the organic materials (cellulose, lignin) and the development of vascular bundles or channel structures [55].

The physicochemical properties of biochar is affected by the pyrolysis temperature that may determine its elemental components, pore structure, surface area, and functional groups [56]. After the carbonization, the $S_{BET}$ and $P_v$ of DB decreased from 6.00 $m^2/g$ to 4.33 $m^2/g$ and 0.008 $cm^3/g$ to 0.007 $cm^3/g$, respectively. The average $P_z$ of DB increased from 5.40 nm to 6.07 nm. DBB has a higher $P_z$ as compared to FBB. This result is correlated with lower carbon char yield, higher ash value, and inorganic elements such as K, Mg, and Ca in DBB as compared to FBB. This is maybe due to the higher carbonization temperature for DBB.

Carbonization temperature <500 °C could produce biochar with a more significant number of functional groups, higher cation exchange capacity (CEC), and enhanced yield. However, due to insufficient thermal degradation, such biochar would often have a small surface area caused by the structural stability of lignocellulosic molecules and volatiles at a lower temperature [57]. DBB has a higher $P_v$ and $S_{BET}$ than FBB. The surface area and pore volume increased with pyrolytic temperature [58]. The increasing of pyrolysis temperature, pore-blocking substances such as inorganic elements such as (K, Mg, and Ca) or –COOH and –CO are decomposed or are thermally cracked, increasing the biochar surface area [59]. Hence, it is highly suspected that DBB was carbonized higher temperature than FBB.

## 3.5 Effect of Contact Time on $NH_4^+$ Adsorption

The effect of contact time on the adsorption of $NH_4^+$ using FBB and DBB for 24 h is shown in Fig. 3. The increase in contact time leads to an increase in removal efficiency until equilibrium adsorption was established. As shown in Fig. 3, the equilibrium for FBB and DBB was reached within 7 h, while DBB has not reached the equilibrium even until 24 h; DBB may need a longer time to reach equilibrium.

Based on Fig. 3, the percentage removal of $NH_4^+$ of FBB and DBB are 65.5% and 61.7%, respectively. It is shown that FBB was more effective and required the shortest time in the removal of $NH_4^+$ solution and high adsorption capacity of $NH_4^+$ compared to DBB. DBB has a higher $S_{BET}$ and $P_V$ as compared to FBB, but FBB has a larger pore size (10.1 nm) compared to DBB (6.07 nm). The same condition was observed by Yu et al. [60] and Takaya et al. [61], where a larger surface area but lower adsorption capacity was observed. They have concluded that the surface area

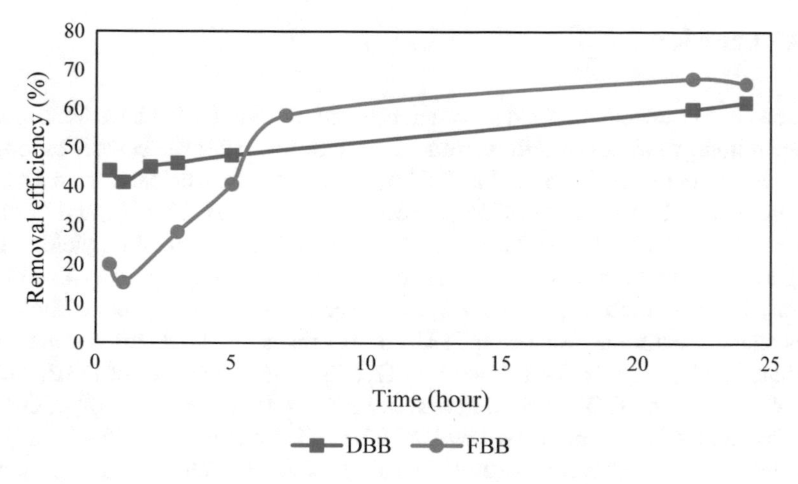


**Fig. 3** Percentage of removal efficiency against time using FBB and DBB

has no direct correlation with surface area, and physisorption is not the dominant mechanism for $NH_4^+$ adsorption. $NH_4^+$ adsorption capacity is a function of oxygen-containing functional groups, and physisorption is not the dominant mechanism [61]. The presence of hydroxyls (–OH) and C = O detected in FBB may indicate the presence of active sites carboxyls (–OOH). –OH and –OOH are active surface functional groups in most biochar that most binding with $NH_4^+$ likely to occur [19].

Table 3 shows the comparison of the adsorption capacity of $NH_4^+$ ions on different sources of biochar with biochar that was used in this study. The adsorption capacity of DBB and FBB is higher than the same precursor (2.44 mg/g [62], 1.75 mg/g [63]) and rice straw 2.9 mg/g [64]. Thus, it can be summarized that both FBB and DBB can remove $NH_4^+$, and FBB is shown as a better adsorbent as it has an 18% higher $NH_4^+$ removal efficiency than DBB.

**Table 3** Comparison of FBB and DBB with other biochar

| Biochar | Equilibrium contact time (h) | Adsorption capacity (mg/g) | Initial concentration (mg/L) | References |
|---|---|---|---|---|
| FBB | 7 | 7.29 | 50 | This study |
| | 40 | 2.44 | 200 | [62] |
| | 8 | 1.75 | 40 | [63] |
| DBB | 22 | 5.98 | 50 | This study |
| Rice straw | 1.5 | 2.9 | 12 | [64] |

## 4 Conclusion

DBB and FBB were synthesized and characterized. DBB and FBB have honeycomb-like morphology structures with various sizes of pores. After the carbonization, the $S_{BET}$ FB has increased from 0.33 to 0.62 m$^2$/g, the $P_z$ has decreased from 13.67 to 10.1 nm. Meanwhile, the $S_{BET}$ of DB decreased from 6.00 to 4.33 m$^2$/g, and $P_z$ of DB has increased from 5.40 to 6.07 nm after the carbonization. The carbon yield of FBB is higher than DBB that may be due to some lignin has been degraded. The higher carbon content and ash content in DBB are suspected due to the higher carbonization temperature as carbonization using TLUD brick kiln cannot be controlled manually. The functional groups of O–H, C = O, C–H, bonds were detected for DBB, and C–O, C–H, –OH, and –OOH were detected in FBB due to the presence of –OOH and –OH that are likely to form a bond with $NH_4^+$. N–H stretching at 2665.49 cm$^{-1}$ was indicated after the adsorption for DBB but absent for FBB. The adsorption capacity of FBB and DD was 7.29 mg/g, 65.5%, and 5.98 mg/g, 61.7%, respectively. Thus, it can be summarized that both FBB and DBB can remove $NH_4^+$, and FBB is showed as a better adsorbent as it has 18% higher $NH_4^+$ removal than DBB.

**Acknowledgements** The authors acknowledge Universiti Malaysia Sabah for financial aid through a research grant SDK0044-2018, SGI0064-2018 and SPB0001-2020.

## References

1. Pérez, J., Isanta, E., Carrera, J.: Would a two-stage N-removal be a suitable technology to implement at full scale the use of anammox for sewage treatment? Water Sci. Technol. **72**, 858–864 (2015)
2. Raboni, M., Torretta, V., Urbini, G.: Influence of strong diurnal variations in sewage quality on the performance of biological denitrification in small community wastewater treatment plants (WWTPs). Sustain. **5**, 3679–3689 (2013)
3. Mery, C., Guerrero, L., Alonso-Gutiérrez, J., Figueroa, M., Lema, J.M., Montalvo, S., Borja, R.: Evaluation of natural zeolite as microorganism support medium in nitrifying batch reactors: Influence of zeolite particle size. **47**, 420–427 (2012). Doi: https://doi.org/10.1080/10934529.2012.646129
4. Choy, K.K.H., Barford, J.P., McKay, G.: Production of activated carbon from bamboo scaffolding waste—process design, evaluation and sensitivity analysis. Chem. Eng. J. **109**, 147–165 (2005)
5. Tran, V.H.: Growth and quality of indigenous bamboo species in the mountainous regions of Northern Vietnam. (2010)
6. Scurlock, J.M.O., Dayton, D.C., Hames, B.: Bamboo: An overlooked biomass resource? Biomass Bioenerg. **19**, 229–244 (2000)
7. Daily Express Online: Sabah bamboo industry has huge potential, https://www.dailyexpress.com.my/news/155947/sabah-bamboo-industry-has-huge-potential/, (2020)
8. Hinde, O., Kaba, G.: Bamboo utilization practices and challenges of cottage industries: the case of selected towns in Ethiopia. Int. J. Adv. Res. Publ. **2**, 50–55 (2018)
9. Desalegn, G., Tadesse, W.: Resource potential of bamboo, challenges and future directions towards sustainable management and utilization in Ethiopia. For. Syst. **23**, 294 (2014)

10. Liese, W., Kumar, S.: Bamboo Preservation Compendium. Centre for Indian Bamboo Resource and Technology, India (2003)
11. Wiwoho, M.S., Machicky, M., Nawir, R., M., S.I.: Bamboo waste as part of the aggregate pavement the way green infrastructure in the future. In: The 6th International Conference of Euro Asia Civil Engineering Forum (EACEF 2017). pp. 1–9. EDP Sciences (2017)
12. Nath, S.K., Chawla, V.K.: Wood substitution: recent developments in India. J. Indian Acad. Wood Sci. **8**(2), 68–71 (2012)
13. Sahoo, S.S., Vijay, V.K., Chandra, R., Kumar, H.: Production and characterization of biochar produced from slow pyrolysis of pigeon pea stalk and bamboo. Clean. Eng. Technol. **3**, 100101 (2021)
14. Maican, E., Murad, E., Cican, G., Duţu, I.-C.: Analysis of a top lit updraft gasification system designed for greenhouses and hothouses. In: 16th International Multidisciplinary Scientific GeoConference SGEM 2016, Bulgaria (2016)
15. Howell, N., Pimentel, A., Bhattacharia, S.: Material properties and environmental potential of developing world-derived biochar made from common crop residues. Environ. Challenges **4**, 100137 (2021)
16. Nsamba, H.K., Hale, S.E., Cornelissen, G., Bachmann, R.T.: Designing and performance evaluation of biochar production in a top-lit updraft up-scaled gasifier. J. Sustain. Bioenergy Syst. **05**, 41–55 (2015)
17. Tangmankongworakoon, N.: An approach to produce biochar from coffee residue for fuel and soil amendment purpose. Int. J. Recycl. Org. Waste Agric. **8**, 37–44 (2019)
18. Yang, H.I., Lou, K., Rajapaksha, A.U., Ok, Y.S., Anyia, A.O., Chang, S.X.: Adsorption of ammonium in aqueous solutions by pine sawdust and wheat straw biochars. Environ. Sci. Pollut. Res. **25**, 25638–25647 (2018)
19. Boopathy, R., Karthikeyan, S., Mandal, A.B., Sekaran, G.: Adsorption of ammonium ion by coconut shell-activated carbon from aqueous solution: kinetic, isotherm, and thermodynamic studies. Environ. Sci. Pollut. Res. **20**, 533–542 (2012)
20. Satayeva, A.R., Howell, C.A., Korobeinyk, A.V., Jandosov, J., Inglezakis, V.J., Mansurov, Z.A., Mikhalovsky, S.V.: Investigation of rice husk derived activated carbon for removal of nitrate contamination from water. Sci. Total Environ. **630**, 1237–1245 (2018)
21. Asada, T., Ohkubo, T., Kawata, K., Oikawa, K.: Ammonia adsorption on bamboo charcoal with acid treatment. J. Heal. Sci. **52**, 585–589 (2006)
22. Zheng, Y., Wang, B., Wester, A.E., Chen, J., He, F., Chen, H., Gao, B.: Reclaiming phosphorus from secondary treated municipal wastewater with engineered biochar. Chem. Eng. J. **362**, 460–468 (2019)
23. Nadir, N., Ismail, N.L., Hussain, A.S.: Fungal pretreatment of lignocellulosic materials. Biomass Bioenergy Recent Trends Future Challenges (2019)
24. Romo, O..: Making biochar and charcoal with the brick chimney kiln, https://www.youtube.com/watch?v=NrTaISI9fm4
25. Gonzaga, M.I.S., Mackowiak, C.L., Comerford, N.B., da Moline, E.F.V., Shirley, J.P., Guimaraes, D.V.: Pyrolysis methods impact biosolids-derived biochar composition, maize growth and nutrition. Soil Tillage Res. **165**, 59–65 (2017)
26. Masís-Meléndez, F., Segura-Chavarría, D., García-González, C.A., Quesada-Kimsey, J., Villagra-Mendoza, K.: Variability of physical and chemical properties of TLUD Stove derived biochars. Appl. Sci. **10**, 507 (2020)
27. Hussin, M.H.A., Abidin, M.S.Z.: Electrical characterization of gold contact on porous silicon layers. In: Proceeding of 2017 IEEE Regional Symposium on Micro and Nanoelectronics RSM 2017. 104–107 (2017)
28. ASTM: ASTM D1762–84 (2013) Standard test method for chemical analysis of wood charcoal. ASTM International, West Conshohocken, PA (2013)
29. Dai, Z., Meng, J., Muhammad, N., Liu, X., Wang, H., He, Y., Brookes, P.C., Xu, J.: The potential feasibility for soil improvement, based on the properties of biochars pyrolyzed from different feedstocks. J. Soils Sediments. **13**, 989–1000 (2013)

30. Gregg, S.J., Sing, K.S.W.: Adsorption, surface area, and porosity. Academic Press, London; New York (1982)
31. Negara, D.N.K.P., Nindhia, T.G.T., Surata, I.W., Hidajat, F., Sucipta, M.: Nanopore structures, surface morphology, and adsorption capacity of tabah bamboo-activated carbons. Surfaces and Interfaces. **16**, 22–28 (2019)
32. Zhuang, J., Li, M., Pu, Y., Ragauskas, A.J., Yoo, C.G.: Observation of potential contaminants in processed biomass using fourier transform infrared spectroscopy. Appl. Sci. **10**, 4345 (2020)
33. Safie, N.N., Zahrim Yaser, A., Hilal, N.: Ammonium ion removal using activated zeolite and chitosan. Asia-Pacific J. Chem. Eng. 1–9 (2020)
34. USEPA: Standard method for the examination of water and wastewater. United State environmental health science & engineering, Washington D.C. (1992)
35. Metcalf & Eddy: Wastewater engineering : Treatment and reuse. McGraw-Hill, Boston (2003)
36. Fang, W., Yang, S., Wang, X.-L., Yuan, T.-Q., Sun, R.-C.: Manufacture and application of lignin-based carbon fibers (LCFs) and lignin-based carbon nanofibers (LCNFs). Green Chem. **19**, 1794–1827 (2017)
37. Nguyen, B.T., Lehmann, J., Hockaday, W.C., Joseph, S., Masiello, C.A.: Temperature sensitivity of black carbon decomposition and oxidation. Environ. Sci. Technol. **44**, 3324–3331 (2010)
38. Domingues, R.R., Trugilho, P.F., Silva, C.A., de Melo, I.C.N.A., Melo, L.C.A., Magriotis, Z.M., Sánchez-Monedero, M.A.: Properties of biochar derived from wood and high-nutrient biomasses with the aim of agronomic and environmental benefits. PLoS One **12**, (2017)
39. Bourke, J., Merilyn, M.-H., Chihiro, F., Kiyoshi, D., Teppei, N., Michael Jerry Antal, J.: Do all carbonized charcoals have the same chemical structure? 2. A model of the chemical structure of carbonized charcoal†. Ind. Eng. Chem. Res. **46**, 5954–5967 (2007)
40. Ye, L., Zhang, J., Zhao, J., Luo, Z., Tu, S., Yin, Y.: Properties of biochar obtained from pyrolysis of bamboo shoot shell. J. Anal. Appl. Pyrolysis. **114**, 172–178 (2015)
41. Fan, R., Chen, C.L., Lin, J.Y., Tzeng, J.H., Huang, C.P., Dong, C., Huang, C.P.: Adsorption characteristics of ammonium ion onto hydrous biochars in dilute aqueous solutions. Bioresour. Technol. **272**, 465–472 (2019)
42. Hernandez-mena, L., Pecora, A., Beraldo, A.: Slow pyrolysis of bamboo biomass : analysis of biochar properties slow pyrolysis of bamboo biomass : analysis of biochar properties. (2014)
43. Xu, G., Wang, L., Liu, J., Wu, J.: FTIR and XPS analysis of the changes in bamboo chemical structure decayed by white-rot and brown-rot fungi. Appl. Surf. Sci. **280**, 799–805 (2013)
44. Al-Taliby, W.H.A.: Evaluation of methylene blue removal from wastewater by adsorption onto different types of adsorbent beds. (2009)
45. Zhang, Y., Ma, Z., Zhang, Q., Wang, J., Ma, Q., Yang, Y., Luo, X., Zhang, W.: Comparison of the physicochemical characteristics of bio-char pyrolyzed from Moso bamboo and rice husk with different pyrolysis temperatures. BioResources **12**, (2017)
46. Ahmad, M., Lee, S.S., Dou, X., Mohan, D., Sung, J.-K.K., Yang, J.E., Ok, Y.S.: Effects of pyrolysis temperature on soybean stover- and peanut shell-derived biochar properties and TCE adsorption in water. Bioresour. Technol. **118**, 536–544 (2012)
47. Okolo, G.N., Neomagus, H.W.J.P., Everson, R.C., Roberts, M.J., Bunt, J.R., Sakurovs, R., Mathews, J.P.: Chemical–structural properties of South African bituminous coals: insights from wide angle XRD–carbon fraction analysis, ATR–FTIR, solid state 13C NMR, and HRTEM techniques. Fuel **158**, 779–792 (2015)
48. Armynah, B., Tahir, D., Tandilayuk, M., Djafar, Z., Piarah, W.H.: Potentials of biochars derived from bamboo leaf biomass as energy sources: effect of temperature and time of heating. Int. J. Biomater. **2019**, 3526145 (2019)
49. Mahata, B.K., Chung, K.L., Chang, S. min: Removal of ammonium nitrogen (NH4+-N) by Cu-loaded amino-functionalized adsorbents. Chem. Eng. J. **411**, 128589 (2021)
50. Liu, H., Dong, Y., Wang, H., Liu, Y.: Adsorption behavior of ammonium by a bioadsorbent—Boston ivy leaf powder. J. Environ. Sci. **22**, 1513–1518 (2010)
51. Viglašová, E., Galamboš, M., Danková, Z., Krivosudský, L., Lengauer, C.L., Hood-Nowotny, R., Soja, G., Rompel, A., Matík, M., Briančin, J.: Production, characterization and adsorption

studies of bamboo-based biochar/montmorillonite composite for nitrate removal. Waste Manag. **79**, 385–394 (2018)
52. Feng, N., Guo, X., Liang, S.: Adsorption study of copper (II) by chemically modified orange peel. J. Hazard. Mater. **164**, 1286–1292 (2009)
53. Guerrero, M., Ruiz, M.P., Alzueta, M.U., Bilbao, R., Millera, A.: Pyrolysis of eucalyptus at different heating rates: studies of char characterization and oxidative reactivity. J. Anal. Appl. Pyrol. 307–314. Elsevier (2005)
54. Nartey, O.D., Zhao, B.: Biochar preparation, characterization, and adsorptive capacity and its effect on bioavailability of contaminants: an overview. Adv. Mater. Sci. Eng. **2014**, 715398 (2014)
55. Li, X., Shen, Q., Zhang, D., Mei, X., Ran, W., Xu, Y., Yu, G.: Functional groups determine biochar properties (pH and EC) as studied by two-dimensional 13C NMR correlation spectroscopy. PLoS One. 8, e65949 (2013)
56. Dhyani, V., Bhaskar, T.: A comprehensive review on the pyrolysis of lignocellulosic biomass. Renew. Energy. **129**, 695–716 (2018)
57. Uroić Štefanko, A., Leszczynska, D.: Impact of biomass source and pyrolysis parameters on physicochemical properties of biochar manufactured for innovative applications. Front. Energy Res. **0**, 138 (2020)
58. Hammes, K., Smernik, R.J., Skjemstad, J.O., Schmidt, M.W.I.: Characterisation and evaluation of reference materials for black carbon analysis using elemental composition, colour, BET surface area and 13C NMR spectroscopy. Appl. Geochemistry. **23**, 2113–2122 (2008)
59. Rafiq, M.K., Bachmann, R.T., Rafiq, M.T., Shang, Z., Joseph, S., Long, R.: Influence of pyrolysis temperature on physico-chemical properties of corn stover (Zea mays L.) biochar and feasibility for carbon capture and energy balance. PLoS One. **11**, e0156894 (2016)
60. Yu, Q., Xia, D., Li, H., Ke, L., Wang, Y., Wang, H., Zheng, Y., Li, Q.: Effectiveness and mechanisms of ammonium adsorption on biochars derived from biogas residues. RSC Adv. **6**, 88373–88381 (2016)
61. Takaya, C.A., Fletcher, L.A., Singh, S., Anyikude, K.U., Ross, A.B.: Phosphate and ammonium sorption capacity of biochar and hydrochar from different wastes. Chemosphere **145**, 518–527 (2016)
62. Chen, L., Chen, X.L., Zhou, C.H., Yang, H.M., Ji, S.F., Tong, D.S., Zhong, Z.K., Yu, W.H., Chu, M.Q.: Environmental-friendly montmorillonite-biochar composites: facile production and tunable adsorption-release of ammonium and phosphate. J. Clean. Prod. **156**, 648–659 (2017)
63. Van Hien, N., Valsami-Jones, E., Vinh, N.C., Phu, T.T., Tam, N.T.T., Lynch, I.: Adsorption of ammonium (NH4+) ions onto various Vietnamese biomass residue-derived biochars (wood, rice husk and bamboo). Biochar Prod. Charact. Appl. (2017)
64. Khalil, A., Sergeevich, N., Borisova, V.: Removal of ammonium from fish farms by biochar obtained from rice straw: isotherm and kinetic studies for ammonium adsorption. Adsorpt. Sci. Technol. **36**, 1294–1309 (2018)

# Waste and Health: Sewage Sludge and Its Hazard to Human

**Azam Muzafar Bin Ahmad Mokhtar, Muaz Mohd Zaini Makhtar, and Ana Masara Ahmad Mokhtar**

**Abstract** Sewage worker is routinely exposed to sewage or sewage sludge containing various biological and chemical irritants such as toxic gases, pathogens, genotoxic agents and harmful organic chemicals. Acute and prolonged contact with these harmful substances may cause several health-related issues such as infection, inflammation, skin irritation, heavy metal poisoning, pulmonary diseases, and even cancer. These symptoms and diseases are not only reported among sewage workers but also the residents who stay near the wastewater treatment plants (WWTPs). Nevertheless, due to the lack of knowledge of risk agents' symptoms and their association with diseases, the workers are continuously exposed to these risk agents. Thus, the review discussed the possible health risk or hazards associated with sewage sludge in more detail.

**Keywords** Sewage · Infectious disease · Inflammation · Cancer · Irritation · Pulmonary disease

## 1 Introduction

Throughout mankind history, various methods of waste disposal have been implemented. From the first ancient Romans, primitive residential municipal sewage and latrines systems [1] to a modern-day dewatered sludge management system, the evolution of wastes disposal has come a long way. To fully understand the proper concept of the waste management system, it is necessary to understand the types of waste. The types of waste can be characterized into solid, liquid and gaseous waste and these wastes can come from various sources such as household, industrial, agricultural, demolition and construction, commercial, and mining [2]. Some of the

A. M. B. A. Mokhtar
Department of Biology, Faculty of Science and Mathematics, Universiti Pendidikan Sultan Idris, 35900 Tanjong Malim, Perak, Malaysia

M. M. Z. Makhtar · A. M. A. Mokhtar (✉)
Bioprocess Technology Division, School of Industrial Technology, Universiti Sains Malaysia, 11800 Gelugor, Penang, Malaysia
e-mail: anamasara@usm.my

 
A. Z. Yaser et al. (eds.), *Waste Management, Processing and Valorisation*,
https://doi.org/10.1007/978-981-16-7653-6_8

waste produced from these sources can be hazardous to the environment and humans due to its hazardous contents such as toxic, irritant, carcinogenic, flammable, explosive and oxidizing agents [3]. Therefore, waste disposal management should be one of the top priorities today as it can leave impacts not only on the environment but also on our society by disrupting our quality of life [4].

Generally, sewage sludge is considered a by-product obtained from wastewater treatment. It is stated by the United Nations Educational, Scientific and Cultural Organization (UNESCO) that wastewater consisting of 99% water and 1% suspended, colloidal and dissolved solids is called sewage sludge [5]. The sewage sludges will be treated further for them to be used effectively, for instance, in the agricultural field. It has been stated that sewage sludge has been considered a vital biological resource for viable agriculture management. Jamil et al. have mentioned that with the use of sewage sludge in agriculture, three times increase in straw and grain production were observed [6]. Similarly, in Jordan, Mohammad and Athamneh also found that adding ~40 t $ha^{-1}$ of sewage sludge can increase the growth of lettuce [7]. This proved that the use of sewage sludge as fertilizer can improve the production of crop yields and ultimately give benefits to the society and economy [8].

As sewage sludge becomes more relevant to many countries, quick and precise information regarding its origin, composition, treatment method, disposal uses should be reviewed to ensure the readers can fully grasp the concept easily. Besides, the review also discussed the risks or hazards that stemmed from improper handling of sewage sludge. This would not only improve the wastewater treatments in line with today's application but also educate society with the knowledge of health hazards associated with sewage sludge.

## *1.1 Definition of Sewage Sludge*

Before the term sewage sludge was discovered, people around the world disposed their wastewater quite differently. Before the 1950s, most industrial factories and municipal residents in the United States (US) disposed their wastewater into streams and rivers without any proper treatment [9]. As time goes by, the quality of their water source starts to wane. To combat this predicament from becoming worse, people have come up with a better method of wastewater disposal. Between the 1950s and 1960s, the concept of sewage sludge was introduced to US society to promote proper waste disposal while keeping the environment safe [10]. This solution eventually helped societies in the US maintain a safe and clean water supply. As mentioned before, the term sewage sludge comes from the treatment process of wastewater or sewage that comes from municipal residents and industrial factories. As the name suggests, sludge is a 1% suspended, colloidal and dissolved solid, a by-product formed as a result of wastewater treatment. The obtained sewage sludge is hazardous not only to the environment but also to human health due to the various harmful contaminants it holds such as pathogens, genotoxicants and heavy metal substances [11]. Therefore, people have designed a proper and safer way to use this 1% by-product efficiently,

Besides the term sewage sludge, this 1% by-product can also be called biosolids. This term was first introduced recently by the wastewater treatment industry. The purpose of this renaming was to differentiate between raw sewage sludge that came directly from the wastewater treatment and the treated sewage sludge that has undergone a proper treatment process, ranging from alkaline stabilization, acid oxidation/disinfection, aerobic or anaerobic digestion, composting and thermal drying [12, 13]. The term biosolid is appropriate when describing the treated sewage sludges so that people can differentiate between the untreated sewage sludge to biosolid, which is safer to be applied in various land applications [14]. Furthermore, the reason why treated sewage sludge is called biosolid is due to its treatment process that yields a more stable, safe and pathogen-reduced substance. However, it still does not imply that biosolid is completely free from harmful substances as some of them are resistant to wastewater treatments.

## *1.2 Production and Composition of Sewage Sludge*

The amount of sewage sludge produced depends on the density of the population, the advancement of the technology and its management at a given time. The major countries that produce a high amount of sewage sludge are mainly located in the US, Europe and East Asia [15, 16]. For instance, in 2010, ~1.96 million tons of sewage sludge (as dry substance, d.s.) were produced and collected from urban wastewater in Europe (14 over 38 countries) and the number was shown to increase to ~2.3 million tons in 2018 [17]. However, the data from the most important sludge producer, including Italy, Spain, Germany, and the UK were missing, suggesting that the reported number were far less than the actual number. Consistently, Campo et al. mentioned that in 2017, the annual sludge production in the whole European countriesmight be more than 9 million tons of d.s. instead of ~3.5 million tons (as reported by Eurostat [17]) based on the published figures and interpolation of the missing data [18].

Reasons behind the great amount of sewage sludge produced by the aforementioned countries are probably due to their high population density and the high-quality machinery they possess. Interestingly, with the overproduction of sewage sludge every year, people start to wonder about its effect on the environment and human health. Before understanding the production and the contents present within the sewage sludges, full details on the wastewater treatment process are needed. Sewage sludges are often characterized into three types: (1) primary sewage sludge, (2) secondary sewage sludge and (3) tertiary sewage sludge.

(1) Primary sewage sludge is the settable solids and floating materials resulting after the primary sedimentation process of raw wastewater (Fig. 1) [12]. This treatment process involves gravity sedimentation and flotation processes that will help to eliminate insoluble matters such as grit, grease, and scum and; reduce half of the solid in raw wastewater [19]. The remaining solid materials,

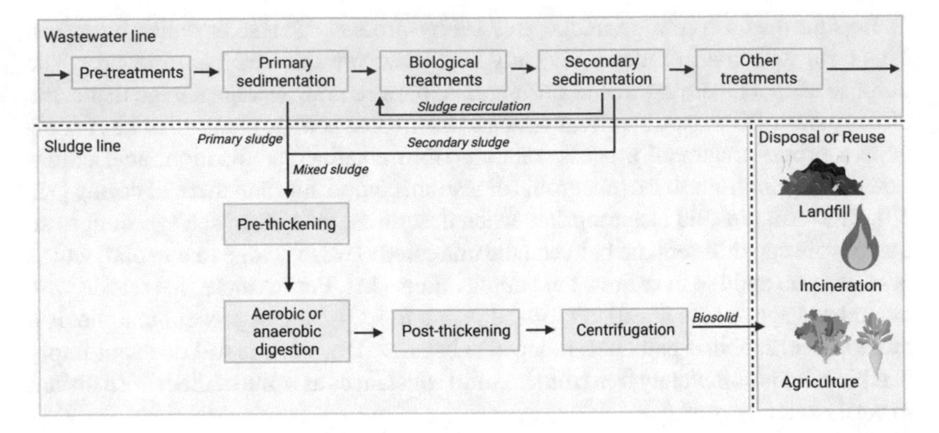


**Fig. 1** Example of a standard wastewater treatment process. The standard wastewater treatment process includes sedimentation, thickening, biological digestion and centrifugation process. The biosolid produced will then be disposed or reused in agriculture. Adapted from [12]. Created with BioRender.com

organic and inorganic, will be separated from the raw wastewater by accumulating at the bottom. Usually, primary sewage sludge is distinguished by its high total solid count and putrescibility [20].

(2) Secondary sludge is the sediment that resulted from a biological treatment (activated sludge process or biofilm systems) of wastewater [12]. This treatment process involves an intricate heterogeneous mixture of microorganisms, bacterial constituents, undigested organic and inorganic materials and water [21]. Naturally occurring microorganisms are used to break down the dissolved organic material in the wastewater and change it into carbon dioxide that will be released to the atmosphere and into microbial cell mass.

(3) Tertiary sewage sludge is the most uncommon of all three. In tertiary sewage sludge production, chemicals such as alum, ferric chloride, or lime will be inserted into the wastewater. This is to further decrease the amount of plant nutrients such as nitrogen and phosphorus, suspended solids, and biological oxygen demand of wastewater [22]. The reason why this is labelled as uncommon is that the treatment procedure only occurred in some WWTPs.

After the wastewater has gone through several treatment processes, a rejuvenated and pathogen-reduced water source is produced. The freshly rejuvenated water will then be released into streams, rivers and in some cases, sprayed over huge areas of land and soil.

As mentioned before, most wastewater produced from municipal and industrial sources contains several hazardous materials that can be dangerous to human health and the environment. Usually, most sewage sludges contain organic and inorganic materials, plant nutrients, trace elements and pathogens [23]. Following the wastewater treatment process, these hazardous materials are generally reduced, forming a much safer sewage sludge. However, the degree of reduction depends on the initial composition of the wastewater and the treatment procedure conducted [24]. Thus, to ensure safer treated wastewater and biosolids are produced, some countries, including Malaysia, the US and Europe have enforced a regulation concerning the total limits of these harmful substances produced and released in wastewater systems [25].

## 1.3 Sewage Sludge Treatment

Upon sewage sludges formation, all three types of sewage sludge (primary, secondary and tertiary) are often mixed, forming a substance called raw sewage sludge [16]. This mixture of sewage sludges is quite unstable and contained many hazardous materials such as pathogens, irritants and genotoxicants that is bad for human health. To overcome this problem, the raw sewage sludge often undergoes another treatment process to make it safer and usable for application and disposal. When it comes to sewage sludge treatments, there are several different methods available (Table 1). Each method has its pros and cons as well as a different final product.

As mentioned before, sewage sludge treatment is crucial for a variety of reasons, including agricultural development and land restoration. It has been confirmed by many researchers that sewage sludge can improve the yield in various farms [26]. Besides, by reusing treated sewage sludge in abandoned opencast mining areas, it can essentially boost its overall organic matter, microorganism population, available nitrogen, phosphorus and soil biological fertility, while preventing soil erosion [27].

## 1.4 Sewage Sludge Disposal

Currently, there are three types of sewage sludge disposal methods, including incineration, landfill and land application [28]. These three methods are commonly applied by several major countries such as the US, Europe, Asia and Africa [29].

(1) The incineration method is the disposal of sewage sludge using a burning method. It is considered the most effective and safer way of sewage disposal due to its controlled discharges to air, water and soil [30]. This method helps to reduce the pathogens, decomposes most organic chemicals and reduces the volume of sewage sludges [16]. In Japan, ~70% of sewage sludge is disposed through this method [15]. However, the ash produced through this method is dangerous to the environment due to its chemical content [31].
(2) The landfill method is the disposal of sewage sludge by dumping it on a specific open field. This method is believed to be the simplest and cheapest way compared to the other two [31]. Nevertheless, this method may lead to various complications for the environment such as overproduction of methane, a greenhouse gas from the anaerobic decomposition that may cause global warming [16]. Besides, the landfill method can also cause secondary emissions (heavy

**Table 1** Treatment methods for raw sewage sludge and its effects [16]

| Treatment method | Description | Effects on sludge |
|---|---|---|
| *Thickening* | Concentrated sludge solids settle and float with the help of gravity or after being introduced to the air | About 5–6% increase of solid contents can be observed. However, the sludge maintains the properties of the liquid |
| *Dewatering* | This process mostly involves air drying on sand beds, centrifugation and belt pressing (filtration) | Air drying decreased the content of pathogens inside the sludge. At the same time, centrifugation and belt pressing will cause some nutrient loss<br>This method will increase the total solid contents (15–30%) |
| *Anaerobic digestion* | Sewage sludge is left without any air at the temperature of 20–55 °C for 15–60 days while anaerobic bacteria consume the sludge creating carbon dioxide and methane gas. In some cases, the burning of the sludge is helped by the methane gas produced. This method has been considered the most widely used method for sewage sludge treatment | This method has been said to increase the total solids content of sewage sludge while decreasing its odours, volatile solids and viable pathogens. Furthermore, it also preserves the plant nutrients present inside the sewage sludge |
| *Aerobic digestion* | Sewage sludge is mixed with air or oxygen at 15–20 °C for 40–60 days while aerobic bacteria consume the sludge, producing only carbon dioxide | This method has been said to increase the total solids content of sewage sludge while decreasing its odours, volatile solids and viable pathogens. Besides, there may be some reduction of nitrogen content in the sewage sludge produced |
| *Alkaline stabilization* | The pH of sewage sludge is manipulated. First, the optimum pH for the sewage sludge would be maintained at 12 for at least 2 h, followed by 11.5 for the next 22 h. This is done by adding an appropriate amount of alkaline metal such as lime (CaO) into the sewage sludge | This method has been said to decrease the odours and viable pathogens of the sewage sludge produced. However, phosphorus is converted to a form that is not suitable to plants and loss of ammonia ($NH_3$) may also occur |
| *Composting* | This method involves a dewatered process, a mixture of high-carbon organic material, and aerobic composting. Initially, the sewage sludge undergoes the dewatered process, and the produced sludge is then mixed with high-carbon organic material such as sawdust. Finally, the mixture undergoes aerobic composting at 55 °C for several days | This method has been said to decrease the volume, odours, plant nutrients, volatile solids and pathogen. Besides, the organic matter is mostly stabilized |

metals) to water and soil resulting in contamination [28]. Even with its high risks, some countries still apply this method, which includes Africa and South Korea [29].

(3) The land application method is where sewage sludge is used in the land improvement process. This method focuses on the content present within the treated sewage sludge or biosolids that are known to be beneficial for agriculture and forestry [32]. With this method, the plant nutrients, organic and inorganic matter contained within the sewage sludge are used as fertilizer for crops enhancement or land restoration [16]. Many studies have described this method as one of the most efficient and practical ways rather than being discarded into the environment or incinerated, which can promote pollution [30]. Furthermore, this land application can help lower Greenhouse Gas (GHG) emissions and subsequently reduce global warming [33]. Because of this, most countries, including the US, Europe and China are now implementing this method [29]. However, some studies also have reported the health risk associated with this land application due to the contents of organic pollutants and improper handling [34, 35].

Besides these three methods, there is also ocean dumping, but this practice has mostly been terminated because of its implications to the environment. It started with the US in the 1980s, but 8 years later, a new regulation has been established to cease this dumping method. South Korea also has implemented this method but discontinued it in 2012 [29]. This indicates that the disposal method must be carefully planned and complemented with proper regulations so that it will not cause additional problems in the future.

## 2 Potential Hazard: Health-Related Issues

As mentioned in the previous section, even though sewage sludge has undergone sufficient treatment before being applied to the soil, it still does not completely remove the pathogen, heavy metals organic chemical or chemical irritant presence in the sewage sludge. For instance, while sewage treatment is known to reduce the pathogen, there are still some pathogen remains in the treated sewage sludge such as *C. perfringens*, which are resistant to several disinfection methods [24, 36]. Besides, hepatitis A virus (HAV) is very stable in the environment for a long period of time and is resistant to the current wastewater treatment [37]. The survived pathogen would then be transmitted to humans or animals through ingestion of contaminated vegetables or water, inhalation or direct skin contact.

Continuous or prolonged exposure to these substances could adversely affect human health, especially the sewage worker and the resident who live near the WWTPs. This is because some of them who stay 100–500 m near the WWTP were reported to experience gastrointestinal symptoms and skin diseases [38, 39], common symptoms seen with the sewage workers. Other symptoms include heavy metal toxicity, inflammation, skin irritation, pulmonary diseases and cancer.

## 2.1 Infection

Although sewage sludge is known to contain several nutrients and organic matters that are useful to improve soil structural, chemical and biological properties [40], it also contains several pathogens which include bacteria, viruses, helminths and fungi (Table 2). All these pathogens are the source of infection and may elicit an immune response, causing inflammation. This will then lead to various diseases as described in Table 2.

Besides all the named pathogens in Table 2, we are expected to see more serotypes as advances in analytical methods and changes in society lifestyles gradually occur through time. Additionally, since microorganisms are subjected to mutation and evolution, it allows them to adapt to the environment and eventually will promote resistance to certain antibiotic treatments. This is consistent with some studies showing the presence of antibiotic-resistance strains such as *E. coli* against cephalexin, ciprofloxacin and ampicillin, compared to *Salmonella* in the treated sewage produced from WWTPs based in Penang [24, 42]. These antibiotic-resistance strains were mainly originated from hospital sewage due to the common hospital practice in using antibiotics and also partly derived from municipal sewage as a result of human or animal excretion [43].

Another common virus-induced disease occurred among sewage workers is hepatitis. Even though the incidence rate is low, the workers who work with both untreated sewage and sewage sludge are still exposed to both hepatitis A and hepatitis E viruses, leading to a community spreading [44]. Poor wastewater facilities and hygienic conditions may also play causative roles in the development of the disease, but it can be prevented through vaccination [45, 46]. Another infection is Weil's disease or also known as Leptospirosis, caused by a bacterium known as Leptospira. The infected patient usually caught the disease upon contact with rat or cattle urine through contaminated freshwater [47].

Recently, some studies also have found a potential transmission of the SARS-CoV-2 virus, a virus known to cause COVID-19, to the environment and humans through sewage sludge (Fig. 2). This can occur not only through infected human secretion and insufficient hand sanitization but also via improper face masks disposal, without disinfection, into water or wastewater [48]. Consistent with this, the SARS-CoV-2 virus has been reported in the inflowing wastewater collected from WWTPs in North America, the Netherlands and Japan [49–51]. The same scenarios have been observed for other viruses such as Middle East respiratory syndrome coronavirus (MERS-CoV), ebola viruses, and severe acute respiratory syndrome coronavirus (SARS-CoV), but these enveloped viruses are not causing a significant threat to the sewage industries due to their insignificant amount in wastewater and poor survivability in

**Table 2** Example of pathogens found in sewage and potential associated diseases [24, 41]

| Pathogen type | Strain | Disease |
|---|---|---|
| *Bacteria* | *Salmonella* | Typhoid, paratyphoid, salmonellosis |
| | *Shigella* | Bacillary dysentery |
| | *Escherichia coli O157:H7* | Bloody diarrhea, cramping and abdominal pain, hemolytic uremic syndrome |
| | *Yersinia enterocolitica* | Gastroenteritis |
| | *Campylobaeterjejuni* | Gastroenteritis |
| | *Vibrio cholerae* | Cholera |
| | *Leptospira* | Weil's disease |
| | *Clostridium perfringens* | Food poisoning |
| *Virus* | Adenovirus | Pharyngitis, conjunctivitis, vomiting, diarrhea |
| | Astrovirus | Vomiting, diarrhea |
| | Calicivirus | Vomiting, diarrhea |
| | Coronavirus | Vomiting, diarrhea |
| | Poliovirus | Paralysis, meningitis, fever |
| | Echovirus | Meningitis, encephalitis, rash, diarrhea, fever |
| | Hepatitis A virus | Hepatitis |
| | Hepatitis E virus | Hepatitis |
| | Rotavirus | Vomiting, diarrhea |
| *Fungi* | *Aspergillus/umigatus* | Respiratory otomycosis |
| | *Candida albicans* | Candidiasis |
| | *Cryptococcus neoformans* | Subacute chronic meningitis |
| | *Trichosporon* | Infection of hair follicles |
| | *Phialophora* | Deep tissue infections |
| *Helminths* | *Ascaris lumbricoides* | Ascariasis |
| | *Ancyclostoma duodenale* | Anaemia |
| | *Necatoramerieanus* | Anaemia |
| | *Trichuris* | Abdominal pain, diarrhea |
| | *Toxocara* | Fever, abdominal pain |
| | *Strongyloides* | Abdominal pain, nausea, diarrhea |

aqueous condition [52]. Compared to these aforementioned viruses, SARS-CoV-2 can survive in wastewater for several days [53], suggesting the urgent need for adequate wastewater disinfection. Current disinfection technologies such as free chlorine and dry heat pasteurization are efficient in removing or inactivating the SARS-CoV-2 virus [49, 54]. However, leaky treatments may contribute to the release of the SARS-CoV-2 virus to the environment and increase the cases of infected people.

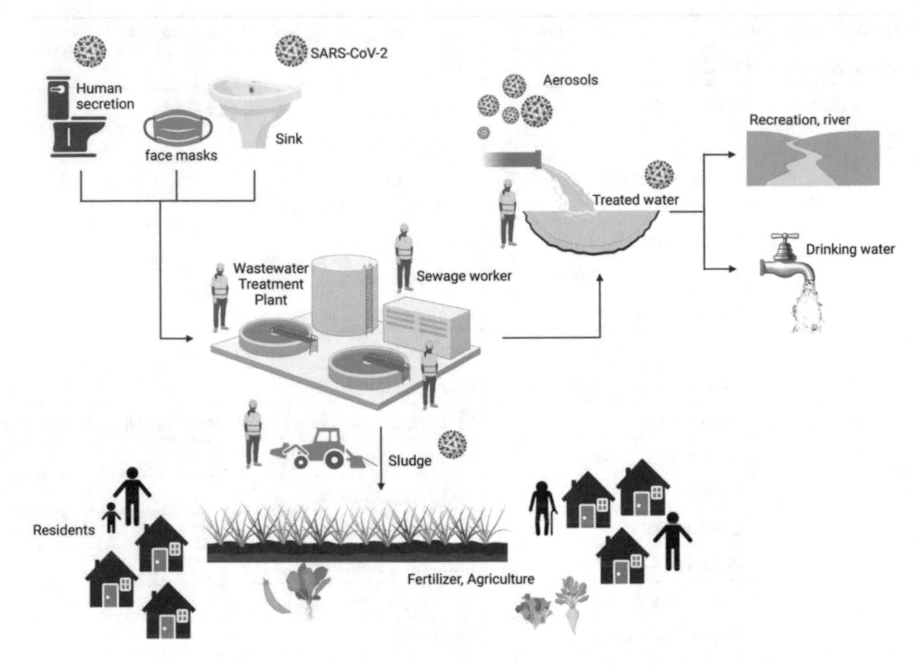


**Fig. 2** Overview of potential SARS-CoV-2 transmission via sewage. SARS-CoV-2 is considered bioaerosol and can also be transmitted to the sewage system through infected human secretion, improper face masks disposal (without disinfection) and insufficient hand sanitization. The virus will then transfer to the WWTP. Improper attire worn by the sewage worker or leaky treatments may contribute to virus transmission among the sewage worker and residents who live near the WWTP, either through drinking infected water or consuming infected vegetables. Created with BioRender.com

Additionally, the efficient disinfection method is necessary, especially for the Middle East and Mediterranean countries as they are known to use their treated sewage sludge for irrigation due to the shortage of water experienced by these countries [24, 55]. Nevertheless, this approach may promote the transfer of viruses, including SARS-CoV-2, and some antibiotic-resistance pathogens to humans, for instance via consumption of effluent-irrigated vegetables [56].

Collectively, sewage sludge treatments are urgently needed as constant or prolonged exposure to the pathogen may lead to life-threatening diseases such as those mentioned in Table 2. Interestingly, the common diseases seen with pathogen infections are due to excessive inflammation such as meningitis and encephalitis. Continuous or prolonged exposure to these inflammatory events may also contribute to cancer progression and this will be explained later in Sect. 2.5.

**Table 3** The content of heavy metals in sludge collected from different countries [59]

| Heavy metals | Land application | | Countries | | | | | | |
|---|---|---|---|---|---|---|---|---|---|
| | A | B | UK | France | Slovenia | Poland | China | Sweden | Japan |
| *Pb* | 750 | 2000 | 221.5 | 119.9 | 128 | 211.8 | 69.33 | 48.2 | 53 |
| *Cu* | 1000 | 1200 | 562 | 322 | 436 | 237.5 | 1467 | 522 | 255 |
| *Zn* | 2500 | 3500 | 778 | 837 | 2049 | 3641 | 7027 | 620.5 | 979 |
| *Cd* | 20 | 25 | 3.5 | 4.1 | 2.78 | 9.93 | 36.81 | 1.5 | 2.3 |
| *Cr* | 500 | 1000 | 159.5 | 69.4 | 841 | 144.2 | 210 | 38.4 | 69 |
| *Ni* | 300 | 400 | 58.5 | 35.5 | 622 | 41.1 | 214 | 19.3 | 40 |

All values in mg·kg$^{-1}$/d.m.; A: Agricultural purpose; B: Non-agricultural purpose

## 2.2 Heavy Metal Poisoning

Besides pathogen, sewage sludge has also been shown to compose greatly of heavy metals such as zinc (Zn), nickel (Ni), lead (Pb), cadmium (Cd) and chromium (Cr). Their concentration varied according to the types and origins of sewage sludge with some of them containing negligible levels of certain heavy metals while others containing higher levels. The concentrations of heavy metals (mg·kg$^{-1}$/dry matter (d.m.)) in sludge collected from different countries are described in Table 3. Usually, these heavy metals, which are mainly derived from industrial or domestic waste are treated before they can be disposed of. Amongst the treatments or remediation technology to remove heavy metals are precipitation, adsorption bio-sorption, coagulation and filtration [57]. However, these methods require large amounts of sludge and are very expensive [58].

Insufficient treatment may lead to human poisoning even at low concentrations and worse, bioaccumulation in the human body [60]. This would then contribute to several diseases including cancer, kidney damage, gastroenteritis, anaemia and jaundice [60–63]. Other diseases observed because of heavy metals poisoning are described in Table 4. Thus, it is necessary to maintain the concentration of discharged heavy metals within the accepted range, according to the Department of Environmental (DOE), Malaysia [25]. Besides, it is necessary to ensure that the heavy metals concentrations in sewage sludge did not exceed the ceiling concentration for land application as described in Table 3, and if it does, it must be used or disposed of in some other way.

## 2.3 Skin Irritation

The next common adverse effect experienced by a human upon contact with sewage sludge is skin irritation [64, 65]. Again, the presence of biological or chemical irritants is lower compared to the untreated sewage but it does not eliminate the possibility

**Table 4** Acceptable conditions for heavy metals discharge of industrial or mixed effluent and their toxic effects on humans

| Heavy metals | Standard [25] | | Human toxicity effect [61] |
|---|---|---|---|
| | A | B | |
| Cadmium | 0.01 | 0.02 | Kidney damage, fragile bones |
| Chromium | 0.05 | 0.05 | Allergy, increased cancer risk |
| Arsenic | 0.05 | 0.10 | Increased cancer risk |
| Lead | 0.10 | 0.50 | Neurological problem, kidney damage |
| Nickel | 0.20 | 1.00 | Allergy, lung fibrosis, kidney and cardiovascular damage |
| Zinc | 2.00 | 2.00 | Vomiting, diarrhea, nausea, neutropenia, gastrointestinal irritation |
| Copper | 0.2 | 1.00 | Hepatic necrosis, kidney damage |

A: Any inland water within the catchment areas; B: Any other inland waters or Malaysian water according to [25]

for the symptoms to occur if exposed in a longer term. For instance, skin irritation has been reported by some residents who reside within 1 km of land-sewage sludge application sites and this is potentially after being exposed to the wind blowing from the treated lands [66].

Compared to the other organ, skin is the largest and is the common organ being exposed to chemical or biological irritants. However, it is difficult to measure the effect and degree of human skin exposure to irritants, making the absence of dermal exposure standard established. These hence suggest the importance to identify the risk factor and developing the assessment method [67]. Amongst the common skin irritation conditions that result from exposure to sewage or sewage sludge are dermatitis, folliculitis and erythema [64]. These conditions possibly rise due to the direct contact of chemical irritants including perfluorinated chemicals, polychlorinated alkanes, polybrominated diphenyl ethers, triclosan, polychlorinated naphthalenes, benzothiazoles and organotins [34, 35]. Besides, poor hygiene and airborne irritant such as dust may also promote skin irritation [68].

Usually, the concentration of organic contaminants in wastewater is not affected with wastewater treatments due to poor biodegradation of the compounds [69]. Previous studies have identified several factors that may influence their biodegradation which includes the physicochemical properties of the compounds (hydrophobicity, molecular weight and pKa), sludge characteristics (organic matter and pH) and the operational parameters of WWTPs (primary sedimentation, hydraulic and sludge residence time) [34].

## 2.4 Pulmonary Disease

Some sewage workers and residents were also reported to experience several respiratory symptoms, ranging from a mild condition such as loss of odour recognition to severe chronic pulmonary diseases, including asthma and chronic obstructive pulmonary diseases (COPD), that needed special attention [70, 71]. Besides these two diseases, some of the residences, especially those that stay near the Class B sludged land (~1.6 km) were also reported to have pulmonary fibrosis, bronchitis, giardiasis and upper respiratory infection [72]. Excessive and prolonged inhalation of noxious gases, oil mists and bioaerosol are among the main contributors to the aforementioned diseases.

An example of noxious gas is hydrogen sulfide ($H_2S$), a rotten eggs-like odour, produced by the breakdown of waste material. It is extremely hazardous and inhalation of $H_2S$ may lead to varied clinical presentation especially in the cardiovascular, nervous and respiratory systems [73]. Acute exposure of $H_2S$ was shown to cause alveolar injury, leading to pulmonary oedema [74]. Meanwhile, repeated exposure, even at a low level was reported to promote chronic respiratory symptoms [75] and damage the upper respiratory tract [73], leading to respiratory failure. At high concentration, $H_2S$ was found to cause cognitive impairment, severe neurotoxicity and bronchospasms [76] (Table 5).

The mechanism of $H_2S$ primarily related to its cytotoxic effect and ability to inhibit enzyme activity and metabolic process, resulting in a decrease in ATP synthesis and subsequently protein phosphorylation [77, 78]. Upon inhalation, $H_2S$ will dissociate in blood as free sulfide and hydrogen ions. The free sulfide will then deliver to the tissue by blood circulation and bind with several macromolecules such as cytochrome oxidase, preventing oxidative phosphorylation. This will then lead to hypoxia and anoxia at the cellular levels (Fig. 3) [78, 79].

**Table 5** Health effect of hydrogen sulfide inhalation at different exposure [74]

| Concentration (Ppm) | Health effect |
|---|---|
| *3–5* | Moderate offensive 'rotten-egg' smell. May be associated with nausea, headaches or insomnia |
| *10–20* | Very strong offensive smell. Conjunctivitis may occur |
| *20–100* | Conjunctivitis, lung irritation, eye damage, digestive upset and loss of appetite |
| *100–150* | Olfactory fatigue; can no longer smell hydrogen sulfide odour |
| *150–250* | Severe eye and lung irritation, nausea, headache, vomiting, dizziness |
| *250–500* | Severe respiratory irritation, pulmonary oedema |
| *500–1000* | May cause death (4–8 h exposure), CNS stimulation, seizures, hyperpnoea, apnea, coma |
| *1000 and above* | Death from respiratory paralysis, immediate collapse |

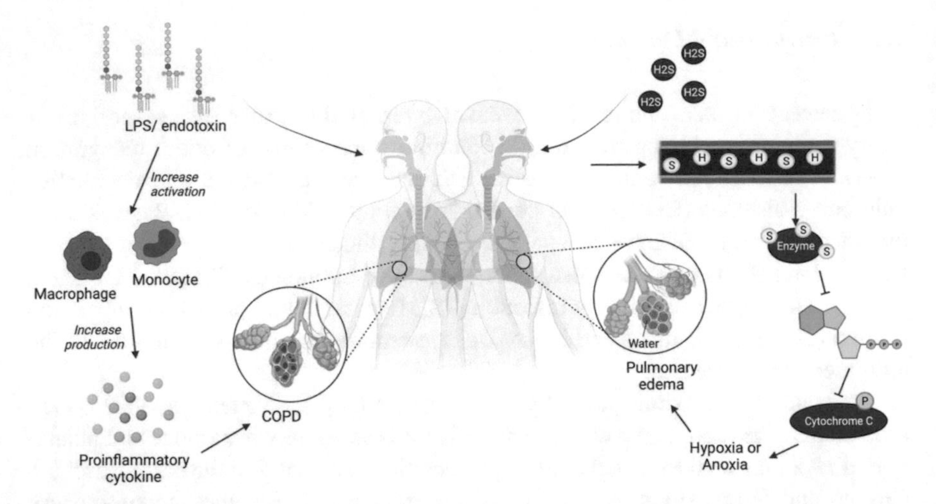


**Fig. 3** Potential pathophysiology of $H_2S$-induced pulmonary oedema and LPS-induced COPD. Exposure to bioaerosol such as endotoxin may promote macrophage and monocyte activation leading to increase production of pro-inflammatory cytokine. This will then contribute to COPD (left). Upon inhalation, $H_2S$ will dissociate into sulfide and hydrogen in the bloodstream. Free sulfide will then deliver to the tissue and inhibit enzyme activities, causing a decrease in ATP production and cytochrome C oxidative phosphorylation. As a result, this will lead to hypoxia and anoxia at the cellular levels and subsequent development of lung oedema (right). Created with BioRender.com

Another contributor is bioaerosol, an airborne collection of biological material such as allergens, endotoxin and mycotoxin. They were shown to promote inflammation and altered lung function [80]. For instance, a low amount of endotoxin or also known as lipopolysaccharides (LPS) can cause acute bronchial obstruction [81], airway irritation and cough among sewage workers [82]. These symptoms are due to endotoxin ability to trigger an inflammatory reaction, macrophages and monocytes activation and; overproduction of pro-inflammatory cytokines such as tumour necrosis factor-α (TNF-α), interleukin (IL)-23 and IL-6 [83]. Continuous inflammatory events will then lead to the development of COPD and also impaired lung function [84, 85] (Fig. 3).

## 2.5 Cancer

Among many substances in sewage, there are carcinogens and mutagens that may induce DNA mutagenesis and promote cancer progression [86–88]. Cancer incidence does not only happen among sewage workers but also involves residents who lived near the WWTPs, occurring through inhalation or dermal contact [89]. Besides, since exposure to sewage and sewage sludge were shown to cause various inflammatory diseases, it is possible therefore that it may also contribute to cancer progression as inflammation and cancer cross-talked (Fig. 4) [90].

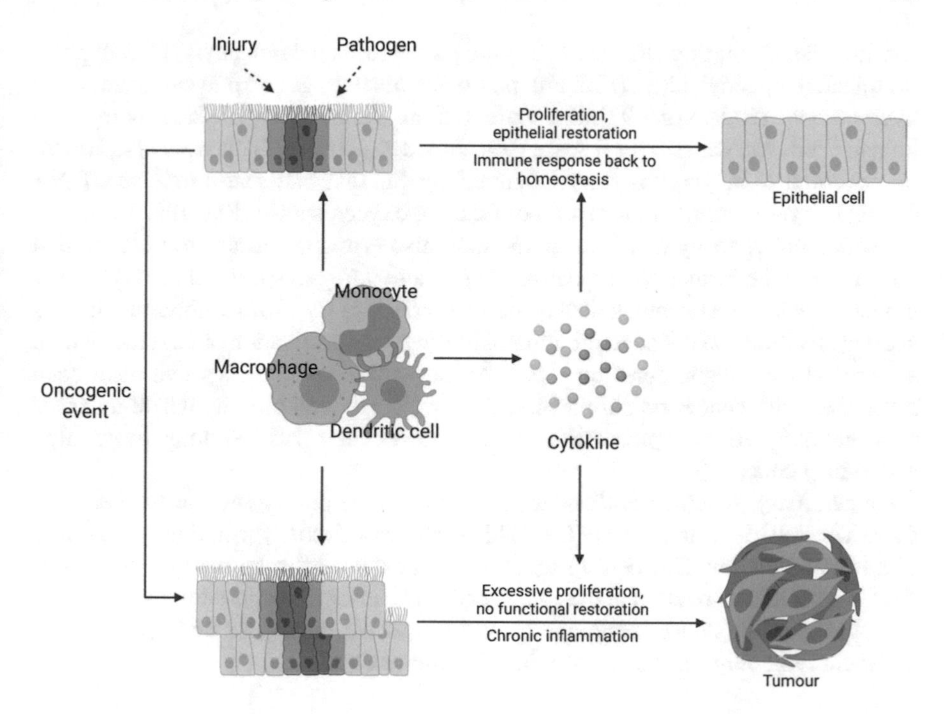


**Fig. 4** Similarities between inflammation in cancer and inflammation during infection. Cell injury, infection and an oncogenic event will promote some inflammatory cells such as monocyte, macrophage and dendritic cell activations. This will then stimulate pro-inflammatory cytokine or chemokine production, leading to restoration of cell homeostasis or tumour formation. Adapted from [90]. Created with BioRender.com

Inflammation is not only important to eliminate pathogens but also essential for maintaining tissue homeostasis such as tissue repair, regeneration and remodelling [91]. This event required a complex interaction of immunocompetent cells and inflammatory mediators like cytokine and chemokine. They communicate through a series of cascades and feedback circuits to maintain normal physiologic inflammation and immune surveillance [92]. Thus, it is expected that deregulation or prolonged inflammatory events may promote cancer progression (Fig. 4), such as seen with immunocompromised patients [93]. Almost 20% of cancer cases are preceded by infection and chronic inflammation at the same inflammation site [90, 94]. For instance, some patients who were diagnosed with inflammatory bowel disease (IBD), *Helicobacter*-induced gastritis and chronic hepatitis were considered to be at high risk of colorectal cancer (CRC), stomach cancer and liver cancer, respectively [95].

This is also apparent for the sewage industry as some of the workers were shown to develop cancer especially colorectal, bronchial and prostate cancers [87]. Additionally, the previous study also shows the risk of long-term used sewage sludge as fertilizer with the occurrence of liver cancer [96]. This is because sewage sludge is known to contain a mixture of environmental chemicals at low concentrations such as endocrine-disrupting chemicals (EDCs), polychlorinated biphenyls (PCBs), polybrominated diphenyl ethers (PBDEs), pharmaceutical drugs and polycyclic aromatic hydrocarbons (PAHs) [96–98]. Exposure to them, for instance, EDCs, is harmful to humans and may cause liver disease [99]. Besides, some of them were also known or suspected to be genotoxicants, chemical agents that may cause oxidative DNA damage by promoting the formation of reactive oxygen species [98, 100, 101].

Additionally, heavy metal poisoning may also promote cancer progression, and this is due to the formation of reactive oxygen and nitrogen species that may lead to oxidative stress and eventually, DNA damages (Fig. 5) [60, 102]. Subsequently, this genetic instability will lead to protein misfolding and inactivation of enzymes which are crucial for cells to function [103]. Amongst the heavy metals that have been associated with cancer risk are lead, cadmium, nickel and arsenic, wherein excess exposure or uptake may promote the development of skin, bladder, lung, liver, colon and kidney cancers [60].

In summary, several health issues have been reported by sewage workers and those who reside near the WWTPs. This includes infection, irritation, pulmonary diseases and cancer. This hence suggests an urgent need for the sewage industry or the Government to overcome the issues and this can be done possibly by including an enhanced sanitization technique or wastewater treatments; or by imposing new legislation not only for the industry but also the worker.

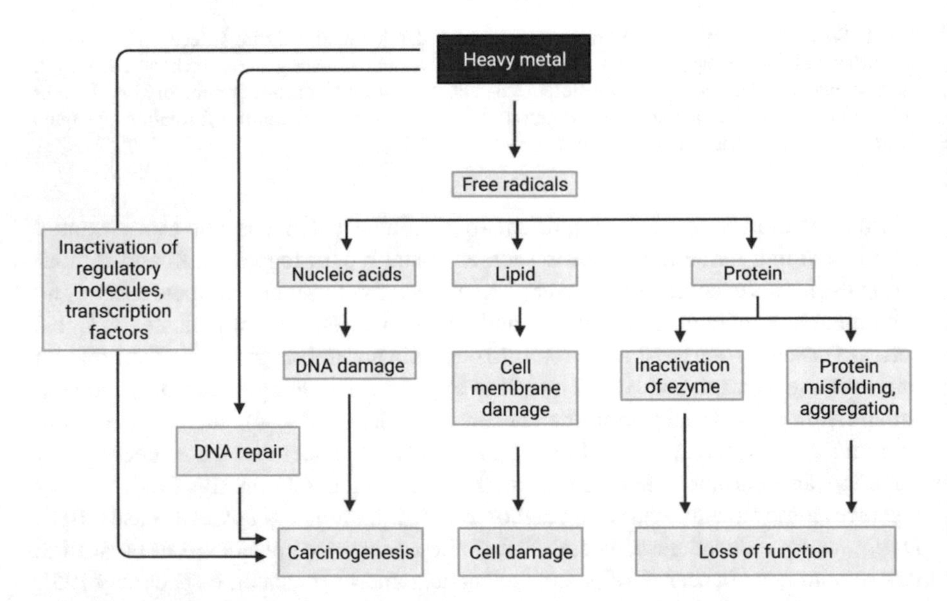


**Fig. 5** Mechanism of heavy metal intoxication and its association with cancer in humans. Heavy metal intoxification may cause DNA damage and cell membrane damage, resulting in tumour formation and cell damage, respectively. Besides, it may also cause protein misfolding and aggregation, leading to loss of protein function. Inactivation and deregulation of some regulatory molecules, transcription factors and DNA repair mechanisms due to heavy metal poisoning may also promote cancer progression in humans. Adapted from [60]. Created with BioRender.com

## 3 Approaches to Reduce the Health-Related Risk

As mentioned earlier, sewage sludge still contains some contaminants or pathogens that can affect human health. Unfortunately, until today there is no additional method has been designed to completely remove the contaminants from sewage sludge. However, numerous methods have been applied to reduce the contaminants, including wastewater treatments, which is known to reduce, eliminate or inactivate some of the contaminants, pollutants or disease-causing microorganisms in wastewater. Generally, the solid material will first be separated through filtration or sedimentation from the rest of the wastewater. The solid or sludge will then be treated with some chemical or biological agents for a certain period of time to allow the treatment process to occur. Next, the sludge is subjected to lime stabilizing, air drying, heat drying, conditioning, dewatering, or composting, before being disposed of or applied to the land (Fig. 1) [64]. The current approach also uses sewage sludge to produce renewable energy such as through microbial fuel cell (MFC) technology that utilizes the use of electrogenic bacteria (EB) to catalyst the degradation of chemical oxygen demand (COD) and promote the electrical generation [104].

Recent technology also includes Soil Biotechnology (SBT), a low-cost and eco-friendly technology that utilizes the use of soil, sand, gravel and biological media like earthworms, bacteria and plants to treat wastewater. This is because some of the pollutants can be removed, inactivated or adhered to soil's materials before reaching groundwater. This approach is also practical and handy in the case of no sewer transportation available [105]. That is probably the reason why land disposal is the most common method to dispose of sewage sludge.

Additionally, disinfection methods using gamma irradiation, ozone, chlorine and ultraviolet (UV) light can also be used to inactivate microorganisms [106]. Some studies show the detrimental effect of the by-product released through these disinfection methods such as neurological impairment and cancer [107, 108]. Amongst the concern disinfection byproducts (DPBs) are halogenated acetic acid, trihalomethanes and halogenated acetonitriles [105], wherein they are frequently associated with chlorination methods [109]. Compared to chlorination, UV radiation produced a relatively low concentration of DBPs, but some microorganisms can recover from the UV damage and sometimes result in an extensive bacterial regrowth through photoreactivation processes and/or dark repair [110, 111]. Currently, there is a growing interest in the use of gamma irradiation to treat wastewater as they are shown to be effective in damaging organic contaminants and inactivates pathogenic microorganisms [106, 112]. It is more efficient than UV radiation since it is not dependent on seasonal variation and has a lower bacterial regrowth rate. This high disinfection efficiency is probably due to its ability to promote reactive species formation that may cause irreversible injury to the microorganism [106].

However, the best way to prevent the occurring of diseases among sewage workers is by wearing proper attire and avoiding any confined and/or non-ventilated areas. This is more practical and can be applied to all. Besides, Gormley et al. also have suggested several ways to mitigate pathogen transmission via wastewater plumbing system which include: (1) report any unexplained foul smell in bathrooms, kitchens and wash areas; (2) ensure all water appliances are well fitted with functioning U-bend; (3) seal the crack or leak in pipework and (4) routinely monitor all system performances [113].

## 4 Conclusion

Collectively, the exposure of sewage and sewage sludge to humans especially sewage worker and residents who stay near the WWTPs involve various forms including consumption of infected food and water, inhalation of polluted air and skin contact. Although wastewater treatments are efficient in removing pathogens, heavy metals and organic chemicals, there is residual left in the treated water and sewage sludge which can cause delirious health effects to humans. These include infection by bacteria or viruses, heavy metal intoxification, skin irritation, pulmonary disease and cancer. Thus, new disinfection or wastewater treatment technologies and extra precautions are needed when handling sewage sludge.

## References

1. Havlíček, F., Morcinek, M.: Waste and pollution in the ancient roman empire. J. Landscape Ecol. **9**, 33–49 (2016). https://doi.org/10.1515/jlecol-2016-0013
2. Amasuomo, E., Baird, J.: The concept of waste and waste management. J. Manage. Sustain. **6**, 88 (2016). https://doi.org/10.5539/jms.v6n4p88
3. Kummer, K.: International Management of Hazardous Wastes: The Basel Convention and Related Legal Rules. Oxford University Press, Oxford (1999)
4. Grobelak, A., Czerwińska, K., Murtaś, A.: General considerations on sludge disposal, industrial and municipal sludge. In: Industrial and Municipal Sludge, pp. 135–153. Elsevier (2019)
5. Wastewater: the untapped resource. UNESCO, Paris (2017)
6. Jamil, M., Qasim, M., Umar, M.: Utilization of sewage sludge as organic fertiliser in sustainable agriculture. J. Appl. Sci. **6**, 531–535 (2006). https://doi.org/10.3923/jas.2006.531.535
7. Mohammad, M.J., Athamneh, B.M.: Changes in soil fertility and plant uptake of nutrients and heavy metals in response to sewage sludge application to calcareous soils. J. Agron. **3**, 229–236 (2004). https://doi.org/10.3923/ja.2004.229.236
8. Usman, K., Khan, S., Ghulam, S., Khan, M.U., Khan, N., Khan, M.A., Khalil, S.K.: Sewage sludge: an important biological resource for sustainable agriculture and its environmental implications. Am. J. Plant Sci. **03**, 1708–1721 (2012). https://doi.org/10.4236/ajps.2012.312209
9. Louis, G.E.: A historical context of municipal solid waste management in the United States. Waste Manage. Res. J. Sustain. Circular Econ. **22**, 306–322 (2004). https://doi.org/10.1177/0734242X04045425
10. Lofrano, G., Brown, J.: Wastewater management through the ages: a history of mankind. Sci. Total Environ. **408**, 5254–5264 (2010). https://doi.org/10.1016/j.scitotenv.2010.07.062

11. Smith, S.R.: Organic contaminants in sewage sludge (biosolids) and their significance for agricultural recycling. Philos. Trans. R. Soc. A Math. Phys. Eng. Sci. **367**, 4005–4041 (2009). https://doi.org/10.1098/rsta.2009.0154
12. Collivignarelli, M.C., Canato, M., Abbà, A., Carnevale Miino, M.: Biosolids: what are the different types of reuse? J. Cleaner Prod. **238**, 117844 (2019). https://doi.org/10.1016/j.jclepro.2019.117844
13. Ukwatta, A., Mohajerani, A., Setunge, S., Eshtiaghi, N.: Possible use of biosolids in fired-clay bricks. Constr. Build. Mater. **91**, 86–93 (2015). https://doi.org/10.1016/j.conbuildmat.2015.05.033
14. Ashekuzzaman, S.M., Forrestal, P., Richards, K., Fenton, O.: Dairy industry derived wastewater treatment sludge: generation, type and characterization of nutrients and metals for agricultural reuse. J. Cleaner Prod. **230**, 1266–1275 (2019). https://doi.org/10.1016/j.jclepro.2019.05.025
15. Spinosa, L.: Wastewater sludge: a global overview of the current status and future prospects. Water Intell. Online **6**, 9781780402154–9781780402154 (2015). https://doi.org/10.2166/9781780402154
16. Stehouwer, R.: What is sewage sludge and what can be done with it? https://extension.psu.edu/what-is-sewage-sludge-and-what-can-be-done-with-it (2010)
17. Eurostat: sewage sludge production and disposal from urban wastewater (in dry substance (d.s)). https://ec.europa.eu/eurostat/databrowser/product/view/ENV_WW_SPD%0A (2018)
18. Campo, G., Cerutti, A., Lastella, C., Leo, A., Panepinto, D., Zanetti, M., Ruffino, B.: Production and destination of sewage sludge in the Piemonte region (Italy): the results of a survey for a future sustainable management. Int. J. Environ. Res. Public Health **18** (2021). https://doi.org/10.3390/ijerph18073556
19. Rosendahl, L.: Direct thermochemical liquefaction for energy applications (2017)
20. Safoniuk, M.: Wastewater Engineering: Treatment and Reuse (Book) (2004)
21. Bianchini, A., Bonfiglioli, L., Pellegrini, M., Saccani, C.: Sewage sludge drying process integration with a waste-to-energy power plant. Waste Manag. **42**, 159–165 (2015). https://doi.org/10.1016/j.wasman.2015.04.020
22. Tezel, U., Tandukar, M., Pavlostathis, S.G.: Anaerobic biotreatment of municipal sewage sludge. In: Comprehensive Biotechnology, pp. 447–461. Elsevier (2011)
23. Bień, J., Nowak, D.: Biological composition of sewage sludge in the aspect of threats to the natural environment. Arch. Environ. Prot. **40** (2014). https://doi.org/10.2478/aep-2014-0040
24. Al-Gheethi, A.A., Efaq, A.N., Bala, J.D., Norli, I., Abdel-Monem, M.O., Ab Kadir, M.O.: Removal of pathogenic bacteria from sewage-treated effluent and biosolids for agricultural purposes. Appl. Water Sci. **8**, 74 (2018). https://doi.org/10.1007/s13201-018-0698-6
25. Environmental Quality Act: Environmental quality industrial effluent regulations 2009 (2009)
26. Sigua, G., Adjei, M., Rechcigl, J.: Cumulative and residual effects of repeated sewage sludge applications: forage productivity and soil quality implications in South Florida, USA (9 pp). Environ. Sci. Pollut. Res.—Int. **12**, 80–88 (2005). https://doi.org/10.1065/espr2004.10.220
27. Li, S., Di, X., Wu, D., Zhang, J.: Effects of sewage sludge and nitrogen fertilizer on herbage growth and soil fertility improvement in restoration of the abandoned opencast mining areas in Shanxi. China Environ. Earth Sci. **70**, 3323–3333 (2013). https://doi.org/10.1007/s12665-013-2397-9
28. Nazari, L., Sarathy, S., Santoro, D., Ho, D., Ray, M.B., Xu, C. (Charles): Recent advances in energy recovery from wastewater sludge. In: Direct Thermochemical Liquefaction for Energy Applications. pp. 67–100. Elsevier (2018)
29. Shaddel, S., Bakhtiary-Davijany, H., Kabbe, C., Dadgar, F., Østerhus, S.: Sustainable sewage sludge management: from current practices to emerging nutrient recovery technologies. Sustainability. **11**, 3435 (2019). https://doi.org/10.3390/su11123435
30. Twardowska, I., Schramm, K.-W., Berg, K.: Sewage sludge. In: Waste Management Series, pp. 239–295. Elsevier (2004)
31. Kwarciak-Kozłowska, A.: Co-composting of sewage sludge and wetland plant material from a constructed wetland treating domestic wastewater. In: Industrial and Municipal Sludge, pp. 337–360. Elsevier (2019)

32. Pritchard, D.L., Penney, N., McLaughlin, M.J., Rigby, H., Schwarz, K.: Land application of sewage sludge (biosolids) in Australia: risks to the environment and food crops. Water Sci. Technol. **62**, 48–57 (2010). https://doi.org/10.2166/wst.2010.274
33. Miller-Robbie, L., Ulrich, B.A., Ramey, D.F., Spencer, K.S., Herzog, S.P., Cath, T.Y., Stokes, J.R., Higgins, C.P.: Life cycle energy and greenhouse gas assessment of the co-production of biosolids and biochar for land application. J. Cleaner Prod. **91**, 118–127 (2015). https://doi.org/10.1016/j.jclepro.2014.12.050
34. Fijalkowski, K., Rorat, A., Grobelak, A., Kacprzak, M.J.: The presence of contaminations in sewage sludge—the current situation. J. Environ. Manage. **203**, 1126–1136 (2017). https://doi.org/10.1016/j.jenvman.2017.05.068
35. Clarke, B.O., Smith, S.R.: Review of 'emerging' organic contaminants in biosolids and assessment of international research priorities for the agricultural use of biosolids. Environ. Int. **37**, 226–247 (2011). https://doi.org/10.1016/j.envint.2010.06.004
36. Alonso, E., Santos, A., Riesco, P.: Micro-organism re-growth in wastewater disinfected by UV radiation and ozone: a micro-biological study. Environ. Technol. **25**, 433–441 (2004). https://doi.org/10.1080/09593332508618452
37. Ouardani, I., Turki, S., Aouni, M., Romalde, J.L.: Detection and molecular characterization of hepatitis a virus from Tunisian wastewater treatment plants with different secondary treatments. Appl. Environ. Microbiol. **82**, 3834–3845 (2016). https://doi.org/10.1128/AEM.00619-16
38. Vantarakis, A., Paparrodopoulos, S., Kokkinos, P., Vantarakis, G., Fragou, K., Detorakis, I.: Impact on the quality of life when living close to a municipal wastewater treatment plant. J. Environ. Public Health **2016**, 8467023 (2016). https://doi.org/10.1155/2016/8467023
39. Jaremków, A., Szałata, Ł., Kołwzan, B., Sówka, I., Zwoździak, J., Pawlas, K.: Impact of a sewage treatment plant on health of local residents: gastrointestinal system symptoms. Pol. J. Environ. Stud. **26**, 127–136 (2017). https://doi.org/10.15244/pjoes/64793
40. Gubišová, M., Horník, M., Hrčková, K., Gubiš, J., Jakubcová, A., Hudcovicová, M., Ondreičková, K.: Sewage sludge as a soil amendment for growing biomass plant Arundo donax L. (2020)
41. Straub, T.M., Pepper, I.L., Gerba, C.P.: Hazards from pathogenic microorganisms in land-disposed sewage sludge. In: Reviews of Environmental Contamination and Toxicology, pp. 55–91 (1993)
42. AL-Gheethi, A.A.S., Ismail, N., Lalung, J., Talib, A., Efaq, A.N., Kadir, M.O.A.: Susceptibility for antibiotics among faecal indicators and pathogenic bacteria in sewage treated effluents. Water Pract. Technol. **8**, 1–6 (2013). https://doi.org/10.2166/wpt.2013.001
43. Pazda, M., Kumirska, J., Stepnowski, P., Mulkiewicz, E.: Antibiotic resistance genes identified in wastewater treatment plant systems—a review. Sci. Total Environ. **697**, 134023 (2019). https://doi.org/10.1016/j.scitotenv.2019.134023
44. Prado, T., Gaspar, A.M.C., Miagostovich, M.P.: Detection of enteric viruses in activated sludge by feasible concentration methods (2014)
45. Jacobsen, K.H., Koopman, J.S.: The effects of socioeconomic development on worldwide hepatitis A virus seroprevalence patterns. Int. J. Epidemiol. **34**, 600–609 (2005). https://doi.org/10.1093/ije/dyi062
46. Ouardani, I., Manso, C.F., Aouni, M., Romalde, J.L.: Efficiency of hepatitis A virus removal in six sewage treatment plants from central Tunisia. Appl. Microbiol. Biotechnol. **99**, 10759–10769 (2015). https://doi.org/10.1007/s00253-015-6902-9
47. Haake, D.A., Levett, P.N.: Leptospirosis in humans. Curr. Top. Microbiol. Immunol. **387**, 65–97 (2015). https://doi.org/10.1007/978-3-662-45059-8_5
48. Tran, H.N., Le, G.T., Nguyen, D.T., Juang, R.-S., Rinklebe, J., Bhatnagar, A., Lima, E.C., Iqbal, H.M.N., Sarmah, A.K., Chao, H.-P.: SARS-CoV-2 coronavirus in water and wastewater: a critical review about presence and concern. Environ. Res. **193**, 110265 (2021). https://doi.org/10.1016/j.envres.2020.110265
49. Sherchan, S.P., Shahin, S., Ward, L.M., Tandukar, S., Aw, T.G., Schmitz, B., Ahmed, W., Kitajima, M.: First detection of SARS-CoV-2 RNA in wastewater in North America: a study

in Louisiana, USA. Sci. Total Environ. **743**, 140621 (2020). https://doi.org/10.1016/j.scitotenv.2020.140621
50. Medema, G., Heijnen, L., Elsinga, G., Italiaander, R., Brouwer, A.: Presence of SARS-coronavirus-2 RNA in sewage and correlation with reported COVID-19 prevalence in the early stage of the epidemic in The Netherlands. Environ. Sci. Technol. Lett. **7**, 511–516 (2020). https://doi.org/10.1021/acs.estlett.0c00357
51. Haramoto, E., Malla, B., Thakali, O., Kitajima, M.: First environmental surveillance for the presence of SARS-CoV-2 RNA in wastewater and river water in Japan. Sci. Total Environ. **737**, 140405 (2020). https://doi.org/10.1016/j.scitotenv.2020.140405
52. Naddeo, V., Liu, H.: Editorial perspectives: 2019 novel coronavirus (SARS-CoV-2): what is its fate in urban water cycle and how can the water research community respond? Environ. Sci. Water Res. Technol. **6**, 1213–1216 (2020). https://doi.org/10.1039/D0EW90015J
53. Bogler, A., Packman, A., Furman, A., Gross, A., Kushmaro, A., Ronen, A., Dagot, C., Hill, C., Vaizel-Ohayon, D., Morgenroth, E., Bertuzzo, E., Wells, G., Kiperwas, H.R., Horn, H., Negev, I., Zucker, I., Bar Or, I., Moran-Gilad, J., Balcazar, J.L., Bibby, K., Elimelech, M., Weisbrod, N., Nir, O., Sued, O., Gillor, O., Alvarez, P.J., Crameri, S., Arnon, S., Walker, S., Yaron, S., Nguyen, T.H., Berchenko, Y., Hu, Y., Ronen, Z., Bar-Zeev, E.: Rethinking wastewater risks and monitoring in light of the COVID-19 pandemic. Nat. Sustain. **3**, 981–990 (2020). https://doi.org/10.1038/s41893-020-00605-2
54. Xiang, Y., Song, Q., Gu, W.: Decontamination of surgical face masks and N95 respirators by dry heat pasteurization for one hour at 70 °C. Am. J. Infect. Control. **48**, 880–882 (2020). https://doi.org/10.1016/j.ajic.2020.05.026
55. Kalavrouziotis, I., Arslan-Alaton, I.: Reuse of urban wastewater and sewage sludge in the Mediterranean countries: case studies from Greece and Turkey. Fresenius Environ. Bull. **17**, 625–639 (2008)
56. Rai, P., Tripathi, B.: Microbial contamination in vegetables due to irrigation with partially treated municipal wastewater in a tropical city. Int. J. Environ. Health Res. **17**, 389–395 (2007). https://doi.org/10.1080/09603120701628743
57. Masindi, V., Muedi, K.L.: Environmental contamination by heavy metals. In: Aglan, K.L.M.E.-H.E.-D.M.S.E.-R.F. (ed.) Heavy Metals, p. Ch. 7. InTech, Rijeka (2018)
58. Nleya, Y., Simate, G.S., Ndlovu, S.: Sustainability assessment of the recovery and utilisation of acid from acid mine drainage. J. Cleaner Prod. **113**, 17–27 (2016). https://doi.org/10.1016/j.jclepro.2015.11.005
59. Milik, J., Pasela, R., Lachowicz, M., Chalamoński, M.: The concentration of trace elements in sewage sludge from wastewater treatment plant in Gniewino. J. Ecol. Eng. **18**, 118–124 (2017). https://doi.org/10.12911/22998993/74628
60. Azeh Engwa, G., Udoka Ferdinand, P., Nweke Nwalo, F., N. Unachukwu, M.: Mechanism and health effects of heavy metal toxicity in humans. In: Ferdinand, P.U. (ed.) Poisoning in the Modern World—New Tricks for an Old Dog? p. Ch. 5. IntechOpen, Rijeka (2019)
61. e Silva, A.R.B., Camilotti VIII, F.: Risks of heavy metals contamination of soil-pant system by land application of sewage sludge: a review with data from Brazil. In: Environmental Risk Assessment of Soil Contamination. InTech. (2014). https://doi.org/10.5772/58384
62. Jaishankar, M., Tseten, T., Anbalagan, N., Mathew, B.B., Beeregowda, K.N.: Toxicity, mechanism and health effects of some heavy metals. Interdiscip. Toxicol. **7**, 60–72 (2014). https://doi.org/10.2478/intox-2014-0009
63. Järup, L.: Hazards of heavy metal contamination. Br. Med. Bull. **68**, 167–182 (2003). https://doi.org/10.1093/bmb/ldg032
64. Garg, N.K.: Multicriteria assessment of alternative sludge disposal methods Neeraj Kumar Garg (2009)
65. Lundholm, M., Rylander, R.: Work related symptoms among sewage workers. Br. J. Ind. Med. **40**, 325–329 (1983). https://doi.org/10.1136/oem.40.3.325
66. Lewis, D.L., Gattie, D.K., Novak, M.E., Sanchez, S., Pumphrey, C.: Interactions of pathogens and irritant chemicals in land-applied sewage sludges (biosolids). New Solutions **12**, 409–423 (2002). https://doi.org/10.2190/LHRY-90EH-HT21-VPH7

67. De Craecker, W., Roskams, N., De Beeck, R.O.: Occupational skin diseases and dermal exposure in the European Union (EU-25). Policy Pract. Overview (2008)
68. Nethercott, J.R.: Airborne irritant contact dermatitis due to sewage sludge. J. Occup. Med. Official Publ. Ind. Med. Assoc. **23**, 771–774 (1981). https://doi.org/10.1097/00043764-198111000-00012
69. Stasinakis, A.S.: Review on the fate of emerging contaminants during sludge anaerobic digestion. Bioresour. Technol. **121**, 432–440 (2012). https://doi.org/10.1016/j.biortech.2012.06.074
70. Chandra, K., Arora, V.K.: Occupational lung diseases in sewage workers: A systematic review. J, Indian Acad. Clin. Med. **19**, 121–132 (2018)
71. Douwes, J., Thorne, P., Pearce, N., Heederik, D.: Bioaerosol health effects and exposure assessment: progress and prospects. Ann. Occup. Hyg. **47**, 187–200 (2003). https://doi.org/10.1093/annhyg/meg032
72. Khuder, S., Milz, S.A., Bisesi, M., Vincent, R., McNulty, W., Czajkowski, K.: Health survey of residents living near farm fields permitted to receive biosolids. Arch. Environ. Occup. Health **62**, 5–11 (2007). https://doi.org/10.3200/AEOH.62.1.5-11
73. Mousa, H.A.L.: Short-term effects of subchronic low-level hydrogen sulfide exposure on oil field workers. Environ. Health Prev. Med. **20**, 12–17 (2015). https://doi.org/10.1007/s12199-014-0415-5
74. Rumbeiha, W., Whitley, E., Anantharam, P., Kim, D.-S., Kanthasamy, A.: Acute hydrogen sulfide–induced neuropathology and neurological sequelae: challenges for translational neuroprotective research. Ann. N. Y. Acad. Sci. **1378**, 5–16 (2016). https://doi.org/10.1111/nyas.13148
75. Richardson, D.B.: Respiratory effects of chronic hydrogen sulfide exposure. Am. J. Ind. Med. **28**, 99–108 (1995). https://doi.org/10.1002/ajim.4700280109
76. Austigard, Å.D., Svendsen, K., Heldal, K.K.: Hydrogen sulphide exposure in waste water treatment. J. Occup. Med. Toxicol. **13**, 10 (2018). https://doi.org/10.1186/s12995-018-0191-z
77. Kfir, H., Rimbrot, S., Markel, A.: Toxic effects of hydrogen sulfide: experience with three simultaneous patients. QJM **108**, 977–978 (2015). https://doi.org/10.1093/qjmed/hcv108
78. Guidotti, T.L.: Hydrogen sulfide: advances in understanding human toxicity. Int. J. Toxicol. **29**, 569–581 (2010). https://doi.org/10.1177/1091581810384882
79. Lindenmann, J., Matzi, V., Neuboeck, N., Ratzenhofer-Komenda, B., Maier, A., Smolle-Juettner, F.-M.: Severe hydrogen sulphide poisoning treated with 4-dimethylaminophenol and hyperbaric oxygen. Diving Hyperb. Med. **40**, 213–217 (2010)
80. Sigsgaard, T., Bonefeld-Jørgensen, E.C., Hoffmann, H.J., Bønløkke, J., Krüger, T.: Microbial cell wall agents as an occupational hazard. Toxicol. Appl. Pharmacol. **207**, 310–319 (2005). https://doi.org/10.1016/j.taap.2004.12.031
81. Cyprowski, M., Sobala, W., Buczyńska, A., Szadkowska-Stańczyk, I.: Endotoxin exposure and changes in short-term pulmonary function among sewage workers. Int. J. Occup. Med. Environ. Health **28**, 803–811 (2015). https://doi.org/10.13075/ijomeh.1896.00460
82. Heldal, K.K., Madsø, L., Huser, P.O., Eduard, W.: Exposure, symptoms and airway inflammation among sewage workers. Ann. Agric. Environ. Med. **17**, 263–268 (2010)
83. Kang, M., Edmundson, P., Araujo-Perez, F., McCoy, A.N., Galanko, J., Keku, T.O.: Association of plasma endotoxin, inflammatory cytokines and risk of colorectal adenomas. BMC Cancer **13**, 91 (2013). https://doi.org/10.1186/1471-2407-13-91
84. Rylander, R.: Endotoxin and occupational airway disease. Curr. Opin. Allergy Clin. Immunol. **6**, 62–66 (2006). https://doi.org/10.1097/01.all.0000202356.83509.f7
85. Heldal, K.K., Barregard, L., Larsson, P., Ellingsen, D.G.: Pneumoproteins in sewage workers exposed to sewage dust. Int. Arch. Occup. Environ. Health **86**, 65–70 (2013). https://doi.org/10.1007/s00420-012-0747-7
86. Malakahmad, A., Abd Manan, T.S., Hilmin, N.: Biological procedures to detect carcinogenic compounds in domestic wastewater. WIT Trans. Ecol. Environ. **182**, 345–356 (2014). https://doi.org/10.2495/WP140301

87. Nasterlack, M., Messerer, P., Pallapies, D., Ott, M.G., Zober, A.: Cancer incidence in the wastewater treatment plant of a large chemical company. Int. Arch. Occup. Environ. Health **82**, 851 (2009). https://doi.org/10.1007/s00420-009-0397-6
88. Scarlett-Kranz, J.M., Babish, J.G., Strickland, D., Goodrich, R.M., Lisk, D.J.: Urinary mutagens in municipal sewage workers and water treatment workers. Am. J. Epidemiol. **124**, 884–893 (1986). https://doi.org/10.1093/oxfordjournals.aje.a114478
89. Friis, L., Edling, C., Hagmar, L.: Mortality and incidence of cancer among sewage workers: a retrospective cohort study. Br. J. Ind. Med. **50**, 653–657 (1993). https://doi.org/10.1136/oem.50.7.653
90. Greten, F.R., Grivennikov, S.I.: Inflammation and cancer: triggers, mechanisms, and consequences. Immunity **51**, 27–41 (2019). https://doi.org/10.1016/j.immuni.2019.06.025
91. Medzhitov, R.: Origin and physiological roles of inflammation. Nature **454**, 428–435 (2008). https://doi.org/10.1038/nature07201
92. Nicholson, L.B.: The immune system. Essays Biochem. **60**, 275–301 (2016). https://doi.org/10.1042/EBC20160017
93. Ahmad Mokhtar, A.M., Hashim, I.F., Mohd Zaini Makhtar, M., Salikin, N.H., Amin-Nordin, S.: The Role of RhoH in TCR signalling and its involvement in diseases. Cells **10**, 950 (2021). https://doi.org/10.3390/cells10040950
94. Grivennikov, S.I., Greten, F.R., Karin, M.: Immunity, inflammation, and cancer. Cell **140**, 883–899 (2010). https://doi.org/10.1016/j.cell.2010.01.025
95. Trinchieri, G.: Cancer and inflammation: an old intuition with rapidly evolving new concepts. Annu. Rev. Immunol. **30**, 677–706 (2012). https://doi.org/10.1146/annurev-immunol-020711-075008
96. Filis, P., Walker, N., Robertson, L., Eaton-Turner, E., Ramona, L., Bellingham, M., Amezaga, M.R., Zhang, Z., Mandon-Pepin, B., Evans, N.P., Sharpe, R.M., Cotinot, C., Rees, W.D., O'Shaughnessy, P., Fowler, P.A.: Long-term exposure to chemicals in sewage sludge fertilizer alters liver lipid content in females and cancer marker expression in males. Environ. Int. **124**, 98–108 (2019). https://doi.org/10.1016/j.envint.2019.01.003
97. Rhind, S.M., Kyle, C.E., Ruffie, H., Calmettes, E., Osprey, M., Zhang, Z.L., Hamilton, D., McKenzie, C.: Short- and long-term temporal changes in soil concentrations of selected endocrine disrupting compounds (EDCs) following single or multiple applications of sewage sludge to pastures. Environ. Pollut. **181**, 262–270 (2013). https://doi.org/10.1016/j.envpol.2013.06.011
98. Jureczko, M., Kalka, J.: Cytostatic pharmaceuticals as water contaminants. Eur. J. Pharmacol. **866**, 172816 (2020). https://doi.org/10.1016/j.ejphar.2019.172816
99. Al-Eryani, L., Wahlang, B., Falkner, K.C., Guardiola, J.J., Clair, H.B., Prough, R.A., Cave, M.: Identification of environmental chemicals associated with the development of toxicant-associated fatty liver disease in rodents. Toxicol. Pathol. **43**, 482–497 (2014). https://doi.org/10.1177/0192623314549960
100. Olinski, R., Rozalski, R., Gackowski, D., Foksinski, M., Siomek, A., Cooke, M.S.: Urinary measurement of 8-OxodG, 8-OxoGua, and 5HMUra: a noninvasive assessment of oxidative damage to DNA. Antioxid. Redox Signal. **8**, 1011–1019 (2006). https://doi.org/10.1089/ars.2006.8.1011
101. Al Zabadi, H., Ferrari, L., Sari-Minodier, I., Kerautret, M.-A., Tiberguent, A., Paris, C., Zmirou-Navier, D.: Integrated exposure assessment of sewage workers to genotoxicants: an urinary biomarker approach and oxidative stress evaluation. Environ. Health **10**, 23 (2011). https://doi.org/10.1186/1476-069X-10-23
102. Valko, M., Morris, H., Cronin, M.: Metals, toxicity and oxidative stress. Curr. Med. Chem. **12**, 1161–1208 (2005). https://doi.org/10.2174/0929867053764635
103. Nagaraj, N.S., Singh, O.V., Merchant, N.B.: Proteomics: a strategy to understand the novel targets in protein misfolding and cancer therapy. Expert Rev. Proteomics **7**, 613–623 (2010). https://doi.org/10.1586/epr.10.70
104. Muaz, M.Z.M., Abdul, R., Vadivelu, V.M.: Recovery of energy and simultaneous treatment of dewatered sludge using membrane-less microbial fuel cell. Environ. Prog. Sustain. Energy **38**, 208–219 (2019). https://doi.org/10.1002/ep.12919

105. Kamble, S.J., Chakravarthy, Y., Singh, A., Chubilleau, C., Starkl, M., Bawa, I.: A soil biotechnology system for wastewater treatment: technical, hygiene, environmental LCA and economic aspects. Environ. Sci. Pollut. Res. **24**, 13315–13334 (2017). https://doi.org/10.1007/s11356-017-8819-6
106. Lee, O.-M., Kim, H.Y., Park, W., Kim, T.-H., Yu, S.: A comparative study of disinfection efficiency and regrowth control of microorganism in secondary wastewater effluent using UV, ozone, and ionizing irradiation process. J. Hazard. Mater. **295**, 201–208 (2015). https://doi.org/10.1016/j.jhazmat.2015.04.016
107. Ahmed, A.E., Campbell, G.A., Jacob, S.: Neurological impairment in fetal mouse brain by drinking water disinfectant byproducts. Neurotoxicology **26**, 633–640 (2005). https://doi.org/10.1016/j.neuro.2004.11.001
108. Richardson, S.D., Plewa, M.J., Wagner, E.D., Schoeny, R., DeMarini, D.M.: Occurrence, genotoxicity, and carcinogenicity of regulated and emerging disinfection by-products in drinking water: a review and roadmap for research. Mutat. Res. **636**, 178–242 (2007). https://doi.org/10.1016/j.mrrev.2007.09.001
109. Bull, R.J., Birnbaum, L., Cantor, K.P., Rose, J.B., Butterworth, B.E., Pegram, R.E.X., Tuomisto, J.: Water chlorination: essential process or cancer hazard? Toxicol. Sci. **28**, 155–166 (1995). https://doi.org/10.1093/toxsci/28.2.155
110. Liltved, H., Landfald, B.: Effects of high intensity light on ultraviolet-irradiated and non-irradiated fish pathogenic bacteria. Water Res. **34**, 481–486 (2000). https://doi.org/10.1016/S0043-1354(99)00159-1
111. Oguma, K., Katayama, H., Mitani, H., Morita, S., Hirata, T., Ohgaki, S.: Determination of pyrimidine dimers in *Escherichia coli* and *Cryptosporidium parvum* during UV light inactivation, photoreactivation, and dark repair. Appl. Environ. Microbiol. **67**, 4630 LP–4637 (2001). https://doi.org/10.1128/AEM.67.10.4630-4637.2001
112. Yuan, S., Zheng, Z., Mu, Y., Yu, X., Zhao, Y.: Use of gamma-irradiation pretreatment for enhancement of anaerobic digestibility of sewage sludge. Front. Environ. Sci. Eng. China **2**, 247–250 (2008). https://doi.org/10.1007/s11783-008-0041-9
113. Gormley, M., Aspray, T.J., Kelly, D.A.: COVID-19: mitigating transmission via wastewater plumbing systems. Lancet Glob. Health **8**, e643–e643 (2020). https://doi.org/10.1016/S2214-109X(20)30112-1

# Microalgae Mediated Sludge Treatment

**Julfequar Hussain, Kaveri Dang, Shruti Chatterjee, and Ekramul Haque**

**Abstract** Microalgae are truly versatile in nature. Their application ranges from food supplements to the bioremediation of sludge and wastewater. Integration of microalgae in the wastewater treatment plants not only increases the efficiency of the removal of pollutants but also significantly brings down the operational cost of the facility, which is due to the fact that microalgae are able to utilize the sunlight and uptake the inorganic carbon, nitrogen and phosphorus abundant in wastewater. Along with the rapid growth and multiplication of the microalgae population, there is a simultaneous increase in the amount of biomass produced, which can be harvested to generate energy (biohydrogen, biogas, biodiesel, etc.); to obtain pigments, lipids and fatty acids to be used in the cosmetic industry; and as feed for livestock and aquaculture. Furthermore, the fate of inorganic carbon after it enters microalgae, as well as its interactions with the Calvin cycle, Citric acid cycle and fatty acid production, has been briefly discussed. It is interesting to note that the species belonging to the genera *Chlorella* and *Scenedesmus* have excellent Endocrine Disrupting Chemicals (EDCs) removal capacity (e.g. *Chlorella fusca* can degrade >95% of bisphenol A, *Chlorella pyrenoidosa* can remove >75% of triclosan and 3,4-dichloroaniline, *Scenedesmus dimorphus* can remove up to 95% of total estrogens from 17α-Estradiol and *Scenedesmus obliquus* can remove about 89% of nonylphenol). Apart from that, microalgae also have excellent nutrient removal (C, N and P) and heavy metal removal capacity.

J. Hussain
Department of Biotechnology, School of Life Sciences, Pondicherry University, Puducherry 605014, India

K. Dang
Department of Microbiology, Kasturba Medical College, Manipal MAHE, Karnataka 576104, India

S. Chatterjee
APBD, CSIR-CSMCRI, Bhavnagar, Gujarat 364002, India

E. Haque (✉)
AESD & CIF, CSIR-CSMCRI, Bhavnagar, Gujarat 364002, India



A. Z. Yaser et al. (eds.), *Waste Management, Processing and Valorisation*,
https://doi.org/10.1007/978-981-16-7653-6_9

**Keywords** Microalgae · Bioremediation · Sludge · Biodiesel · Anaerobic fermentation

# 1 Introduction

Microalgae or microscopic algae are one of the oldest living organisms on earth. They are unicellular organisms with sizes ranging in the order of micrometres and are found both in marine and freshwater environments. Their applications range from the production of food supplements, pigments, antibiotics, vitamins, cosmetics, biofuel, biofertilizers and aquaculture feeds to bioremediation of sludge and wastewater [1–3]. The earliest reported use of microalgae dates back approximately 2000 years when the Chinese used the cyanobacteria *Nostoc* as a source of food during a famine. Microalgae mediated sludge treatment has become one of the most sought-after methods for the treatment of sludge and wastewater, the main reason being their superior photosynthetic efficiency, case of management (once conditions are standardized) and higher nutrient removal capability. In addition, they have excellent potential for the up-gradation of the quality of biogas by mitigating the amounts of $CO_2$ (10 times more efficient as compared to forests) and $H_2S$ produced during the process [4–6]. Moreover, one of the major impediments in the production of biogas is the accumulation of volatile fatty acids during anaerobic fermentation, which can be overcome by their conversion to acetyl-CoA using microalgae [7]. Apart from this, the usage of microalgae in the wastewater plants may reduce the cost of the residual sludge disposal, which may sometimes account for >50% of the total expenditure of running a wastewater treatment plant [8].

# 2 Mechanisms of Microalgae for Sewage Treatment

## *2.1 Strain Selection*

Various strains of microalgae are selected depending upon their characteristics and capabilities to treat wastewater [9]. Scendesmus obliquus, a strain responsible for biological elimination of carbonaceous and nitrogenous pollutants, has recently acquired prominence. In addition, *Chlorella vulgaris* has proven to be an efficient candidate for this process [10]. Factors to consider in the selection of suitable species would include: type of wastewater that needs to be treated, level of treatment required, cost requirement of harvesting biomass and its application [9, 11].

## 2.2 Pretreatment

The performance of microalgal species depends upon the wastewater composition as microalgae work on three essential nutrients—Carbon, Nitrogen and Phosphorous [10]. An optimal ratio of these elements which reflects the stoichiometry of microalgae is required for an efficient result. Besides, elemental nutrients and trace amounts of other nutrients like calcium, zinc and manganese are also required which are present in abundance in the wastewater bodies [8, 12].

There can be a presence of certain substances leading to a change in pH that hamper the growth of microalgal species. To avoid this, pretreatment of wastewater is among the first few steps leading to treatment. Treating wastewater physicochemically to reduce the strength of wastewater is an essential step [13]. In a study done by Lin et al. in 2017, a series of strategies have been adopted to treat wastewater generated by the textile industry [14]. Firstly, activated carbon helps in the adsorption of toxic compounds in the wastewater, which is then allowed for anaerobic digestion, in order to generate electricity and decrease the load of Treated Wastewater (TW). Further, incompletely treated wastewater is cultivated with microalgal species. Microalgal strains responsible for this process require a specific amount of ammonia to supplement their need for elemental nitrogen but the concentration of ammonia in wastewater is found to be significantly high, making growth of microalgal strains difficult [14, 15]. In order to reduce ammonia concentration, pretreatment is necessary for which it is set to have the optimum ammonia concentration. Another way to do it is to dilute the wastewater to such an extent that the ammonium present gets diluted to a concentration suitable for microalgae to grow [12, 14]. In some wastewater treatment plants of paper, oil and grease, electrocoagulation is performed in the pretreatment step to remove chemical pollutants and to reduce turbidity.

# 3 Wastewater Treatment

## 3.1 Uptake of Nutrients

Organic nutrients are often added as supplements to the wastewater where C/N ratio is lower than the required amount to improve removal efficiency [10, 16]. Several strains have specific requirements in treatment plants, like sunlight utilization, inorganic carbon dioxide and nitrogen which can range anywhere between 3 and 10% to increase cell numbers [17]. Different amounts of nitrogenous content can be assimilated by microalgal strains from both organic and inorganic sources depending upon the strain type and its capability. Inorganic sources include ammonium, nitrate, nitrite and atmospheric nitrogen whereas organic sources comprise urea and glycine [12, 16].

## 3.2 Carbon Uptake/Removal

Microalgal biomass utilizes inorganic carbon, primarily as $CO_2$, as carbon source. Around 1–6% of carbon is considered optimum for the growth of microalgae and the removal of nutrients. $CO_2$ (g) is dissociated into bicarbonate and carbonate ions in aqueous mode at equilibrium depending upon the pH, concentration of cation and salinity conditions [11]. Because of the non-polar nature of carbon dioxide, it diffuses across plasma membrane of microalgae but bicarbonate is transported actively with energy requirement. Upon reaching chloroplast, bicarbonate catalyzes carbon dioxide with the help of carbonic anhydrase that fixes inorganic carbon [11, 18]. Microalgae converts this inorganic carbon to organic in Calvin cycle using NADPH (Nicotinamide Adenine Dinucleotide Phosphate) which is a reductant, along with energy utilized from hydrolysis of ATP which is produced in photosynthetic electron transport chain. Inorganic carbon is fixed to ribulose 1, 5-bisphosphate which is eventually reduced to glyceraldehyde-3-phosphate (G3P) with the help of a series of reactions. Additional glyceraldehyde-3-phosphate is either sent for storage or metabolized to pyruvate through glycolysis. Pyruvate is then converted to acetyl-CoA, which either goes for lipid synthesis or enters the Citric acid cycle and forms the intermediate 2-oxoglutarate (2-OG), which is transported to the chloroplast wherein it combines with ammonium ions to form glutamic acid (GLU) and finally to form glutamine (GLN), which is involved in the biosynthesis of proteins, nucleic acids and mucopolysaccharides [11, 17]. A simple illustration of the uptake and assimilation of the inorganic carbon is given in Fig. 1.

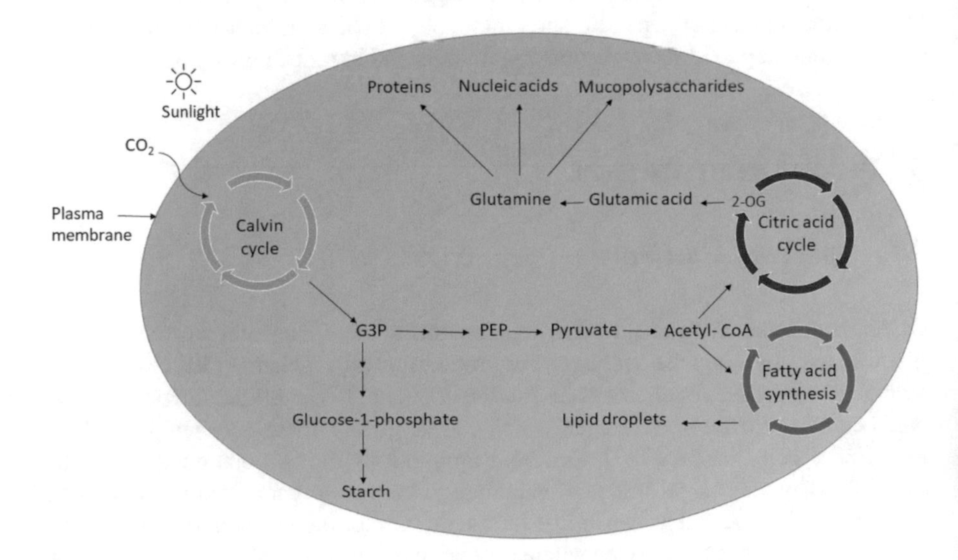


**Fig. 1** Fate of inorganic carbon inside a mixotrophic microalgae. *Footnotes: G3P: Glyceraldehyde-3-phosphate, PEP: Phosphoenol pyruvate, 2-OG: 2-Oxoglutarate

### 3.3 Nitrogen Removal

Microalgae utilize nitrogen from inorganic sources (ammonia, nitrites, nitrates and atmospheric nitrogen) as well as organic sources (urea, glycines, purines). Preference for ammonia as a nitrogen source for microalgae is attributed to its efficient assimilation and energetic incorporation. Species like *Chlorella vulgaris* has shown a preference for NH3+ as its nitrogen source as compared to any other source [19, 20].

### 3.4 Phosphorus Removal

Phosphorous is an essential element because of its role in multiple metabolic pathways [9]. There are various forms of inorganic phosphorus that are present in the wastewater, including $H_3PO_4^-$, $H_2PO_4^-$, $HPO_4^{2-}$ which are pH specific [10]. These forms are preferred by microalgal species for assimilation, especially hydrogen phosphate ($HPO_4^{2-}$) and dihydrogen phosphate ($H_2PO_4^-$) [18]. Phosphate ion enters the cells of algae with the help of active transport and the movement is symport with H+ or Na+ acting as driving force [18, 21]. This movement is assisted by plasma membrane H+-ATPase pump. Apart from inorganic phosphorous, microalgae also need organic phosphorus and it derives that either from membrane bound or free phosphates that hydrolyze bound phosphate groups [22]. With enough energy input, ADP is phosphorylated and the phosphorous gets incorporated into the organic compounds from Adenosine triphosphate (ATP). The newly generated ATP at the substrate level allows transferring phosphate groups to organic compounds [18, 22]. The microalgal strains are cultivated in various treatment plants for municipal, industrial, pharmaceutical and agro-industries wastewater [23]. Depending upon the need, concentration of microalgal biomass can be approximately 0.5 g $L^{-1}$and nitrogen, phosphorous content can vary depending on cellular makeup of the strain. N, P composition in microalgal biomass can be anywhere in the range of 5.4–8.7% and 0.7–1.1%, respectively [14, 24]. A list of microalgae with excellent nitrogen, phosphorous and carbon removal properties has been depicted in Table 1.

### 3.5 Removal of Heavy Metals

Although heavy metals are naturally occurring elements, an increase in their concentration due to natural (volcanic eruptions, soil erosion, etc.) or anthropogenic (industrial, agricultural, etc.) activities may lead to soil pollution and water pollution which in turn affects the entire marine ecosystem. Due to biomagnification, the risk of heavy metal poisoning increases at each trophic level. As a result of this, humans are one of the most affected species. The greatest harm to humans among the heavy metals is brought about by cadmium, mercury, lead and arsenic that cause itai-itai

**Table 1** Microalgae with their removal percentage of nitrogen, phosphorous and carbon

| Microalgae | Removal percentage of | | | References |
|---|---|---|---|---|
| | Nitrogen (%) | Phosphorous (%) | Carbon (%) | |
| Mixed culture of *Woronichinia sp., Aulacoseira sp., Actuodesmus sp., Desmodesmus quadricaudatus, Nitzschia sp., Limnothrix redekei* and *Gomphonema parvulum* | 92 | 96 | 89 | [25] |
| *Chlorella* and *Phormidium* sp. | 94 | 90 | 69 | [26] |
| *Chlorella pyrenoidosa* and microbial community normally found in wastewater | 95 | 84 | 78 | [27] |
| *Scenedesmus obliquus* and bacteria | 98 | 96 | 59 | [20, 28] |
| Primarily microalgae from the genera *Coelastrum* | 94 | 58 | 62 | [29] |
| *Scenedesmus obliquus* and microbial community normally found in wastewater | 86 | 88 | 24 | [30, 31] |
| *Chlorella vulgaris* and microbial community normally found in wastewater | 70 | 52 | 58 | [32] |

disease, minimata disease, mental retardation and arsenicosis, respectively [33, 34]. Microalgae have huge potential for the removal of heavy metals in wastewater treatment plants. In a study, it was reported that the microalga *Pseudochlorococcum typicum* had a removal efficiency of 97.75%, 86.24% and 70.06% for mercury, cadmium and lead, respectively, after an incubation of 30 min. Aside from that, the tolerance of *Scenedesmus quadricauda, Phormidium ambigum* and *Pseudochlorococcum typicum* against heavy metals (mercury, cadmium and lead) were also examined by the determination of chlorophyll content as an indicator of their growth and it was found that mercury aggressively inhibits the growth of microalga in both *P. ambigum* and *P. typicum. P. ambigum* showed almost no growth at concentrations starting from 5 μg/ml. *P. typicum* was slightly better, which exhibited a 25% decrease in growth at 5 μg/ml as compared to control; however, upon increasing the concentration to 10 μg/ml, the growth drastically reduced by 85% as compared to control. In the case of cadmium and lead, the three microalgae showed good tolerance capacity (comparable to control up to a concentration of 80–100 μg/ml except for cadmium against *P. ambigum*, where the tolerable concentration was up to 20 μg/ml). In fact, it spiked the algal growth at lower concentrations (up to 20 μg/ml for *P. typicum* and *S. quadricauda*; up to 10 μg/ml for *P. ambigum*). The increase in the growth may be attributed to the heavy metals being linked with the regulation of carbohydrate and nitrogen metabolism [35]. Another study found several microalgal species that had an excellent heavy metal removal capability when each of them was incubated at 200 rpm; 28 °C for 30 min. A summary of the results with the species having the best

**Table 2** **a** Microalgae with their removal percentage of mercury [36]. **b** Microalgae with their removal percentage of cadmium [36]. **c** Microalgae with their removal percentage of lead [36]

| (a) | |
|---|---|
| Microalgae | Removal percentage of mercury (%) |
| *Scenedesmus sp.* | 97 |
| *Lyngbya spiralis* | 96 |
| *Chlorococcum sp.* | 96 |
| *Chlorella vulgaris var. vulgaris* | 94 |
| *Tolypothrix tenuis* | 94 |
| (b) | |
| Microalgae | Removal percentage of cadmium (%) |
| *Lyngbya heironymusii* | 97 |
| *Gloeocapsa sp.* | 96 |
| *Phormidium molle* | 95 |
| *Nostoc sp.* | 94 |
| *Oscillatoria jasorvensis* | 94 |
| (c) | |
| Microalgae | Removal percentage of lead (%) |
| *Nostoc punctiforme* | 98 |
| *Gloeocapsa sp.* | 96 |
| *Oscillatoria agardhii* | 96 |
| *Nostoc piscinale* | 94 |
| *Nostoc commune* | 94 |

removal percentage (ranging from 98% to 94%) of mercury, cadmium and lead are shown in Table 2a–c [36]. Sometimes, immobilizing the microalgae may increase the efficiency by which it uptakes heavy metals. Many other factors such as the pH of the medium, initial concentration of the heavy metal and biosorbent dose affects the heavy metal removal efficiency. For example, *Chlorella minutissima* entrapped in calcium-alginate beads results in 100% removal of chromium (VI) within 30 min at pH 1.0. However, when the pH is increased to 5, the Cr (VI) removal rate is drastically reduced by 80% [37].

### 3.6 Removal of Endocrine Disrupting Chemicals (EDCs)

Endocrine Disrupting Chemicals (EDCs) are those chemicals that interfere with or disrupt the endocrine system, which is composed of ductless glands (hypothalamus, pituitary, adrenal and gonads, etc.) that produce hormones and receptors that detect and subsequently transduce a cellular response. EDCs are most commonly found in industrial chemicals, pesticides, herbicides, pharmaceuticals, cosmetics and plastic contaminants. Some of the existing technologies for the removal of EDCs in the wastewater treatment plants include sand filtration coupled with ozonation, activated sludge process in conjunction with sequencing batch biofilter granular reactor, competitive ligand binding, microfiltration along with reverse osmosis and catalytic photodegradation [38–42]. These methods, however, require a sizeable initial investment to set up the facility and subsequent high energy requirements, which can be overcome by employing microalgae as they are based on the principle of biological degradation of pollutants while deriving their energy from the organic carbon, inorganic carbon, nitrogen, phosphorous and sunlight. In a study, it was found that the microalgae *Chlorella fusca* was able to degrade >95% of bisphenol A (BPA) up to a concentration of 80 $\mu$M when cultured at 27.5 °C, aerated with 1% $CO_2$ for 168 h in a continuous light condition with an energy flux of 18 W/m$^2$. It also resulted in the disappearance of the estrogenic activity associated with BPA [43]. In another study, it was found that thc microalgae *Scenedesmus obliquus* had excellent nonylphenol and octylphenol bioremoval capabilities. It exhibited 100%, 96.3%, 90.3% and 89.2% removal efficiency of nonylphenol (NP) at 0.5, 1.0, 2.0 and 4.0 mg $L^{-1}$ respectively. Similarly, it exhibited 96.6%, 97.7%, 97.9% and 98.3% removal efficiency of 4-t-octylphenol (OP) at 0.5, 1.0, 2.0 and 4.0 mg $L^{-1}$ at 150 rpm, 25 $\pm$ 1 °C and a continuous supply of cool white light at an intensity of 3000 lx with a 12 h/12 h of light/dark cycle for 5 days [44]. A couple of marine microalgal species are able to degrade phthalate acid esters such as diethyl phthalate (DEP) and di-n-butyl phthalate (DBP) with the help of both their intracellular as well as extracellular enzymes. These microalgae include *Cylindrotheca closterium, Dunaliella salina* and *Chaetoceros mulleri.* Among them, *C. closterium* is the most potent with a biodegradation efficiency of 93.1% against DBP and 81.2% against DEP [45]. A summary of EDCs and their bioremediation efficiency using microalgae has been depicted in Table 3.

### 3.7 Microalgal Cultivation and Configuration of Bioreactor for Wastewater Treatment

The process is quite efficient performance-wise, albeit having its fair share of challenges that needs to be overcome. With different compositions of wastewater to large volumes that require microalgal treatment, numerous cultivation techniques are taken into consideration [51]. A perfect technique is a concoction of several aspects that include optimal productivity of microalgae, high performance in nutrient removal, and accommodation of large volumes of wastewater that needs to be treated with cost efficiency [24, 51]. Bioreactors are designed in such a way that affects the efficiency and performance with control on factors like temperature and light for a better influence on the assimilation of nutrients and their removal by microalgal growth.

**Table 3** EDCs and their bioremediation strategies employing microalgae

| EDCs | Microalgae | Removal efficiency | References |
|---|---|---|---|
| Bisphenol A (BPA) | *Chlorella fusca* var. *vacuolata* | >95% degradation of BPA to monohydroxybisphenol A and other intermediate compounds having no estrogenic activity | [43] |
| Nonylphenol (NP) and octylphenol (OP) | *Scenedesmus obliquus* | At 4 mg $L^{-1}$ of NP & OP, ~89% of NP and ~58% of OP were removed within 5 days | [44] |
| Diethyl phthalate (DEP) and Di-n-butyl phthalate (DBP) | *Cylindrotheca closterium* | Biodegradation efficiency of 93.1% against DBP and 81.2% against DEP at 25 ± 1 °C,16:8-h light: dark with 3000 lx of light intensity for 168 h | [45] |
| Triclosan | *Chlorella pyrenoidosa* | ~77% of triclosan removal was achieved at 800 ng/mL when incubated for 96 h | [46] |
| 3,4-Dichloroaniline (3,4-DCA) | *Chlorella pyrenoidosa* | ~78% removal of 3,4-DCA at 25 °C, 16:8-h light: dark cycle with 4,800 lx of light intensity | [47] |
| 17α-Estradiol, 17β-estradiol, estriol and estrone | *Scenedesmus dimorphus* | ~85–95% of total estrogens were removed from 17α-Estradiol, 17β-estradiol, estriol and estrone within 8 days | [48] |
| β-Estradiol (E2) and 17α-ethinylestradiol (EE2) | *Selenastrum capricornutum* and *Chlamydomonas reinhardtii* | >90% removal of E2 within 24 h and up to 100% in 7 days. For EE2, up to 95% removal by *C. reinhardtii* in 7 days | [49] |
| 4-(1,1,3,3-tetramethylbutyl) phenol (OP), technical-nonylphenol (t-NP) and Bisphenol A (BPA) | Microalgae inoculum, primarily consisting of cyanobacteria | Total removal ratios for alkylphenols (op and t-NP) and BPA ranged from ~50 to 80% without aeration, which increased up to 96% upon aeration and further up to 98.8% upon inclusion of microalgae | [50] |

Cultivation of microalgae is largely done either as suspended or attached growth [52].

**Suspended growth** This method is the most common adapted method for microalgal cultivation [52]. In a suspended growth system, there can be either an open or closed system. The closed systems are also called photobioreactors (PBRs). The major advantage of PBRs is that they reduce unwanted residuals, contamination and water loss by evaporation [53]. Different types of PBRs include biocoil, horizontal tubular and vertical PBRs [53, 54]. Despite such advantages, the cost of PBRs is high and they can surpass the calorific value of produced biomass [54, 55]. On the other hand, open systems, despite challenges like loss of water by evaporation, are economically feasible. The open systems include High-Rate Algal Pond or Corrugated Raceway Pond [56]. There should be an optimum depth for open cultivation system varying between 0.15m and 0.45 m for efficient light absorption by microalgae, where the top layers of cells absorbs maximum light and as we move down, light penetration decreases exponentially, thereby leaving the cells in the bottom layers in the dark [57].

**Attached/Immobilized growth** This method is considered useful because of cost reduction linked with separating out biomass from treated water before discharging it [58]. In a study done by Christenson and Sims in 2011, it was reported that immobilization resulted in a higher concentration of cells that could be maintained in water as compared to suspended growth systems [59]. In immobilized systems, 3.3 g $L^{-1}$ dry weight of cells was maintained as compared to 1.5–1.7 g $L^{-1}$ dry weight in tubular ponds and 0.25–1 g $L^{-1}$ in raceway ponds [59, 60]. Once the wastewater is treated, it is essential to remove the biomass from it [11, 61]. Many types of techniques like sedimentation, auto- flocculation, coagulation-flocculation, filtration are used once bioremediation is over. These techniques are applied depending upon biomass application and requirement of energy per unit of produced biomass [62]. A list of microalgae in different types of wastewater with their biomass productivity has been shown in Table 4.

**Table 4** Biomass productivity of various microalgae in different types of wastewater

| Microalgae | Wastewater type | Biomass productivity | References |
|---|---|---|---|
| Consortium of *Scenedesmus, Chlorella, Pediastrum, Nitzschia, Cosmarium* and others | Municipal wastewater | 9.1–13.5 g $m^{-2}$ $d^{-1}$ | [63] |
| Primarily consisting of *Microspora willeana* along with *Ulothrix zonata, Ulothrix aequalis* and others | Undigested dairy manure | 5.35 g $m^{-2}$ $d^{-1}$ | [64] |
| *Arthrospira* sp. | Swine wastewater | 11.8 g $L^{-1}$ $d^{-1}$ | [65] |
| *Chroococcus* sp. 1 | Dairy cattle wastewater | 2.22 g $L^{-1}$ $d^{-1}$ (4.44 g $L^{-1}$ 48 $h^{-1}$) | [66] |
| *Chlamydomonas sp. TAI-2* | Industrial wastewater | 1.34 g $L^{-1}$ $d^{-1}$ | [67] |
| *Chlorella sorokiniana* | Urea supplemented domestic wastewater | 0.22 g $L^{-1}$ $d^{-1}$ | [68] |

***Footnote: sp.: species,** *TAI-2:* Taiwan-2 (as it was isolated from a pond in Taiwan)

# 4 Applications of Algal Biomass from Wastewater

**Biofuel generation** Microalgal biomass generated from wastewater is anaerobically digested to generate biogas like biomethane, depending upon the composition of microalgal biomass [9, 55]. The microalgal strains having low amounts of lipids are preferred because the conversion of methane from carbohydrates and proteins is higher as compared to lipids [69, 70]. Conversion of methane to electricity produces carbon dioxide in anaerobic digestion, which can be used in the cultivation of microalgae in wastewater [24, 71].

**Chemicals** Microalgal biomass can be used to produce chemicals like bioplastics/polyhydroxyalkonates (PHA), polysaccharides and bioglycerol [11, 71]. When specific cyanobacterial strains are inoculated in wastewater treatment plants, they produce an intracellular substance, PHA having properties like that of plastics. Bioglycerol is also produced from microalgal lipid, as a byproduct of biodiesel [72].

**Cosmetics industry** There are microalgal species like *Chlorella* sp, *Spirulina* sp that produce secondary metabolites and their extracts are used in the cosmetics industry as antioxidants, UV protectants [72]. Some extracts from strains like *Dunaliella* sp, *Noctiluca* sp are used in products like moisturizers and other skincare products [24, 73].

**Biofertilizers** Microalgal biomass generated from wastewater is applied as a soil supplement or biofertilizer for enhancement of nutrient content like N, P and other minerals like calcium, iron and potassium [8]. These biofertilizers have been demonstrated to be useful to grow crops like barley, corn and cucumber. The microalgal biomass has been shown to improve the organic content of soil [74].

**Production of feed** Biomass generated from microalgal strains has nutrients that can enhance the nutritional component of feed for livestock and poultry. The biomass is composed of nutrients like carbohydrates, proteins, lipids and vitamins and antioxidants in trace amounts [24]. When the biomass is added to the feed, it improves the quality of ingredients needed to make a wholesome feed for animals. Besides this, it is cost-efficient as there is no need for additional nutrients, making the cost of feed production using microalgal biomass cheaper [75]. Microalgal biomass is also used for aquaculture feed. For example, Astaxanthin produced from *Haematococcus* microalgae and *Chlorella zofingiensis,* is used as a supplement for carotenoid pigment to culture Salmon as they are unable to synthesize this pigment naturally [76, 77]. The different usages of microalgae in a wastewater treatment plant are summarized in Fig. 2.

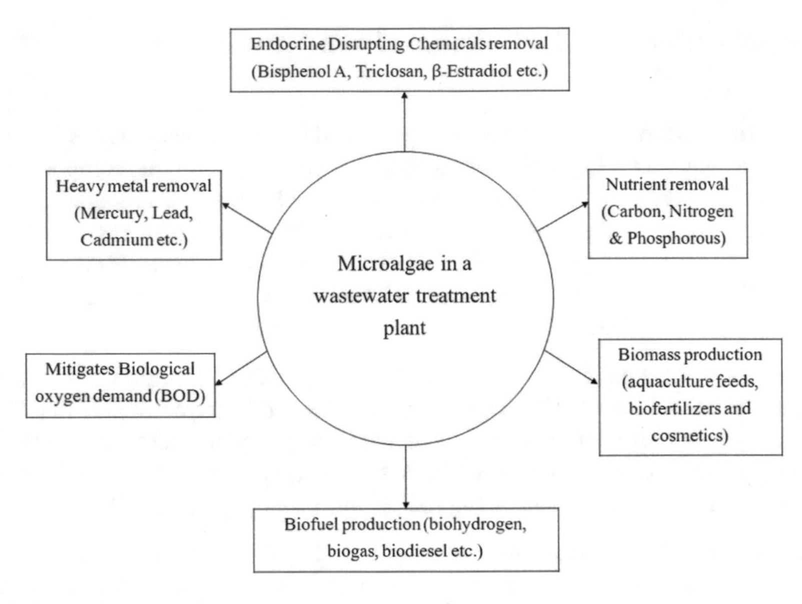


**Fig. 2** Microalgae usage in a wastewater treatment plant

## 5 Conclusion

Microalgae are akin to self-sustaining mini factories for the bioremediation of wastewater. They do not require additional supply of energy or nutrients for their growth and metabolic activities as they are photosynthetic, and at the same time uptake nutrients from the wastewater with minimal maintenance and care such as pH and temperature regulation of the medium. They can thrive and yield biomass which can be harvested for food additives, skincare products, biofuel, biofertilizers, livestock and aquaculture feeds, etc. It is interesting to note that microalgae belonging to the genera *Chlorella* and *Scenedesmus* are excellent candidates when it comes to the removal of any kind of wastewater pollutants. For example, they can efficiently remove several EDCs such as bisphenol A, nonylphenol, octyphenol, triclosan and β-Estradiol (removal efficiency ranging from 75% to 95%); nutrients such as C, N and P (up to 95%); and heavy metals (for instance, up to 97% removal of mercury) from wastewater. Several reports have shown that, when there is a mixture of several microalga species, it leads to an increased nutrient removal as well as better biomass productivity. This could be attributed to the fact that one species could overcome the limitations of the other species and together they show a synergistic effect. For example, if one microalgal species primarily degrades a pollutant by releasing extracellular enzymes and the other one removes the same pollutant primarily by accumulating them in their cell wall, together they would lead to an improvement in the remediation of that pollutant. By understanding the intricacies by which microalgae work, it is possible to design a consortium of microalgae targeted towards the removal of specific pollutants.

CSIR-CSMCRI communication no is 89/2021.

## References

1. Spolaore, P., Joannis-Cassan, C., Duran, E., Isambert, A.: Commercial applications of microalgae. J. Biosci. Bioeng. **101**(2), 87–96 (2006)
2. Priyadarshani, I., Rath, B.: Commercial and industrial applications of micro algae—a review. J. Algal Biomass Utilization **3**(4), 89–100 (2012)
3. Solé-Bundó, M., Cucina, M., Folch, M., Tàpias, J., Gigliotti, G., Garfí, M., Ferrer, I.: Assessing the agricultural reuse of the digestate from microalgae anaerobic digestion and co-digestion with sewage sludge. Sci. Total Environ. **586**, 1–9 (2017)
4. Ho, S.H., Nagarajan, D., Ren, N.Q., Chang, J.S.: Waste biorefineries—integrating anaerobic digestion and microalgae cultivation for bioenergy production. Curr. Opin. Biotechnol. **50**, 101–110 (2018)
5. Bahr, M., Díaz, I., Dominguez, A., Gonzalez Sanchez, A., Muñoz, R.: Microalgal-biotechnology as a platform for an integral biogas upgrading and nutrient removal from anaerobic effluents. Environ. Sci. Technol. **48**(1), 573–581 (2014)
6. Brune, D.E., Lundquist, T.J., Benemann, J.R.: Microalgal biomass for greenhouse gas reductions: potential for replacement of fossil fuels and animal feeds. J. Environ. Eng. **135**(11), 1136–1144 (2009)
7. Ramanan, R., Kim, B.H., Cho, D.H., Ko, S.R., Oh, H.M., Kim, H.S.: Lipid droplet synthesis is limited by acetate availability in starchless mutant of Chlamydomonas reinhardtii. FEBS Lett. **587**(4), 370–377 (2013)
8. Zhang, G., Yang, J., Liu, H., Zhang, J.: Sludge ozonation: disintegration, supernatant changes and mechanisms. Biores. Technol. **100**(3), 1505–1509 (2009)
9. Al-Jabri, H., Das, P., Khan, S., Thaher, M., AbdulQuadir, M.: Treatment of wastewaters by microalgae and the potential applications of the produced biomass—a review. Water **13**(1), 27 (2020)
10. Lee, J., Sohn, D., Lee, K., Park, K.: Solid fuel production through hydrothermal carbonization of sewage sludge and microalgae Chlorella sp. from wastewater treatment plant. Chemosphere **230**, 157–163 (2019)
11. Mohsenpour, S., Hennige, S., Willoughby, N., Adeloye, A., Gutierrez, T.: Integrating micro-algae into wastewater treatment: a review. Sci. Total Environ. **752**, 142168 (2021)
12. Xiong, J., Kurade, M., Jeon, B.: Can microalgae remove pharmaceutical contaminants from water? Trends Biotechnol. **36**(1), 30–44 (2018)
13. Olsson, J., Feng, X., Ascue, J., Gentili, F., Shabiimam, M., Nehrenheim, E., Thorin, E.: Co-digestion of cultivated microalgae and sewage sludge from municipal waste water treatment. Biores. Technol. **171**, 203–210 (2014)
14. Lin, C.Y., Nguyen, M.L.T., Lay, C.H.: Starch-containing textile wastewater treatment for biogas and microalgae biomass production. J. Clean. Prod. **168**, 331–337 (2017)
15. Cai, T., Park, S.Y., Li, Y.: Nutrient recovery from wastewater streams by microalgae: status and prospects. Renew. Sustain. Energy Rev. **19**, 360–369 (2013)
16. Mahdy, A., Mendez, L., Ballesteros, M., González-Fernández, C.: Algaculture integration in conventional wastewater treatment plants: Anaerobic digestion comparison of primary and secondary sludge with microalgae biomass. Biores. Technol. **184**, 236–244 (2015)
17. Chen, Y., Li, S., Ho, S., Wang, C., Lin, Y., Nagarajan, D., Chang, J., Ren, N.: Integration of sludge digestion and microalgae cultivation for enhancing bioenergy and biorefinery. Renew. Sustain. Energy Rev. **96**, 76–90 (2018)

18. Shen, Q., Jiang, J., Chen, L., Cheng, L., Xu, X., Chen, H.: Effect of carbon source on biomass growth and nutrients removal of Scenedesmus obliquus for wastewater advanced treatment and lipid production. Biores. Technol. **190**, 257–263 (2015)
19. Bolognesi, S., Bernardi, G., Callegari, A., Dondi, D., Capodaglio, A.: Biochar production from sewage sludge and microalgae mixtures: properties, sustainability and possible role in circular economy. Biomass Conversion Bioref. **11**(2), 289–299 (2019)
20. Shen, H., Wang, Z., Zhou, A., Chen, J., Hu, M., Dong, X., Xia, Q.: Adsorption of phosphate onto amine functionalized nano-sized magnetic polymer adsorbents: mechanism and magnetic effects. RSC Adv. **5**(28), 22080–22090 (2015)
21. Falkowski, P., Raven, J.: Aquatic Photosynthesis. Blackwell Science, Malden, Mass (1997)
22. Borowitzka, M.A., Moheimani, N.R. (eds.): Algae for Biofuels and Energy, vol. 5, pp. 133–152. Springer, Dordrecht (2013)
23. Borowitzka, M.A., Beardall, J., Raven, J.A. (eds.): The Physiology of Microalgae, vol. 6. Springer International Publishing, Cham (2016)
24. Williams, P.J.L.B., Laurens, L.M.: Microalgae as biodiesel & biomass feedstocks: review & analysis of the biochemistry, energetics & economics. Energy Environ. Sci. **3**(5), 554–590 (2010)
25. Posadas, E., García-Encina, P.A., Domínguez, A., Díaz, I., Becares, E., Blanco, S., Munoz, R.: Enclosed tubular and open algal–bacterial biofilm photobioreactors for carbon and nutrient removal from domestic wastewater. Ecol. Eng. **67**, 156–164 (2014)
26. Choudhary, P., Prajapati, S.K., Kumar, P., Malik, A., Pant, K.K.: Development and performance evaluation of an algal biofilm reactor for treatment of multiple wastewaters and characterization of biomass for diverse applications. Biores. Technol. **224**, 276–284 (2017)
27. Dahmani, S., Zerrouki, D., Ramanna, L., Rawat, I., Bux, F.: Cultivation of Chlorella pyrenoidosa in outdoor open raceway pond using domestic wastewater as medium in arid desert region. Biores. Technol. **219**, 749–752 (2016)
28. Zamalloa, C., Boon, N., Verstraete, W.: Decentralized two-stage sewage treatment by chemical–biological flocculation combined with microalgae biofilm for nutrient immobilization in a roof installed parallel plate reactor. Biores. Technol. **130**, 152–160 (2013)
29. De Godos, I., Arbib, Z., Lara, E., Rogalla, F.: Evaluation of high rate algae ponds for treatment of anaerobically digested wastewater: effect of CO2 addition and modification of dilution rate. Biores. Technol. **220**, 253–261 (2016)
30. Arbib, Z., Ruiz, J., Álvarez-Díaz, P., Garrido-Pérez, C., Barragan, J., Perales, J.A.: Photo-biotreatment: influence of nitrogen and phosphorus ratio in wastewater on growth kinetics of Scenedesmus obliquus. Int. J. Phytorem. **15**(8), 774–788 (2013)
31. Arbib, Z., Ruiz, J., Álvarez-Díaz, P., Garrido-Pérez, C., Barragan, J., Perales, J.A.: Long term outdoor operation of a tubular airlift pilot photobioreactor and a high rate algal pond as tertiary treatment of urban wastewater. Ecol. Eng. **52**, 143–153 (2013)
32. Choi, H.: Intensified production of microalgae and removal of nutrient using a microalgae membrane bioreactor (MMBR). Appl. Biochem. Biotechnol. **175**(4), 2195–2205 (2015)
33. Briffa, J., Sinagra, E., Blundell, R.: Heavy metal pollution in the environment and their toxicological effects on humans. Heliyon **6**(9), e04691 (2020)
34. Engwa, G.A., Ferdinand, P.U., Nwalo, F.N., Unachukwu, M.N.: Mechanism and health effects of heavy metal toxicity in humans. Poison Modern World-New Tricks Old Dog **10** (2019)
35. Shanab, S., Essa, A., Shalaby, E.: Bioremoval capacity of three heavy metals by some microalgae species (Egyptian Isolates). Plant Signal. Behav. **7**(3), 392–399 (2012)
36. Inthorn, D., Sidtitoon, N., Silapanuntakul, S., Incharoensakdi, A.: Sorption of mercury, cadmium and lead by microalgae. Sci. Asia **28**(3), 253–261 (2002)
37. Ajayan, K.V., Selvaraju, M., Thirugnanamoorthy, K.: Growth and heavy metals accumulation potential of microalgae grown in sewage wastewater and petrochemical effluents. Pak. J. Biol. Sci. **14**(16), 805–811 (2011)
38. Nakada, N., Shinohara, H., Murata, A., Kiri, K., Managaki, S., Sato, N., Takada, H.: Removal of selected pharmaceuticals and personal care products (PPCPs) and endocrine-disrupting chemicals (EDCs) during sand filtration and ozonation at a municipal sewage treatment plant. Water Res. **41**(19), 4373–4382 (2007)

39. Balest, L., Mascolo, G., Di Iaconi, C., Lopez, A.: Removal of endocrine disrupter compounds from municipal wastewater by an innovative biological technology. Water Sci. Technol. **58**(4), 953–956 (2008)
40. Liu, Z.H., Ito, M., Kanjo, Y., Yamamoto, A.: Profile and removal of endocrine disrupting chemicals by using an ER/AR competitive ligand binding assay and chemical analyses. J. Environ. Sci. **21**(7), 900–906 (2009)
41. Al-Rifai, J.H., Khabbaz, H., Schäfer, A.I.: Removal of pharmaceuticals and endocrine disrupting compounds in a water recycling process using reverse osmosis systems. Sep. Purif. Technol. **77**(1), 60–67 (2011)
42. Zhang, Y., Zhou, J.L.: Occurrence and removal of endocrine disrupting chemicals in wastewater. Chemosphere **73**(5), 848–853 (2008)
43. Hirooka, T., Nagase, H., Uchida, K., Hiroshige, Y., Ehara, Y., Nishikawa, J.I., Nishihara, T., Miyamoto, K., Hirata, Z.: Biodegradation of bisphenol A and disappearance of its estrogenic activity by the green alga Chlorella fusca var. vacuolata. Environ. Toxicol. Chem. Int. J. **24**(8), 1896–1901 (2005)
44. Zhou, G.J., Peng, F.Q., Yang, B., Ying, G.G.: Cellular responses and bioremoval of nonylphenol and octylphenol in the freshwater green microalga Scenedesmus obliquus. Ecotoxicol. Environ. Saf. **87**, 10–16 (2013)
45. Gao, J., Chi, J.: Biodegradation of phthalate acid esters by different marine microalgal species. Mar. Pollut. Bull. **99**(1–2), 70–75 (2015)
46. Wang, S., Wang, X., Poon, K., Wang, Y., Li, S., Liu, H., Lin, S., Cai, Z.: Removal and reductive dechlorination of triclosan by Chlorella pyrenoidosa. Chemosphere **92**(11), 1498–1505 (2013)
47. Wang, S., Poon, K., Cai, Z.: Biodegradation and removal of 3, 4-dichloroaniline by Chlorella pyrenoidosa based on liquid chromatography-electrospray ionization-mass spectrometry. Environ. Sci. Pollut. Res. **20**(1), 552–557 (2013)
48. Zhang, Y., Habteselassie, M.Y., Resurreccion, E.P., Mantripragada, V., Peng, S., Bauer, S., Colosi, L.M.: Evaluating removal of steroid estrogens by a model alga as a possible sustainability benefit of hypothetical integrated algae cultivation and wastewater treatment systems. ACS Sustain. Chem. Eng. **2**(11), 2544–2553 (2014)
49. Hom-Diaz, A., Llorca, M., Rodríguez-Mozaz, S., Vicent, T., Barceló, D., Blánquez, P.: Microalgae cultivation on wastewater digestate: β-estradiol and 17α-ethynylestradiol degradation and transformation products identification. J. Environ. Manage. **155**, 106–113 (2015)
50. Abargues, M.R., Ferrer, J., Bouzas, A., Seco, A.: Removal and fate of endocrine disrup tors chemicals under lab-scale postreatment stage. removal assessment using light, oxygen and microalgae. Biores. Technol. **149**, 142–148 (2013)
51. Hoffmann, J.: Wastewater treatment with suspended and nonsuspended algae. J. Phycol. **34**(5), 757–763 (1998)
52. Fazal, T., Mushtaq, A., Rehman, F., Ullah Khan, A., Rashid, N., Farooq, W., Rehman, M., Xu, J.: Bioremediation of textile wastewater and successive biodiesel production using microalgae. Renew. Sustain. Energy Rev. **82**, 3107–3126 (2018)
53. Hossain, L., Sarker, S., Khan, M.: Evaluation of present and future wastewater impacts of textile dyeing industries in Bangladesh. Environ. Devel. **26**, 23–33 (2018)
54. Alkaya, E., Demirer, G.: Sustainable textile production: a case study from a woven fabric manufacturing mill in Turkey. J. Clean. Prod. **65**, 595–603 (2014)
55. Hoh, D., Watson, S., Kan, E.: Algal biofilm reactors for integrated wastewater treatment and biofuel production: a review. Chem. Eng. J. **287**, 466–473 (2016)
56. Craggs, R., McAuley, P., Smith, V.: Wastewater nutrient removal by marine microalgae grown on a corrugated raceway. Water Res. **31**(7), 1701–1707 (1997)
57. Kumar, D., Murthy, G.: Impact of pretreatment and downstream processing technologies on economics and energy in cellulosic ethanol production. Biotechnol. Biofuels **4**(1), 27 (2011)
58. Kesaano, M., Sims, R.: Algal biofilm based technology for wastewater treatment. Algal Res. **5**, 231–240 (2014)
59. Christenson, L., Sims, R.: Production and harvesting of microalgae for wastewater treatment, biofuels and bioproducts. Biotechnol. Adv. **29**(6), 686–702 (2011)

60. Whitton, R., Ometto, F., Pidou, M., Jarvis, P., Villa, R., Jefferson, B.: Microalgae for municipal wastewater nutrient remediation: mechanisms, reactors and outlook for tertiary treatment. Environ. Technol. Rev. **4**(1), 133–148 (2015)
61. Christenson, L.B., Sims, R.C.: Rotating algal biofilm reactor and spool harvester for wastewater treatment with biofuels by-products. Biotechnol. Bioeng. **109**(7), 1674–1684 (2012)
62. Gutiérrez, R., Passos, F., Ferrer, I., Uggetti, E., García, J.: Harvesting microalgae from wastewater treatment systems with natural flocculants: effect on biomass settling and biogas production. Algal Res. **9**, 204–211 (2015)
63. Lee, S.H., Oh, H.M., Jo, B.H., Lee, S.A., Shin, S.Y., Kim, H.S., Lee, S.H., Ahn, C.Y.: Higher biomass productivity of microalgae in an attached growth system, using wastewater. J. Microbiol. Biotechnol. **24**(11), 1566–1573 (2014)
64. Wilkie, A.C., Mulbry, W.W.: Recovery of dairy manure nutrients by benthic freshwater algae. Biores. Technol. **84**(1), 81–91 (2002)
65. Olguín, E.J., Galicia, S., Mercado, G., Pérez, T.: Annual productivity of Spirulina (Arthrospira) and nutrient removal in a pig wastewater recycling process under tropical conditions. J. Appl. Phycol. **15**(2), 249–257 (2003)
66. Prajapati, S.K., Choudhary, P., Malik, A., Vijay, V.K.: Algae mediated treatment and bioenergy generation process for handling liquid and solid waste from dairy cattle farm. Biores. Technol. **167**, 260–268 (2014)
67. Wu, L.F., Chen, P.C., Huang, A.P., Lee, C.M.: The feasibility of biodiesel production by microalgae using industrial wastewater. Biores. Technol. **113**, 14–18 (2012)
68. Ramanna, L., Guldhe, A., Rawat, I., Bux, F.: The optimization of biomass and lipid yields of Chlorella sorokiniana when using wastewater supplemented with different nitrogen sources. Biores. Technol. **168**, 127–135 (2014)
69. Milledge, J.J., Heaven, S.: Energy balance of biogas production from microalgae: effect of harvesting method, multiple raceways, scale of plant and combined heat and power generation. J. Marine Sci. Eng. **5**(1), 9 (2017)
70. Torres, A., Fermoso, F.G., Rincón, B., Bartacek, J., Borja, R., Jeison, D.: Challenges for cost-effective microalgae anaerobic digestion. Biodegradation Eng. Technol. pp.139–159. Intech, Rijeka (2013)
71. ter Veld, F.: Beyond the fossil fuel era: on the feasibility of sustainable electricity generation using biogas from microalgae. Energy Fuels **26**(6), 3882–3890 (2012)
72. Xiao, R., Zheng, Y.: Overview of microalgal extracellular polymeric substances (EPS) and their applications. Biotechnol. Adv. **34**(7), 1225–1244 (2016)
73. Shahid, A., Khan, F., Ahmad, N., Farooq, M., Mehmood, M.A.: Microalgal carbohydrates and proteins: synthesis, extraction, applications and challenges. In: Microalgae Biotechnology for Food, Health and High Value Products, pp. 433–468. Springer, Singapore (2020)
74. Renuka, N., Prasanna, R., Sood, A., Ahluwalia, A.S., Bansal, R., Babu, S., Singh, R., Shivay, Y.S., Nain, L.: Exploring the efficacy of wastewater-grown microalgal biomass as a biofertilizer for wheat. Environ. Sci. Pollut. Res. **23**(7), 6608–6620 (2016)
75. Schwinn, D.E., Dickson, B.H.: Nitrogen and phosphorus variations in domestic wastewater. J. (Water Pollution Control Federation), 2059–2065 (1972)
76. Dominguez, A., Ferreira, M., Coutinho, P., Fábregas, J., Otero, A.: Delivery of astaxanthin from Haematocuccus pluvialis to the aquaculture food chain. Aquaculture **250**(1–2), 424–430 (2005)
77. Del Campo, J.A., Rodriguez, H., Moreno, J., Vargas, M.A., Rivas, J., Guerrero, M.G.: Accumulation of astaxanthin and lutein in Chlorella zofingiensis (Chlorophyta). Appl. Microbiol. Biotechnol. **64**(6), 848–854 (2004)

# Microbial Factory; Utilization of Pectin-Rich Agro-Industrial Wastes for the Production of Pectinases Enzymes Through Solid State Fermentation (SSF)

**Nor Hawani Salikin and Muaz Mohd Zaini Makhtar**

**Abstract** Agro-industrial wastes are abundantly generated every year from the agriculture processing industries. While some of the organic biomasses are used as animal feed, most of the organic materials are either dumped or burnt in the open environment, thus resulting in debilitating pollution and health hazards. However, such wastes are usually composed of organic pectic substances, sugars, minerals, proteins and moisture that may serve as an ideal environment that can support microbial growth. The nutritious property of agro-industrial wastes allows its utilization as the perfect substrate for solid state fermentation (SSF). The SSF is a low-cost technology that enables bioconversion of agro-industrial wastes into various value-added by-products such as pectinases enzymes as well as an eco-friendly approach for the control and waste management system. While SSF technology is less explored compared to submerged fermentation, it has been proven to give higher microbial bioactive compound productivity. This chapter will provide an overview of agro-industrial wastes valorisation via SSF methods and its application in pectinases production. Further, general requirements for efficient SSF bioprocess technology and final productivity are comprehensively described.

## 1 Introduction

The agriculture sector is believed as an efficient tool to stimulate economic growth, combat poverty among the human population worldwide as well as encourage environmental sustainability [1, 2]. The agriculture sector in the United States of America represents the major industry with 2.2 million farms (~922 million acres) of which corn, tomatoes, potatoes, peanuts and sunflower seeds are among the major crops being cultivated. In China, over 300 million farmers are employed in the agriculture industry, endowing this country as the first place in worldwide farm output. Primarily, China is the main producer of rice, wheat, sorghum, potatoes, peanuts, tea, millet,

N. H. Salikin (✉) · M. M. Z. Makhtar
Bioprocess Technology Division, School of Industrial Technology, Universiti Sains Malaysia, 11800 Gelugor, Penang, Malaysia
e-mail: norhawani@usm.my



A. Z. Yaser et al. (eds.), *Waste Management, Processing and Valorisation*,
https://doi.org/10.1007/978-981-16-7653-6_10

barley, oilseed and cotton. Besides, India ranks as the second largest contributor of farm output after China, representing as the largest supplier of many fresh fruits and vegetables, spices, milk, jute, millet and castor oil seeds. India is also the second largest producer of wheat and rice and ranked as one of the five largest contributors of 80% of agricultural products including coffee and cotton in 2010 [3].

Likewise, the agriculture industry plays an imperative role in Malaysia economy, providing 16% of employment to the local citizens. Rubber, palm oil and cocoa are among the foremost commercial crops being cultivated locally while paddy, banana, coconut, durian, pineapple, rambutan and other local fruits and vegetables are also cultivated for the domestic market. In 1999, about 10.55 million metric tons of palm oil was generated endowing Malaysia as the world's largest producer [4]. The gross output of local agricultural sector indicates an 11% of increment in 2017 (~RM 91.2 billion) compared to ~RM 73.9 billion in 2015. This promising inclination is attributed to the increasing productivity derived from crops cultivation followed by livestock, forestry and logging and fisheries sub-sectors. Concurrently, the job opportunities are widening, providing salaries to 835 974 people and increased wages payment from only RM 7.90 billion in 2015 to RM 10.4 billion in 2017 [5].

In the upcoming years, the development of agriculture industry is projected to rise in order to accomplish the worldwide population's feeding demand. It is estimated that the global human population will reach approximately 9.1 billion people in 2050 which may induce global agricultural expansion [6]. While this circumstance may indirectly stimulate the local economic growth as well as improve the human socio-economy, the accumulation of biowastes or agro-industrial residues resulting from the agriculture sector remains a major challenge, particularly to the environment. The inefficient control management and disposal of such organic materials may result in multiple intimidating health hazards as well as detrimental impacts to the ecosystem [7]. However, considering the hidden potential of such carbon and nitrogen-rich agro-industrial wastes that can be wisely manipulated, this chapter will orchestrate the fundamental principles related to the utilization of biowastes and agro-industrial residues for the production of valuable pectinases enzymes through the cost-saving solid state fermentation (SSF) technique. This approach is believed as an efficient method in converting "wastes" into "wealth" while sustaining a green and healthy environment for a better future of the world [8].

## 2 Undesired Accumulation of Agro-Industrial Wastes as a Detrimental Issue

Despite the escalating development of global agricultural industries, the accumulation of biological wastes, i.e. agro-industrial residues remain a major challenge to the public and environment. Approximately, 998 million tonnes of agricultural wastes were produced worldwide within a year [9]. As in the 1990s, around 147.2 million metric tons of fibre were disposed, while 673.3 and 709.2 million metric tons of rice straws and wheat straw residues were abandoned worldwide [8, 10]. In Malaysia, it is estimated that 1.2 million tonnes of agro-wastes are disposed annually in addition to the other waste biomasses produced by the other Asian countries such as Indonesia, Cambodia and Thailand (see Table 1). In general, the main source of these horrendous agro-industrial wastes comes directly either from the agricultural practices or from industries associated with the agriculture sector that include processed animal products and crops [11]. Brewer's waste, maize milling by-product, molasses, oilseed cakes and bagasse are among the common wastes produced by the agro-industrial processing sector. Furthermore, the crop residues such as straw, stem, peel, shell, leaves, husk, seed, pulp, lint and stubble that are derived from cereals (which include rice, maize or corn, millet, barley, wheat and sorghum), legumes (soybean and tomatoes), jute, groundnut, cotton, coffee, cocoa, tea, olive, fruits (banana, mango, cashew) and palm oil are also defined as agricultural wastes. Unfortunately, in most developing countries, those biological wastes are indiscriminately dumped or burnt in public, hence resulting in greenhouse gases (GHG) emission, global warming, health hazards as well as soil, water and air pollution [7, 8].

Indeed, managing agro-industrial waste remains a big task for the underdeveloped and low-income regions around the world due to the lack of awareness and understanding of green renewable technologies [7, 8]. However, with the evolution of optimized fermentation technologies and utilization of highly productive microorganisms, the bioconversion of such wastes to become nutrient-rich raw materials to generate renewable energy (e.g. biomethane, biohydrogen, bioethanol) [12] or valuable microbial by-products such as hydrolytic enzymes (e.g. pectinases, cellulase, protease) can be successfully achieved [13–15].

**Table 1** Agricultural waste generation in Asia and the estimation of increment in the year 2025 [9]

| List of countries | Agricultural waste generation (kg/capita/day) | Estimated agricultural waste generation in 2025 (kg/capita/day) |
|---|---|---|
| Brunei | 0.099 | 0.143 |
| Cambodia | 0.078 | 0.165 |
| Indonesia | 0.114 | 0.150 |
| Laos | 0.083 | 0.135 |
| Malaysia | 0.122 | 0.210 |
| Myanmar | 0.068 | 0.128 |
| Philippines | 0.078 | 0.120 |
| Singapore | 0.165 | 0.165 |
| Thailand | 0.096 | 0.225 |
| Vietnam | 0.092 | 0.150 |
| Nepal | 0.060 | 0.090 |
| Bangladesh | 0.040 | 0.090 |

# 3 Utilization of Agro-Industrial Wastes with Pectic Substances for Pectinases Enzymes Production

Various agro-industrial wastes can be used as a source of energy for microbial growth during fermentation as well as a carbon source for biomass and other useful microbial by-products synthesis [8, 16]. While the bio waste materials serve as an anchorage for the microbial cells, the substrate also provides sufficient nutrients to support microbial growth. However, the substrate should be supplied externally for certain nutrients that are accessible in sub-optimal concentrations or are simply lacking [8, 17]. Given the biological potential of agro-wastes to support the fermentation process and their availability are unseasonal, those biomasses can be comprehensively exploited to produce profitable metabolites (e.g. vitamins, β-glucan, carotenoids, antibiotics and profitable enzymes) instead of being dumped in the open environment [18]. This environmentally friendly approach serves as an alternative for efficient waste disposal management to avoid undesired biowaste accumulation due to industrial development as well as to sustain a natural and pollution-free environment [18].

## *3.1 Pectic Substances*

In nature, various substrates derived from agro-industrial wastes such as peels of pomelo, citrus, apple, cucumber, tomato and papaya, apple pomace, apple bagasse, coffee mucilage and rice husk are applicable for the production of commercially-related by-products such as pectinases enzymes via solid state fermentation (SSF) [15] (see Table 2). In addition to their nutritional value, those organic biomasses are composed of pectic substances that act as a natural inducer for pectinases enzyme production by different microorganisms. Pectic substances are complex, high molecular weight, acidic and negatively charged biomacromolecules made up of seventeen different monosaccharides and at least seven different polysaccharides [19, 20]. Pectic compounds and hemicelluloses operate as a cementing agent for cellulose microfibrils in plant cell walls (Fig. 1), providing strength and stability to the cell wall structure [21]. Pectic substances are ubiquitous to members in the Plant Kingdom and copiously found in the lamella between the primary cell walls of adjacent young plant cells in form of calcium pectate and magnesium pectate that are responsible for the cell cohesion [20]. In young and developing cell walls, pectic substances are synthesized in the Golgi apparatus from the UDP-D-galacturonic acid [22–24].

## *3.2 Classification of Pectic Compounds*

Based on the modification type of its backbone chain, pectic compounds are classified as protopectin, pectic acid, pectinic acid and pectin. As a result, depending on their

**Table 2** Utilization of different agro-industrial wastes in pectinase production via solid state fermentation (SSF)

| Agro-industrial wastes | Microorganisms | References |
|---|---|---|
| Sugar cane bagasse | *Aspergillus niger* | [26] |
| | *Aspergillus awamori* | [27] |
| | *Aspergillus niger* CH4 | [28] |
| | *Moniliella* sp. SB9 | [29] |
| | *Penicillium* sp. EGC5 | [29] |
| Monosodium glutamate wastewater | *Aspergillus niger* | [30] |
| Sugar beet pulp | *Aspergillus niger* | [30] |
| | *Bacillus gibsonii* S2 | [31] |
| | *Penicillium oxalicum* | [32] |
| Wheat bran | *Bacillus safensis* M35 | [33] |
| | *Bacillus altitudinis* J208 | [33] |
| | *Aspergillus niger* | [34] |
| | *Aspergillus sojae* | [35] |
| | *Thermoascus aurantiacus* | [36] |
| | *Bacillus* sp. DT7 | [37] |
| | *Penicillium* sp. EGC5 | [34] |
| | *Streptomyces* sp. RCK-SC | [38] |
| | *Penicillium viridicatum* RFC3 | [39] |
| | *Penicillium expansum* | [40] |
| | *Bacillus pumilus* dcsr1 | [41] |
| | *Alternaria alternata* | [42] |
| *Consumption of different agro-industrial wastes in pectinase production through solid s fermentation (SSF)* | | |
| Rice bran | *Aspergillus niger* | [43] |
| | *Aspergillus awamori* | [27] |
| Orange bagasse | *Thermoascus aurantiacus* | [36] |
| | *Penicillium viridicatum* RFC3 | [39] |
| | *Moniliella* sp. SB9 | [29] |
| | *Penicillium* sp. EGC5 | [29] |
| Apple bagasse | *Penicillium* sp. | [34] |
| | *Aspergillus niger* | [34] |
| Orange pulp | *Trichoderma viridi* | [44] |
| Orange peel | *Aspergillus oryzae* | [45] |
| | *Penicillium atrovenetum* | [45] |
| Coffee pulp | *Aspergillus* sp. VTM5 | [46] |
| | *Aspergillus niger* | [47] |

(continued)

**Table 2** (continued)

| Agro-industrial wastes | Microorganisms | References |
|---|---|---|
| | *Saccharomycopsis fibuligera* | [48] |
| Coffee mucilage | *Erwinia herbicola* | [49] |
| | *Lactobacillus brevis* | [49] |
| Grape pomace | *Aspergillus awamori* | [50] |
| Grape peel | *Saccharomyces cerevisiae* CECT 11783 | [51] |
| Soy bran | *Aspergillus niger* | [52] |
| Citrus peel | *Aspergillus niger* | [53] |
| Citrus waste | *Schizophyllum commune* | [54] |
| | *Aspergillus niger* F3 | [55] |
| Corn cobs | *Alternaria alternata* | [42] |

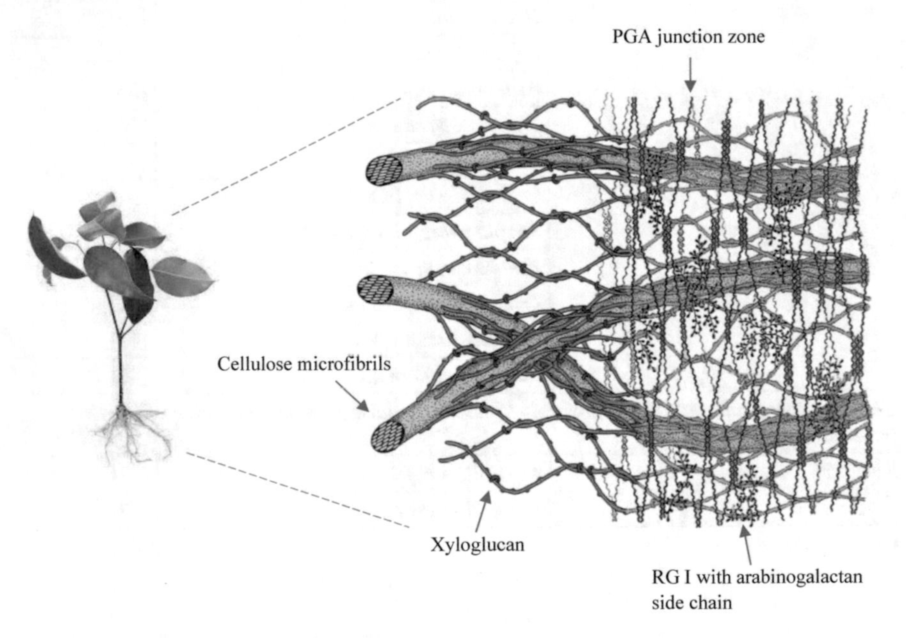


**Fig. 1** The primary cell wall structure of flowering plants. The cellulose microfibrils are interweaved with xyloglucan polymers and are embedded in a matrix composed of pectic polysaccharides, polygalacturonic acid (PGA) and rhamnogalacturonan (RG). (Figure was adapted from [25] with slight modification)

specificity and mode of action, different types of pectinases enzymes have distinct enzymatic reactions to these distinct pectic compounds (MOA) [56].

### 3.2.1 Protopectin

Protopectin is the "parents" of pectic substances which are comprised of high molecular weight methylated galacturonic acid polymer and usually found in immature fruits. Pectin or pectinic acid is produced when protopectin is hydrolysed [20, 57, 58]. This type of pectic substance is a water-insoluble component and is mainly found in the plant tissue [58], particularly in the central lamella between plant cells. Protopectin is insoluble due to its large molecular weight, salt bonding between carboxyl groups and an ester bond structure between protopectin's carboxylic group and the hydroxyl groups of other components in the cell wall [22].

### 3.2.2 Pectic Acid

Pectic acid is a soluble polymer of galacturonans that is essentially free of methoxyl groups [20]. Pectates are pectic acid salts that are either neutral or acidic [22, 57]. Pectic acid is a derivative of pectinic acid and is commonly found in overripe fruits [59, 60] (Fig. 2).

### 3.2.3 Pectinic Acid

Pectinic acid is a colloidal substance and is composed of a polygalacturonan chain that consists of more than 75% of methylated galacturonate units [61, 62]. Pectinates are pectinic acid salts that are either neutral or acidic [20]. Under the right conditions,

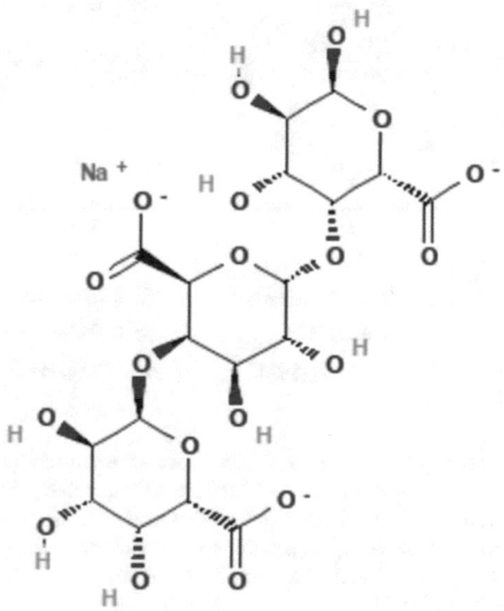

**Fig. 2** The 2D chemical structure of pectic acid, sodium salt ($C_{18}H_{23}NaO_{19}^{-2}$). Picture was derived from PubChem at https://pubchem.ncbi.nlm.nih.gov/compound/86278167

pectinic acid alone has the ability to produce a gel with acid or sugar. If the pectinic acid has a low methoxyl concentration, the gel can also be generated with certain metallic ions or other chemicals such as calcium salts [22].

### 3.2.4 Pectin

Pectin is a complex colloidal acid polysaccharide with a long unbranched galacturonic acid backbone chain and linked together by α, 1–4 linkages [63] (Fig. 3a). Galacturonic acid which is the monomer of this polysaccharide can be methyl esterified at C-6 while the hydroxyl group at C-2 or C-3 can be acetylated [64]. The secondary structure of pectin (Fig. 3b) is composed of two main regions which are "smooth region" (homogalacturonan) and "hairy regions" (rhamnogalacturonan I and II) [63, 65]. Homogalacturonan constitutes the backbone chain of pectin, composed of α, 1–4 linked D-galacturonic acid residues. Rhamnogalacturonan I is a highly branched region which comprised of a huge number of side chain α, 1–2

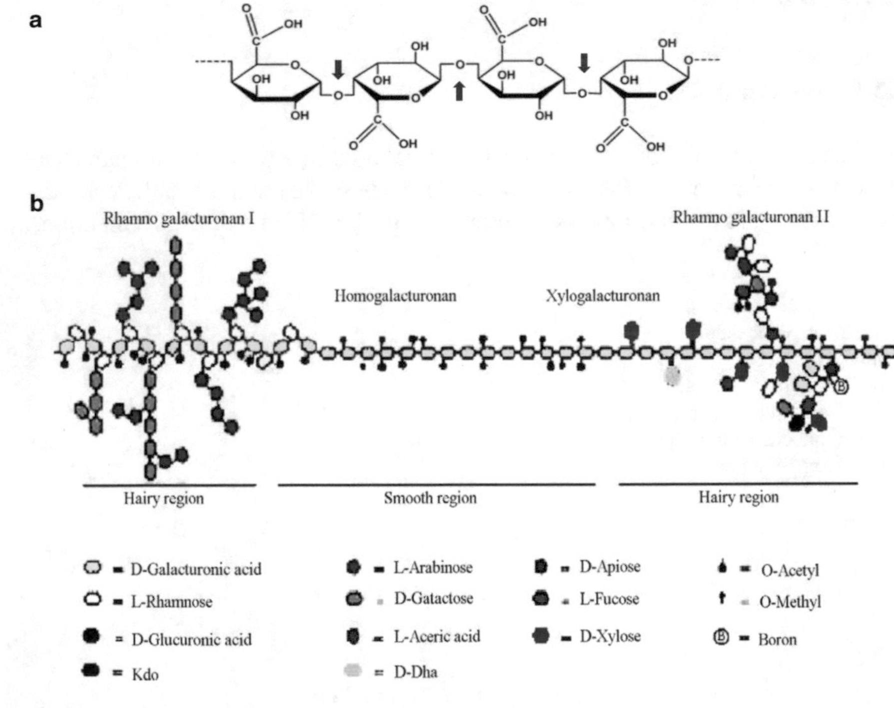


**Fig. 3** The primary and secondary structures of pectin. Galacturonic acids (pectin monomer) are linked together by α, 1–4 linkages (indicated by the red arrows) to form a long unbranched galacturonic acid backbone chain (also known as homogalacturonan) as the primary structure of pectin (**a**). A more complex secondary structure of pectin (**b**) is generally composed of homogalacturonan, xylogalacturonan, rhamnogalacturonan I and rhamnogalacturonan II. Figures were adapted from [66] and [67]

**Table 3** Pectin proportion in different fruits and vegetables [20]

| Fruit or vegetable | Tissue | Pectic substances (%) |
|---|---|---|
| Apple | Fresh | 0.6–1.6 |
| Banana | Fresh | 0.7–1.2 |
| Peach | Fresh | 0.1–0.9 |
| Strawberry | Fresh | 0.6–0.7 |
| Cherry | Fresh | 0.2–0.5 |
| Pea | Fresh | 0.9–1.4 |
| Carrot | Dry matter | 6.9–8.6 |
| Orange pulp | Dry matter | 12.4–28.0 |
| Potato | Dry matter | 1.8–3.3 |
| Tomato | Dry matter | 2.4–4.6 |
| Sugar beet pulp | Dry matter | 10.0–30.0 |

linked L-rhamnopyranose residues while the more complex Rhamnogalacturonan II is predominantly found in the primary cell wall of plants and plays a significant role as a signal molecule in the development of plant cell wall (Fig. 3b).

Polymethyl galacturonate is another name for pectin. It is a soluble polymeric substance that, after interacting with cellulose, gives the cell wall structure stiffness [20]. Methanol is used to esterify 75% of the carboxyl groups of the galacturonate monomer [22]. Pectin is plentiful in plants, especially fruits and vegetables, according to previous research [23]. Pectin makes up about 0.5% to 4.0% of the total fresh weight of a plant [20] (see Table 3). The pectic substances are crucial for plant cell growth and cell differentiation during plant tissue development as new materials will be laid down while the old materials are removed or degraded by the enzymes. Pectin attaches to the cellulose microfibrils that contribute to the cell wall's stability and rigidity in immature fruit. During the ripening stage, however, the primary backbone chain or side chain of pectin changes due to the response of naturally occurring pectinases enzymes [68]. Consequently, the rigid cell wall is softened since the pectin becomes more soluble due to natural chain modification [57, 68].

## 4 Pectinases Enzymes

The pectinases enzymes or pectin-degrading enzymes (i.e. polygalacturonase, polymethyl galacturonase, protopectinase, pectate lyases, pectin lyases, pectin methyl esterase, rhamnogalacturonan rhamnohydrolases and rhamnogalacturonan acetylesterases) are a kind of heterogeneous constituent that degrades pectic substance-rich plant structures [56]. These inducible enzymes can be extensively produced by saprophytic microorganisms either via solid state fermentation (SSF) or submerged fermentation (SmF) utilizing nutrient-rich agro-industrial wastes [26, 48]. More than a century ago in 1886, pectinases were considered as virulence factors causing decomposition of the plant cell walls [69]. Later, the profitable potential of pectinases was applied for home-made preparation of fruit juices and wines and its application at the industrial level was observed in 1930 [57]. Only after three decades, the understanding of chemical nature and enzymatic reaction catalysed by pectinases in plant tissue was improved. With this awareness, research on the application of pectinases enzymes for commercial purposes was implemented extensively which in turn, endowed pectinases as one of the leading enzymes in commercial sector nowadays. Pectinases exhibit an escalating trend in the commercial market [70, 71] while maintaining the average growth rate of 2.86% from $US 27.6 million (in 2013) to $US 30.0 million (in 2016) and projected to rise to $US 35.5 million by 2021 [72]. Kikkoman Shoyu, Societe Rapidase, Clarizyme Wallerstein, Novozymes, Novartis, Roche and Biocon are among the industrial scale manufacturers for commercial pectinases production that are located in Japan, France, the United States of America, Denmark, Switzerland, Germany and India respectively [57, 73].

In addition to saprophytic microorganisms, pectinases are also naturally produced by higher organisms, i.e. plants, insects, nematodes and protozoans [65, 74]. In plants, pectinases play an essential role in supporting the extension of cell walls, facilitating the penetration of pollen tubes [75] and softening the tissue of ripening fruits by altering the pectin backbone chain [20, 68]. In insects, e.g. rice weevil (*Sitophilus oryzae*), pectinases are secreted by the endosymbiotic microbiota to facilitate the host digestive system [76]. Microbial pectinases are also important particularly during the phytopathogenic process, plant–microbe symbiotic interaction and during the natural decomposition process of decaying plant residues. Ultimately, plant materials decomposition by microbial pectinases also contributes to the natural carbon cycle in the ecosystem [65].

## *4.1 Classification of Pectinases Enzymes and Their Mechanism of Actions*

Pectinases enzymes are categorized based on the degradation mechanism towards the pectic substances [77, 78]. More specifically, the different catalytic mechanisms of action (MOAs) against galacturonan region of the targeted pectic substances will distinguish the pectinases enzymes into different classes [29, 79] (see Fig. 4). However, most recently, the primary substrate and the final products resulting from the catalytic activity also can be used to distinguish the pectinases enzymes [79]. In general, pectinases are classified into three main groups including protopectinases, depolymerases (hydrolases and lyases) and esterase [20, 79, 80] (see Fig. 5 for extensive pectinases enzymes classification). Protopectinases degrade the insoluble protopectin to become highly polymerized soluble pectin while depolymerases hydrolyse the α-1,4-glycosidic linkage in the D-galacturonic acid moieties of pectic compounds. The removal of methoxy esters from pectin is catalysed by esterase [20, 80]. Microbial pectinases generated by prokaryotes are generally alkaline in

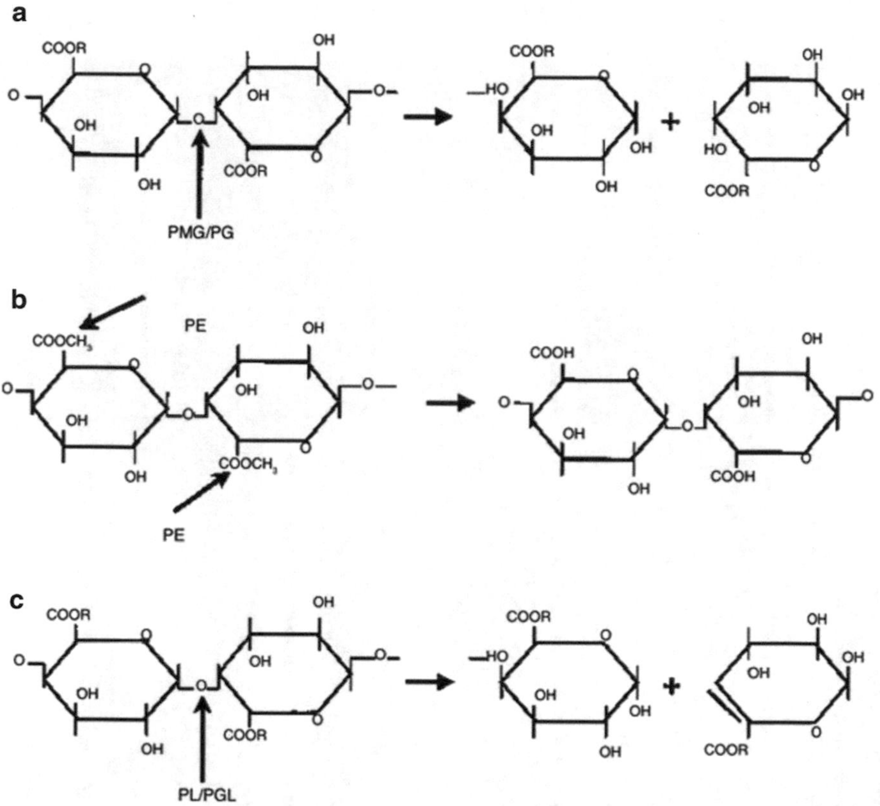

Fig. 4 Different mechanisms of actions (MOAs) of pectinases enzymes. Examples; PMG; polymethylgalacturonases, PG; polygalacturonases, PE; pectinesterase (pectin methyl esterase), PL; pectin lyase. **a** R = H for PG and $CH_3$ for PMG. **b** PE **c** R = H for PGL and $CH_3$ for PL. Pectinases reactions towards the pectic substances are indicated by the arrow [20, 78]

nature while pectinases produced by eukaryotic microorganisms are predominantly acidic [80]. A complex process which comprises the interaction of many pectinolytic enzymes such as polygalacturonase, pectate lyase and pectin methyl esterase is required to achieve a complete degradation of pectin [79, 81].

### 4.1.1 Polygalacturonase

Polygalacturonase hydrolyses the α-1,4-glycosidic linkages in polygalacturonic acid and gives rise to D-galacturonate as the final product. Polygalacturonase is classified under the glycosyl hydrolases group [65]. Acidic endo-polygalacturonase is predominantly applied in reducing the viscosity of fruit juices due to the visibility of pectin substances while alkaline endo-polygalacturonase is utilized vastly in pectic wastewater treatment and paper and textile manufacturing industries [82]. The molecular

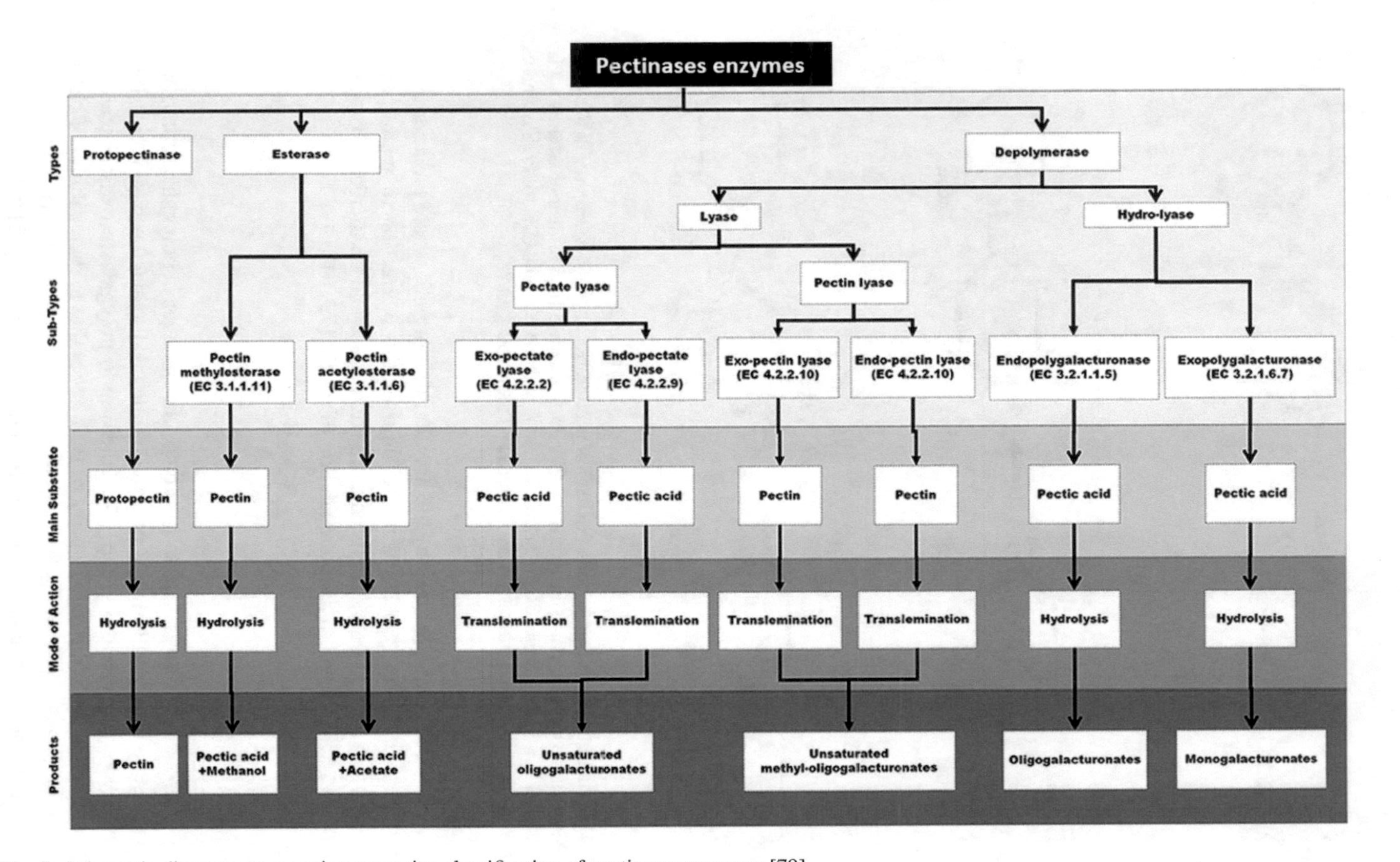


**Fig. 5** Schematic diagram representing extensive classification of pectinases enzymes [79]

weights of endo-polygalacturonase and exo-polygalacturonase often range from 30 to 80 kDa and 30–50 kDa respectively [56].

### 4.1.2 Pectate Lyases

Pectate lyases are categorized into exo- and endo- groups based on the difference of enzymatic MOAs. The exo-pectate lyases cleave polygalacturonic acid from the non-reducing end while the endo-pectate lyases react to the substrate in a random way. Pectate lyases catalyse the cleavage of glycosidic linkages in polygalacturonic acid. The final product of this enzymatic process is unsaturated 4,5-D-galacturonate [65].

### 4.1.3 Pectin Methyl Esterase

Pectin methyl esterase (also known as pectinesterase) catalyses the de-esterification process of methoxyl group from pectin [56, 64], thus giving rise to pectic acid and methanol as the final products. The molecular weight varies between 30 – 50 kDa with optimum reaction temperature from 40 to 60 °C [56]. Pectin methyl esterase reacts before polygalacturonases and pectate lyases given that non-esterified galacturonate units are required for their enzymatic reactions [65]. This enzyme is crucial in controlling the metabolism of the plant's cell wall [82].

## 5 Microorganisms as Potential Pectinase Producer

Microorganisms including bacteria, fungi and yeasts have been endowed with enormous potential in producing numbers of valuable enzymes such as cellulase, mannanase, protease, lipase and pectinases. These enzymes have been vigorously exploited for commercial purposes over the years [20, 83]. Current market trend indicates that almost 50% of industrial pectinases originate from fungi and yeast, 35% from bacteria while the rest of 15% is derived from plants or animals' origin [73, 84]. In previous centuries, crude enzymes preparation involved the utilization of animal tissue (i.e. from stomach or pancreas mucosa) or from plants (i.e. papaya fruits and malt). Only in 1894, Dr. Jokichi Takamine became the pioneer who introduced fungal-derived enzymes production to the industry. Twenty years later, Boidin and Effront introduced the production of bacterial enzymes that further kickstarted industrial growth [85].

Pectinases are one of the imminent enzymes in the current market that can be extensively generated through the exploitation of different microorganisms during the fermentation process [72]. In nature, some of these microorganisms act as saprophytes which depend on other decaying materials such as plant tissues for nutritional purposes [86]. Whereas for some other phytopathogenic microorganisms, abundant pectinolytic enzymes are secreted during the attack against the plant tissues due to the accessibility of pectic constituents compared to the other fibres in the plant [56]. Plant cell wall is the first defence tissue which is constituted of pectin, lignin, hemicelluloses and cellulose. In order to invade, these microorganisms (e.g. *Aspergillus flavus*, *Alternaria citri*, *Agrobacterium tumefaciens* and *Claviceps purpurea*) produce a pool of hydrolytic enzymes including pectinases to succeed during the infection stages [65]. Microbial enzyme producers have gained the main interest in research due to their vast diversity, short life cycle, easy growth and maintenance, cost effective, consistency in enzyme production, capability of enzyme production enhancement through the improvement of culturing condition as well as accessibility for genetic manipulation [87].

### 5.1 Fungi as Potential Pectinase Producer

The Kingdom of Fungi is constituted of four different classes including Zygomycetes, Ascomycetes, Deuteromycetes and Basidiomycetes. The main structure of fungi is composed of septate or aseptate hyphae which play a considerable role in nutrient absorption from the substrate and during the germination process [88]. Catalysing the decomposition of putrefying plants materials and degrading the plant tissues during phytopathogenic attack endowed the fungi a potential in producing a pool of cell wall synergistic degrading enzymes such as pectinolytic, cellulolytic and xylanolytic enzymes [89]. For instance, polygalacturonase as one of the pectinolytic enzymes is abundantly secreted by *Aspergillus flavus* in order to degrade pectin as part of the major constituent in the plant's cell wall [90].

The pectinolytic enzymes produced by filamentous fungi are predominantly acidic in nature [91] while some are alkaline such as pectinases produced by *Penicillium italicum* and *Aspergillus fumigatus* [22]. One of the most studied fungal pectinase producers is *Aspergillus niger* which possesses the status of GRAS (Generally Regarded As Safe) of which the metabolites yielded by this microorganism is considered safe to be utilized [64, 92]. *A. niger* produces a bulk of pectinolytic enzymes such as polymethylgalacturonase, polygalacturonase and pectinesterase. For commercial purposes, solid state fermentation (SSF) and submerged fermentation (SmF) are used to generate the bulk of pectinases employing different fungal strains. In comparison with submerged fermentation, solid state fermentation commonly produces a higher volume of stable pectinase [65] since its condition approximates the natural environment for fungal growth that lacks water activity [93].

### 5.2 Bacteria as Potential Pectinase Producer

Filamentous and non-filamentous bacteria also produce pectinases which are predominantly alkaline in nature [91]. Bacterial strains, i.e. *Bacillus sphaericus*

(MTCC 7542), *Bacillus laterosporus* and *Bacillus firmus* isolated from soil, rotten vegetables and fermented clayed potato peels had been proven as pectinase producers [94, 95]. In general, bacteria from the genus *Bacillus* or any other thermophilic bacteria are predominantly saprophytes or plant pathogens that can produce pectinases to catalyse plant degradation [70].

In industrial scale, bacterial pectinases are produced enormously employing submerged fermentation since the environment with a higher water activity ($a_w$ = 0.95) is more appropriate for bacterial growth [91]. However, there are some bacteria that also can produce pectinases via solid state fermentation. For example, *Bacillus* sp. DT7 produce alkaline and thermotolerant pectinase under solid state fermentation system utilizing wheat bran as the substrate [91]. Studies also confirmed that *Bacillus firmus* could produce polygalacturonase through solid state fermentation using agro-industrial wastes [94].

## 5.3 Yeasts as Potential Pectinase Producer

Studies had shown that yeasts were capable to produce acidic pectinases [96]. For example, *Pichia pinus* produced pectin lyase and β-galactosidase when cultivated in mango paste under semi-solid fermentation [97]. There is also a report that confirmed *Saccharomyces cerevisiae* grown on grape skin can degrade pectin by producing pectinases [51]. Besides, *Rhodotorula sp.* was also found producing pectin methyl esterase and polygalacturonase enzymes during the softening process of olive while endo-polygalacturonase was produced by *Kluyveromyces marxianus* in order to degrade the pectinaceous layer of cocoa beans [98].

In order to maximize pectinase production, strategies such as genetic recombinant and molecular cloning are applied to the potential microbial producers [99, 100]. For example, the pectinase encoding gene in *Aspergillus niger* was cloned and expressed in yeast *Pichia pastoris* as a new host to produce an enormous volume of pectinolytic enzymes. Codon optimization and different promoter inclusion were also tested onto the strain to ensure high pectinase productivity and stability [99]. In another study, the transformation of a recombinant strain with the plasmid pAN52*pgg2*, carrying the gene encoding polygalacturonase belonging to *Penicillium griseoroseum* had successfully resulted in increased polygalacturonase production [100]. This new finding offers an effective alternative for microbial pectinase production employing genetically modified microorganisms through solid state fermentation system. Examples of microorganisms with the capability in producing acidic and alkaline pectinase are shown in the following Table 4.

**Table 4** Examples of acidic and alkaline microbial pectinases producers

| Microbial producer | Type of pectinase | References |
|---|---|---|
| *Acidic pectinase* | | |
| *Aspergillus kawachii* | Endopolygalacturonase | [101] |
| *Penicilloium frequentans* | Endopolygalacturonase | [57] |
| *Kluyveromyces marxianus* | Endopolygalacturonase | [102] |
| *Sclerotium rolfsii* | Endopolygalacturonase | [57] |
| *Aspergillus flavus* | Endopolygalacturonase | [103] |
| *Rhizoctonia solani* | Endopolygalacturonase | [57] |
| *Aspergillus niger* | Polygalacturonase | [104] |
| *Trichoderma harzianum* | Polygalacturonase | [105] |
| *Saccharomyces cerevisiae* | Polygalacturonase | [106] |
| *Aspergillus aculeatus* | Polygalacturonase | [82] |
| *Streptomyces* sp. QG-11-3 | Polygalacturonase | [82] |
| *Wickerhanomyces anomalus* | Polygalacturonase | [107] |
| *Mucor pusilus* | Polygalacturonase | [57] |
| *Mucor circinelloides* ITCC 6025 | Polygalacturonase | [95] |
| *Erwinia carotovora* | Polygalacturonase | [108] |
| *Clostridium thermosaccharolyticum* | Polygalacturonate hydrolase | [57] |
| *Penicillium italicum* | Pectin lyase | [109] |
| *Saccharomyces cerevisiae* | Pectin lyase | [106] |
| *Saccharomyces cerevisiae* | Pectin esterase | [106] |
| *Aspergillus niger* | Pectin methylesterase | [110] |
| *Aspergillus aculeatus* | Rhamnogalacturonase | [111] |
| *Alkaline pectinase* | | |
| *Bacillus* sp. RK9 | Polygalacturonate lyase | [57] |
| *Bacillus* sp. NT-33 | Polygalacturonase | [57] |
| *Bacillus paralicheniformis* CBS32 | Polygalacturonase | [112] |
| *Bacillus polymyxa* | Polygalacturonase | [57] |
| *Bacillus* sp. SMIA-2 | Polygalacturonase | [113] |
| *Bacillus sphaericus* MTCC 7542 | Polygalacturonase | [95] |
| *Penicillium italicum* CECT 229421 | Pectin lyase | [57] |
| *Pseudomonas syringae* | Pectate lyase | [57] |
| *Bacillus clausii* | Pectate lyase | [114] |
| *Bacillus amyloliquefaciens* S6 | Pectate lyase | [115] |
| *Amucola* sp. | Pectate lyase | [57] |
| *Xanthomonas campestris ACCC 10048* | Pectate lyase | [116] |
| *Dickeya dadantii* | Pectate lyase | [117] |
| *Pseudomonas fluorescence* | Pectate lyase | [118] |

# 6 Biotechnological Application of Microbial Pectinases

The current global market indicates an immense number of profits ($US 7 billion) generated from enzymes involved in research, diagnosis, industrial and therapeutic use [119]. The industrial enzymes market itself generates revenue of around $US 4.5 billion and is projected to boost up to $US 7 billion by 2021 with 4% of annual growth rate [120]. Pectinases are among the important products of biotechnological sector and they hold a leading position among the commercially produced industrial enzymes [73]. The application of pectinases in industries varies depending on their optimal acidic or alkaline nature and substrate specificity.

## *6.1 Acidic Pectinases*

### 6.1.1 Fruit Juices Extraction and Clarification

Acidic pectinases play a vital role in fruit juices processing and manufacturing particularly during the extraction and clarification process [15, 91]. Fruits for example apples and oranges are mechanically crushed to release highly viscous juices along with the pectic substances in the fruits that may increase the juice's turbidity and viscosity [20]. During the ultrafiltration process, the pectic substances colloidal polysaccharides cause fouling problems to the membrane filter. Therefore, acidic pectinases are added during the extraction procedure which in turn augments the juice yields, decreasing its viscosity and degrading the gelatinous structure [121]. The incorporation of pectinases during the fruit juice processing successfully improves juice recovery in terms of volume, extraction time as well as the preservation of organoleptic (flavour and colour) and nutritional properties of the juice yielded [65].

### 6.1.2 Industrial Wine Making and Manufacturing

Combination of grape juice with other juices such as apple juice is always preferred by the industrial winemaker to produce different grades of wines. Chronologically, the fruits are crushed, heated to 60–80 °C and filtered. Enzymes including pectinases (i.e. polygalacturonase) and hemicellulase (i.e. arabinogalactanase) are added during the depectinization process. Pectinases are added at the first stage when the fruits are mechanically crushed in order to maximize the juice recovery and reduce the pressing time [73]. Pectinases are then incorporated into the running juice at the second stage before or after the fermentation period to degrade suspended particles and diminish the growth of undesirable microorganisms. At the final stage, pectinases are supplemented to the wine to improve its filtration rate and clarity [57, 91]. As a precautionary approach against the formation of methanol as well as to maintain the aromatic property of the wine beverage, pectin lyase is added during the final stage of wine processing [122].

### 6.1.3 Essential Oil Extraction

Essential oil is extracted from citrus peels and it is commonly constituted of hydrocarbons, i.e. sesquiterpenes and terpenes, aldehydes, esters, alcohols, phenols, ketones and other non-volatile compounds such as fatty acids, waxes and flavonoids. Pectinases and protease are applied to accelerate the processing time and to enhance the quality of the essential oil via hydrolyzation of the pectin–protein substances [57]. Studies also show that the inclusion of pectinases during the bergamot peels (*Citrus bergamia* Risso) essential oil extraction process considerably enhanced the antimicrobial property of the flavonoid compound in the oil against several pathogenic microorganisms [123].

### 6.1.4 Preparation of Dried Animal Feed from Citrus Fruits

Agro-industrial residues including peels, seeds and pulp derived from citrus fruits processing can be exploited for the preparation of dried feeds for dairy cattle and sheep [124]. Pectinases are added to the biowastes to catalyse the demethylation of citrus pectin while cytolase is supplemented to improve the firmness of the mixture. Those residues are subsequently dried leaving the average moisture level to be around 8 to 10%. The final product could be consumed as an alternative nutrient source to feed the animals [57, 73].

## *6.2 Alkaline Pectinases*

### 6.2.1 Retting and Degumming of Plant Fibres

Plant fibres are broadly utilized in developing countries particularly for the manufacturing of bags, ropes and nets [22]. The enzymatic process of retting and degumming of plant fibre such as coir from the coconut husk, kenaff, flax, jute, ramie and hemp is commonly catalysed by the alkaline pectinases [57, 125]. In general, the retting process involves the microbial aerobic or anaerobic decomposition of pectic substances in plant barks to release fibre as the final product [126]. In the aerobic process, a thin layer of plant straw is spread onto the ground followed by fermentation by *Cladosporium* sp., *Penicillium* sp., *Aspergillus* sp. and *Rhodotorula* sp. for two to ten weeks. Whereas anaerobic retting is performed by submerging the plant straw in water tanks with low oxygen supply. This condition is favoured by anaerobic microorganisms such as *Clostridium butyricum*, *Clostridium felsineum*, *Bacillus*, *Pseudomonas* and *Micrococcus* sp. that act as the retting agents that secrete pectinases enzymes [57]. For the degumming process, pectinases are consumed to eliminate the gummy substances in the fibres before being used as the raw materials in the textile industry [73, 91]. The exploitation of microbial pectinases during the degumming process is not only cost-saving but also serves as an environmentally friendly approach against the conservative methods that mainly employs hazardous chemicals.

### 6.2.2 Pectic Wastewater Treatment

Wastewater incorporated with pectic substances is part of a by-product from the citrus and vegetable processing industries [20]. Conventionally, the polluted wastewater is hydrolysed using chemicals, leading to toxic environmental pollution. An eco-friendly alternative is applied to eliminate the pectic wastewater sludge using pectinases produced via alkalophilic microbial fermentation [22]. Studies have shown that extracellular endopectate lyase produced by soil bacteria, *Bacillus* sp. (GIR 621) can be exploited for efficient degradation of pectic residues during wastewater treatment [57, 126].

### 6.2.3 Vegetable Oil Extraction

Conventionally, a toxic organic solvent such as hexane is broadly used during oil extraction from rapeseed (*Canola*), sunflower seed, palm, kernel and olives. However, frequent utilization of hazardous solvents contributes to the risk of cancer and cell abnormalities [65]. The application of cell wall degrading enzyme such as pectinases is considerably a healthy alternative, proven to degrade the cell wall structure of oil-containing crops and unshackle the vegetable oil effectively [22, 73]. For instance, Olivex; a mixture of cell wall degrading enzymes produced by *Aspergillus aculeatus* is added during the extraction of vegetable oil from olives. This multienzymes preparation comprised of pectinases, hemicellulase and cellulase is proven to liberate the vegetable oil efficiently. As a result, optimal oil extraction is accomplished while the recovered vegetable oil also indicates a high proportion of polyphenols and vitamin E [57].

### 6.2.4 Coffee and Tea Fermentation

Pectinases are also used during the fermentation of coffee to dispose of the mucilaginous layer of the coffee beans [22, 91, 127]. Instead of commercial enzymes utilization, microbial pectinases yielded via fungal fermentation on coffee beans mucilage wastes are used as a cost-saving alternative [73]. The fermentation liquid which contains fungal pectinases is washed and filtered before being sprayed to the surface of coffee beans [57]. Pectinases are also used to accelerate the fermentation process of tea leaves as well as to enhance the tea quality and its foam-forming property [22, 73].

# 7 Solid State Fermentation (SSF); A Cost-Saving Tool to Produce Microbial Enzymes

Solid state fermentation is a three-phase heterogeneous process, comprising solid, liquid and gaseous phases that involve a fermentation that occurs in the absence or near absence of free water employing a specific substrate as the main source of energy and carbon supply for the microbial growth [128, 129]. It is an eco-friendly fermentation method and mostly utilizes agro-industrial wastes as the main substrate [128]. Historically, this solid state fermentation was almost neglected in western countries in the 1940s due to the emergence of submerged fermentation technology (SmF) which was extensively conducted in producing penicillin. However, a new profitable finding related to steroid transformation by a fungal isolate through solid state fermentation had effectively fascinated the world from 1950 to 1960 [130]. Solid state fermentation has shown a huge potential for the synthesis of commercially related microbial products including enzymes, feed, biofuel, food, industrial chemicals and pharmaceutical products. Especially in the case of agro-industrial wastes accumulation, solid state fermentation offers an alternative approach for the control management of these often-neglected residues [131]. A better understanding of bioprocess dynamics including substrate selection and multiple parameters optimization has enabled an improved fermentation yield, which in turn appears as a driving force to scale up the solid state fermentation processes. Although there are important challenges yet to be addressed, the advantages of this technology could push the development and implementation of solid state fermentation into the context of a future bioeconomy [129].

## *7.1 Selection of Substrate*

Selection of appropriate substrate is one of the key factors in solid state fermentation as it plays a crucial role as the main source of nutrients and anchorage to support microbial growth [17]. Low cost and the unseasonal availability of raw materials become the determinative factors for substrate selection; two special criteria that have been fulfilled by the utilization of agro-industrial wastes [132–134]. Furthermore, the substrate in use must be porous while possessing sufficient mechanical resistance and highly water absorbent in order to overcome heat and mass transfer problems while supporting microbial growth [135]. An ideal substrate should be indicated by its nutritional properties that are adequate to support microbial growth. However, in some cases, the substrate needs to be supplemented externally due to the lacking or invisibility of particular nutrients. Occasionally, some substrates must be pretreated chemically or mechanically before the fermentation process to enhance their accessibility to the microorganisms [17]. Indeed, the selection of substrate is part of the crucial factors that govern the final yield of solid state fermentation, particularly in microbial pectinase production. While various agro-industrial residues such as apple pomace, mango and banana peels, rice bran and corn cobs are applicable to be used as the substrate for microbial pectinase production (see Table 2), citrus peels has been suggested as the best substrate for the fermentation process due to its higher pectin content of about 24.5% [136].

## 7.2 Screening of Potential Microbial Enzyme Producer

A compatible microorganism with the capability to produce a bulk of enzymes should be considered prior to its incorporation into the solid state fermentation system. The selected industrial strains should exhibit genetic stability, efficient production of the desired final product, limited or no vitamins requirement for growth, capability to utilize a broad range of low-cost carbon sources, applicable for genetic manipulation, non-pathogenic and limited production of other undesired by-products to facilitate purification process [137]. Theoretically, fungi and yeasts are more applicable for solid state fermentation due to the low water activity [138]. However, some bacterial strains such as *Bacillus firmus* and *Bacillus laterosporus* are able to produce enzymes via solid state fermentation [94]. Primarily, potential microorganisms are isolated from different biological sources for example agricultural waste dump soil and rotten fruits or vegetables. Those microbial isolates are subsequently investigated for pectinolytic activity via primary and secondary screenings [139–141]. Viable microbial colonies surrounded with discernible hydrolysis zone on pectin-containing agar during primary screening indicates the existence of pectinases degradation while the activity of microbial crude pectinases produced via small scale fermentation process is determined more precisely during the secondary screening method [139].

## 7.3 Optimal Physical Conditions for Microbial Enzyme Production via SSF

An optimal size of the substrate, initial moisture content, incubation temperature, inoculum size, initial pH, agitation rate and aeration in the fermentation system are among the critical factors that affect the enzyme yield via the SSF technique [130, 134].

In solid state fermentation, close contact of microorganisms to the substrate enables optimal nutrient absorption to support their growth [93, 134]. A smaller particle size of substrate offers a larger surface for microbial attack than the larger one. However, the utilization of the larger size of the substrate allows for better aeration due to the increasing space among the inter particles. Inversely, too small particle size only leads to large agglomerates (clumps) which in turn affects microbial respiration, retarded microbial growth and deteriorated enzyme production [17,

133, 142]. Therefore, a compromised size of substrate needs to be optimized prior to the implementation of the fermentation system.

Initial moisture content is another indispensable factor to be emphasized in solid state fermentation systems as moisture content significantly affects microbial growth, sporulation, enzyme yield and stability [122, 142]. The moisture content contributes to the puffiness of fermented substrate and facilitates the penetration of fungal mycelium for optimal nutrient absorption [52]. Almost 70 to 80% of microbial cell is constituted of water and at the same time, free water is utilized for cellular metabolic process. Decreasing water activity ($a_w$) in solid state fermentation will only result in the decline of pectinases production as has been observed on *Trichoderma viridae* [122]. However, the too high moisture content may cause undesired contamination problems by other invading microorganisms [143].

Incubation temperature also affects the rate of enzyme production via solid state fermentation [129]. The determination of optimal incubation temperature also relies on the type of microbial enzyme producer either it is mesophiles, psychrophiles, or thermophiles. Given that chemical reaction, metabolic networks and signalling in cells are increased in parallel with the rising temperature, consequently, in the case of solid state fermentation, microorganisms will grow more hastily when the incubation temperature is aroused. Furthermore, microbial growth may generate 100–300 kJ of heat per kilogram of cell mass in solid state fermentation system [144]. Excessive heat will contribute to the denaturation of some heat-sensitive macromolecules such as lipid, nucleic acid and proteins. In contrast, the lipids in the cell membrane will not be fluid enough to function accordingly in an environment with too low temperature. Therefore, the incubation temperature needs to be optimized in order to enhance the microbial growth as well as the enzyme biosynthesis [142].

Besides, the pH condition in solid state fermentation should be favoured by the process microorganisms for optimal enzyme yield as well as for microbial growth [134, 142]. Fundamentally, microorganisms can be classified into acidophiles, neutrophiles and alkalophiles. Majority of fungi and yeasts are classified as acidophiles which favour the acidic condition with pH ranges starting from 1.0 to 5.5. Most bacteria are alkalophiles that inhabit the alkaline environment with pH ranges from 8.5 to 11.5. However, microorganisms are predominantly neutrophiles and grow in pH ranges from 5.0 to 9.0. Siddiqui et al. [145] had investigated the production of polygalacturonase by a fungus, *Rhizomucor pusilis* through solid state fermentation, utilizing sugarcane bagasse and fermented in pH 5.5 to 6.0. Pectinase was also massively yielded by *Bacillus* sp. DT7 in alkaline condition after being fermented on wheat bran [37]. A yeast isolate, *Pichia pinus* was also proven in producing pectinase through the fermentation system employing mango wastes with a surrounding pH of 5.5 [97].

An optimized inoculum size of fermenting microorganisms is also crucial in governing the yield of enzymes [17, 142]. The culture to be used as the inoculum should be in a stable condition and capable of producing enzymes during the fermentation period. Inoculum preparation of bacteria and yeast involves the cells while fungal inoculum is prepared using its spores. While a higher size of inoculum may facilitate microbial colonization and amplify the enzyme production [142], nutrient reduction may occur as a result of competitive interaction in dense microbial colonies which eventually may deteriorate the enzyme yield [146]. In contrast, too low inoculum size may result in insufficient microbial biomass and enzyme biosynthesis while promoting the invasion of other microbial contaminants [147].

Heat removal and the control of humidity and aeration are pivotal particularly in industrial scale solid state fermentation [17, 142]. Studies have shown that the inside temperature of the fermented substrate within 5 cm depth is higher (~50 °C) compared to the ambient incubation temperature setting (37 °C) [148]. Therefore, intermittent or continuous mixing is crucial to enable the release of heat [143] even though it may affect the fungal hyphae and the sporulation process [148]. Instead of mixing, evaporative cooling is another practical approach to liberate heat from the fermentation bed. However, in this case, the microbial growth might be retarded due to the drying of the raw substrate after the evaporation process [142, 149]. Therefore, an appropriate heat removal method needs to be optimized accordingly in order to ensure a massive enzyme production by the fermentation system [129].

### *7.4 Optimal Chemical Conditions for Microbial Enzyme Production via SSF*

While substrate with the capability to support all of the nutrient requirements for microbial growth is highly favoured in solid state fermentation [8, 32], in most cases, external supplementation of carbon and nitrogen sources are incorporated into the substrate due to the deficiency of particular nutrient content [17]. Additional carbon sources such as glucose, fructose, maltose, sucrose, lactose, starch and molasses are always incorporated into the fermentation system to ensure maximum energy supply in the microorganisms during the enzyme biosynthesis [150].

In addition, adequate nitrogen source is also crucial to support microbial growth as well as for enzyme production. Organic or inorganic nitrogen sources including industrial by-products such as corn steep liquor, yeast extracts, peptones, urea and soy meal are commonly supplemented into the solid state fermentation system as additional nitrogen sources [151, 152].

## 8 Conclusion and Future Perspective

Malaysia is blessed with abundant fertile agricultural lands and extremely rich and very diverse biological resources that can be manipulated for local socio-economic growth. Despite the excellent achievement depicted by the agriculture sector, the accumulation of undesired agro-industrial wastes may result in unpleasing issues including intimidating environmental pollution, greenhouse gas (GHG) emission, global warming and debilitating health hazards to the community. Solid state fermentation offers an efficient waste valorisation method catalysed by microorganisms that allow conversion of those pectin and nutrient-rich biomass residues into profitable value-added by-products such as pectinases enzymes. While solid state fermentation offers a cost-saving and environmentally friendly fermentation system for commercial or industrial scale enzyme production, several improvements should be considered to enhance the bioprocess system. This includes engineering interventions or soft computing tools using mathematical models that can be applied to replace the "one variable at a time" (OVAT) type optimization method that is more laborious and less cost effective. Furthermore, genetic manipulation using DNA recombinant technique and genome editing involving CRISPR-Cas9 can be applied to transform the microorganism used during the fermentation process and increase the output of the fermentation products. Indeed, agro-industrial residues should be considered as "raw materials" instead of "wastes" of which its wise utilization or manipulation can benefit the human population as well as the environment.

## References


1. Gassner, A., Harris, D., Mausch, K., Terheggen, A., Lopes, C., Finlayson, R.F., Dobie, P.: Poverty eradication and food security through agriculture in Africa: rethinking objectives and entry points. Outlook Agric. **48**, 309–315 (2019). https://doi.org/10.1177/0030727019888513
2. Tian, Z., Wang, J.-W., Li, J., Han, B.: Designing future crops: challenges and strategies for sustainable agriculture. Plant J. **105**, 1165–1178 (2021). https://doi.org/10.1111/tpj.15107
3. FAO: The State of Food and Agriculture: Viale delle Terme di Caracalla, 00153 Rome, Italy (2012)
4. Encyclopedia of the Nation: Malaysia Agriculture, Information about Agriculture in Malaysia. https://www.nationsencyclopedia.com/economies/Asia-and-the-Pacific/Malaysia-AGRICULTURE.html
5. DOS: Department of Statistics Malaysia Official Portal. https://www.dosm.gov.my/v1/index.php?r=column/ctwoByCat&parent_id=45&menu_id=Z0VTZGU1UHBUT1VJMFlpaXRRR0xpdz09
6. FAO: High Level Expert Forum - How to Feed the World in 2050, Viale delle Terme di Caracalla, 00153 Rome, Italy (2009)
7. Adejumo, I.O., Adebiyi, O.A.: Agricultural solid wastes: causes, effects, and effective management. In: Solid Waste Management. IntechOpen (2020). https://doi.org/10.5772/intechopen.93601
8. Sadh, P.K., Duhan, S., Duhan, J.S.: Agro-industrial wastes and their utilization using solid state fermentation: a review. Bioresour. Bioprocess. **5**, 1–15 (2018). https://doi.org/10.1186/s40643-017-0187-z
9. Agamuthu, P.: Inaugural Meeting of First Regional 3R Forum in Asia 11–12, Tokyo, Japan (2009)
10. Belewu, M.A., Babalola, F.T.: Nutrient enrichment of waste agricultural residues after solid state fermentation using *Rhizopus oligosporus*. J. Appl. Biosci. **13**, 695–699 (2009)
11. Javad, S., Akhtar, I., Naz, S.: Nanomaterials and agrowaste. In: Nanoagronomy, pp. 197–207. Springer, Berlin (2020)
12. Basri, M.F., Yacob, S., Hassan, M.A., Shirai, Y., Wakisaka, M., Zakaria, M.R., Phang, L.Y.: Improved biogas production from palm oil mill effluent by a scaled-down anaerobic treatment process. World J. Microbiol. Biotechnol. **26**, 505–514 (2010). https://doi.org/10.1007/s11274-009-0197-x

13. Balachandran, C., Vishali, A., Nagendran, N.A., Baskar, K., Hashem, A., FathiAbd_Allah, E.: Optimization of protease production from *Bacillus halodurans* under solid state fermentation using agrowastes. Saudi J. Biol. Sci. (2021). https://doi.org/10.1016/j.sjbs.2021.04.069
14. Rusdianti, R., Azizah, A., Utarti, E., Wiyono, H.T., Muzakhar, K.: Cheap cellulase production by *Aspergillus* sp. VTM1 through solid state fermentation of coffee pulp waste. In: Key Engineering Materials, pp. 159–164. Trans Tech Publ (2021). https://doi.org/10.4028/www.scientific.net/KEM.884.159
15. Thakur, P., Mukherjee, G.: Utilization of agro-waste in pectinase production and its industrial applications. In: Recent Developments in Microbial Technologies, pp. 145–162. Springer, Berlin (2021)
16. Jacob, N., Prema, P.: Influence of mode of fermentation on production of polygalacturonase by a novel strain of *Streptomyces lydicus*. Food Technol. Biotechnol. **44** (2006)
17. Pandey, A., Selvakumar, P., Soccol, C.R., Nigam, P.: Solid state fermentation for the production of industrial enzymes. Curr. Sci. 149–162 (1999)
18. Kieliszek, M., Piwowarek, K., Kot, A.M., Pobiega, K.: The aspects of microbial biomass use in the utilization of selected waste from the agro-food industry. Open Life Sci. **15**, 787–796 (2020). https://doi.org/10.1515/biol-2020-0099
19. Celus, M., Kyomugasho, C., Van Loey, A.M., Grauwet, T., Hendrickx, M.E.: Influence of pectin structural properties on interactions with divalent cations and its associated functionalities. Compr. Rev. Food Sci. Food Saf. **17**, 1576–1594 (2018). https://doi.org/10.1111/1541-4337.12394
20. Jayani, R.S., Saxena, S., Gupta, R.: Microbial pectinolytic enzymes: a review. Process Biochem. **40**, 2931–2944 (2005). https://doi.org/10.1016/j.procbio.2005.03.026
21. Geetha, M., Saranraj, P., Mahalakshmi, S., Reetha, D.: Screening of pectinase producing bacteria and fungi for its pectinolytic activity using fruit wastes. Int. J. Biochem. Biotech. Sci. **1**, 30–42 (2012)
22. Hoondal, G., Tiwari, R., Tewari, R., Dahiya, N., Beg, Q.: Microbial alkaline pectinases and their industrial applications: a review. Appl. Microbiol. Biotechnol. **59**, 409–418 (2002). https://doi.org/10.1007/s00253-002-1061-1
23. Lara-Espinoza, C., Carvajal-Millán, E., Balandrán-Quintana, R., López-Franco, Y., Rascón-Chu, A.: Pectin and pectin-based composite materials: beyond food texture. Molecules **23**, 942 (2018). https://doi.org/10.3390/molecules23040942
24. Suhaimi, H., Dailin, D.J., Abd Malek, R., Hanapi, S.Z., Ambehabati, K.K., Keat, H.C., Prakasham, S., Elsayed, E.A., Misson, M., El Enshasy, H.: Fungal pectinases: production and applications in food industries. Fungi Sustain. Food Prod. 85–115 (2021)
25. Carpita, N.C., Gibeaut, D.M.: Structural models of primary cell walls in flowering plants: consistency of molecular structure with the physical properties of the walls during growth. Plant J. **3**, 1–30 (1993). https://doi.org/10.1111/j.1365-313x.1993.tb00007.x
26. Díaz, G.V., Coniglio, R.O., Alvarenga, A.E., Zapata, P.D., Villalba, L.L., Fonseca, M.I.: Secretomic analysis of cheap enzymatic cocktails of *Aspergillus niger* LBM 134 grown on cassava bagasse and sugarcane bagasse. Mycologia **112**, 663–676 (2020). https://doi.org/10.1080/00275514.2020.1763707
27. Baladhandayutham, S., Thangavelu, V.: Optimization and kinetics of solid-state fermentative production of pectinase by *Aspergillus awamori*. Int. J. ChemTech Res. **3**, 1758–1764 (2011)
28. Solis-Pereyra, S., Favela-Torres, E., Gutierrez-Rojas, M., Roussos, S., Saucedo-Castaneda, G., Gunasekaran, P., Viniegra-Gonzalez, G.: Production of pectinases by *Aspergillus niger* in solid state fermentation at high initial glucose concentrations. World J. Microbiol. Biotechnol. **12**, 257–260 (1996)
29. Martin, N., de Souza, S.R., da Silva, R., Gomes, E.: Pectinase production by fungal strains in solid-state fermentation using agro-industrial bioproduct. Brazilian Arch. Biol. Technol. **47**, 813–819 (2004)
30. Bai, Z.H., Zhang, H.X., Qi, H.Y., Peng, X.W., Li, B.J.: Pectinase production by *Aspergillus niger* using wastewater in solid state fermentation for eliciting plant disease resistance. Bioresour. Technol. **95**, 49–52 (2004). https://doi.org/10.1016/j.biortech.2003.06.006

31. Li, Z., Bai, Z., Zhang, B., Xie, H., Hu, Q., Hao, C., Xue, W., Zhang, H.: Newly isolated *Bacillus gibsonii* S-2 capable of using sugar beet pulp for alkaline pectinase production. World J. Microbiol. Biotechnol. **21**, 1483–1486 (2005)
32. Neagu, D.A., Destain, J., Thonart, P., Socaciu, C.: Effects of different carbon sources on pectinase production by *Penicillium oxalicum*. Bull. UASVM Agric. **69**, 327–333 (2012)
33. Thite, V.S., Nerurkar, A.S., Baxi, N.N.: Optimization of concurrent production of xylanolytic and pectinolytic enzymes by *Bacillus safensis* M35 and *Bacillus altitudinis* J208 using agro-industrial biomass through response surface methodology. Sci. Rep. **10**, 3824 (2020). https://doi.org/10.1038/s41598-020-60760-6
34. Abbasi, H., Mortazavipour, S.R., Setudeh, M.: Polygalacturonase (PG) production by fungal strains using agro-industrial bioproduct in solid state fermentation. Chem. Eng. Res. Bull. **15**, 1–5 (2011)
35. Heerd, D., Yegin, S., Tari, C., Fernandez-Lahore, M.: Pectinase enzyme-complex production by *Aspergillus* spp. in solid-state fermentation: a comparative study. Food Bioprod. Process. **90**, 102–110 (2012). https://doi.org/10.1016/j.fbp.2011.08.003
36. Martins, E.S., Silva, D., Da Silva, R., Gomes, E.: Solid state production of thermostable pectinases from thermophilic *Thermoascus aurantiacus*. Process Biochem. **37**, 949–954 (2002). https://doi.org/10.1016/S0032-9592(01)00300-4
37. Kashyap, D.R., Soni, S.K., Tewari, R.: Enhanced production of pectinase by *Bacillus* sp. DT7 using solid state fermentation. Bioresour. Technol. **88**, 251–254 (2003). https://doi.org/10.1016/S0960-8524(02)00206-7
38. Kuhad, R.C., Kapoor, M., Rustagi, R.: Enhanced production of an alkaline pectinase from *Streptomyces* sp. RCK-SC by whole-cell immobilization and solid-state cultivation. World J. Microbiol. Biotechnol. **20**, 257–263 (2004). https://doi.org/10.1023/B:WIBI.0000023833.15866.45
39. Silva, D., Tokuioshi, K., da Silva Martins, E., Da Silva, R., Gomes, E.: Production of pectinase by solid-state fermentation with *Penicillium viridicatum* RFC3. Process Biochem. **40**, 2885–2889 (2005). https://doi.org/10.1016/j.procbio.2005.01.008
40. Zhu, M., He, H., Fan, M., Ma, H., Ren, H., Zeng, J., Gao, H.: Application and optimization of solid-state fermentation process for enhancing polygalacturonase production by *Penicillium expansum*. Int. J. Agric. Biol. Eng. **11**, 187–194 (2018). https://doi.org/10.25165/j.ijabe.20181106.3673
41. Sharma, D.C., Satyanarayana, T.: Biotechnological potential of agro residues for economical production of thermoalkali-stable pectinase by *Bacillus pumilus* dcsr1 by solid-state fermentation and its efficacy in the treatment of ramie fibres. Enzyme Res. **2012** (2012). https://doi.org/10.1155/2012/281384
42. Faten, A.M., Abeer, A.A.E.: Enzyme activities of the marine-derived fungus *Alternaria alternata* cultivated on selected agricultural wastes. J. Appl. Biol. Sci. **7**, 39–46 (2013)
43. Chugh, P., Soni, R., Soni, S.K.: Deoiled rice bran: a substrate for co-production of a consortium of hydrolytic enzymes by *Aspergillus niger* P-19. Waste Biomass Valorizat. **7**, 513–525 (2016)
44. Irshad, M., Anwar, Z., Mahmood, Z., Aqil, T., Mehmmod, S., Nawaz, H.: Bio-processing of agro-industrial waste orange peel for induced production of pectinase by *Trichoderma viridi*; its purification and characterization. Turkish J. Biochem. Biyokim. Derg. **39**, (2014). https://doi.org/10.5505/tjb.2014.55707
45. Adeleke, A.J., Odunfa, S.A., Olanbiwonninu, A., Owoseni, M.C.: Production of cellulase and pectinase from orange peels by fungi. Nat. Sci. **10**, 107–112 (2012)
46. Hidayah, A.A., Azizah, Winarsa, R., Muzakhar, K.: Utilization of coffee pulp as a substrate for pectinase production by *Aspergillus* sp. VTMS through solid state fermentation. In: AIP Conference Proceedings, p. 020012. AIP Publishing LLC (2020)
47. Frómeta, R.A.R., Sánchez, J.L., García, J.M.R.: Evaluation of coffee pulp as substrate for polygalacturonase production in solid state fermentation. Emirates J. Food Agric. 117–124 (2020). https://doi.org/10.9755/ejfa.2020.v32.i2.2068
48. Haile, M., Kang, W.H.: Isolation, identification, and characterization of pectinolytic yeasts for starter culture in coffee fermentation. Microorganisms **7**, 401 (2019). https://doi.org/10.3390/microorganisms7100401

49. Avallone, S., Brillouet, J.M., Guyot, B., Olguin, E., Guiraud, J.P.: Involvement of pectolytic micro-organisms in coffee fermentation. Int. J. food Sci. Technol. **37**, 191–198 (2002)
50. Botella, C., Diaz, A., De Ory, I., Webb, C., Blandino, A.: Xylanase and pectinase production by *Aspergillus awamori* on grape pomace in solid state fermentation. Process Biochem. **42**, 98–101 (2007). https://doi.org/10.1016/j.procbio.2006.06.025
51. Arévalo-Villena, M., Fernández, M., López, J., Briones, A.: Pectinases yeast production using grape skin as carbon source. Adv. Biosci. Biotechnol. **2**, 89 (2011). https://doi.org/10.4236/abb.2011.22014
52. Castilho, L.R., Medronho, R.A., Alves, T.L.M.: Production and extraction of pectinases obtained by solid state fermentation of agroindustrial residues with *Aspergillus niger*. Bioresour. Technol. **71**, 45–50 (2000). https://doi.org/10.1016/S0960-8524(99)00058-9
53. Dhillon, S.S., Gill, R.K., Gill, S.S., Singh, M.: Studies on the utilization of citrus peel for pectinase production using fungus *Aspergillus niger*. Int. J. Environ. Stud. **61**, 199–210 (2004). https://doi.org/10.1080/0020723032000143346
54. Mehmood, T., Saman, T., Irfan, M., Anwar, F., Ikram, M.S., Tabassam, Q.: Pectinase production from *Schizophyllum commune* through central composite design using citrus waste and its immobilization for industrial exploitation. Waste Biomass Valorizat. **10**, 2527–2536 (2019). https://doi.org/10.1007/s12649-018-0279-9
55. Rodríguez-Fernández, D.E., Leon, J.A.R., Carvalho, J.C., Karp, S.G., Parada, J.L., Soccol, C.R.: Process development to recover pectinases produced by solid-state fermentation. J. Bioprocess. Biotech. **2** (2012). https://doi.org/10.4172/2155-9821.1000121
56. Sharma, N., Rathore, M., Sharma, M.: Microbial pectinase: sources, characterization and applications. Rev. Environ. Sci. Bio/Technol. **12**, 45–60 (2013). https://doi.org/10.1007/s11157-012-9276-9
57. Kashyap, D.R., Vohra, P.K., Chopra, S., Tewari, R.: Applications of pectinases in the commercial sector: a review. Bioresour. Technol. **77**, 215–227 (2001). https://doi.org/10.1016/S0960-8524(00)00118-8
58. Shiukashvili, V., Vephkhishvili, N., Khositashvili, M.: Quantitative analysis of soluble and insoluble forms of pectic substances in grapes in different phases of vegetation. In: E-Conference Globe, pp. 1–7 (2021)
59. Vaclavik, V.A., Christian, E.W.: Pectins and gums. In: Essentials of Food Science, pp. 53–61. Springer, Berlin (2014). https://doi.org/10.1007/978-1-4614-9138-5_5
60. Susanti, S., Legowo, A.M., Nurwantoro, N., Silviana, S., Arifan, F.: Comparing the chemical characteristics of pectin isolated from various Indonesian fruit peels. Indones. J. Chem. **21**, 1057–1062 (2021). https://doi.org/10.22146/ijc.59799
61. Rehman, H., Baloch, A.H., Nawaz, M.A.: Pectinase: immobilization and applications: a review. Trends Pept. Protein Sci. **6**, 1–16 (2021). https://doi.org/10.22037/tpps.v6i.33871
62. Valdés, A., Burgos, N., Jiménez, A., Garrigós, M.C.: Natural pectin polysaccharides as edible coatings. Coatings **5**, 865–886 (2015). https://doi.org/10.3390/coatings5040865
63. Tsuru, C., Umada, A., Noma, S., Demura, M., Hayashi, N.: Extraction of pectin from Satsuma mandarin orange peels by combining pressurized carbon dioxide and deionized water: a green chemistry method. Food Bioprocess Technol. 1–8 (2021). https://doi.org/10.1007/s11947-021-02644-9
64. Reddy, P.L., Sreeramulu, A.: Isolation, identification and screening of pectinolytic fungi from different soil samples of Chittoor district. Int. J. Life Sci. Biotechnol. Pharma Res. **1**, 186–193 (2012)
65. Pedrolli, D.B., Monteiro, A.C., Gomes, E., Carmona, E.C.: Pectin and pectinases: production, characterization and industrial application of microbial pectinolytic enzymes. Open Biotechnol. J. 9–18 (2009)
66. Yadav, S., Yadav, P.K., Yadav, D., Yadav, K.D.S.: Pectin lyase: a review. Process Biochem. **44**, 1–10 (2009). https://doi.org/10.1016/j.procbio.2008.09.012
67. Tai, C., Bouissil, S., Gantumur, E., Carranza, M.S., Yoshii, A., Sakai, S., Pierre, G., Michaud, P., Delattre, C.: Use of anionic polysaccharides in the development of 3D bioprinting technology. Appl. Sci. **9**, 2596 (2019). https://doi.org/10.3390/app9132596

68. Paniagua, C., Posé, S., Morris, V.J., Kirby, A.R., Quesada, M.A., Mercado, J.A.: Fruit softening and pectin disassembly: an overview of nanostructural pectin modifications assessed by atomic force microscopy. Ann. Bot. **114**, 1375–1383 (2014). https://doi.org/10.1093/aob/mcu149
69. Lang, C., Dörnenburg, H.: Perspectives in the biological function and the technological application of polygalacturonases. Appl. Microbiol. Biotechnol. **53**, 366–375 (2000). https://doi.org/10.1007/s002530051628
70. Kavuthodi, B., Sebastian, D.: Review on bacterial production of alkaline pectinase with special emphasis on *Bacillus* species. Biosci. Biotechnol. Res. Commun. **11**, 18–30 (2018). https://doi.org/10.21786/bbrc/11.1/4
71. Roy, K., Dey, S., Uddin, M., Barua, R., Hossain, M.: Extracellular pectinase from a novel bacterium *Chryseobacterium indologenes* strain SD and its application in fruit juice clarification. Enzyme Res. **2018** (2018). https://doi.org/10.1155/2018/3859752
72. Amin, F., Bhatti, H.N., Bilal, M.: Recent advances in the production strategies of microbial pectinases—a review. Int. J. Biol. Macromol. **122**, 1017–1026 (2019). https://doi.org/10.1016/j.ijbiomac.2018.09.048
73. Garg, G., Singh, A., Kaur, A., Singh, R., Kaur, J., Mahajan, R.: Microbial pectinases: an ecofriendly tool of nature for industries. 3 Biotech. **6**, 47 (2016). https://doi.org/10.1007/s13205-016-0371-4
74. KC, S., Upadhyaya, J., Joshi, D.R., Lekhak, B., Kumar Chaudhary, D., Raj Pant, B., Raj Bajgai, T., Dhital, R., Khanal, S., Koirala, N.: Production, characterization, and industrial application of pectinase enzyme isolated from fungal strains. Fermentation. **6**, 59 (2020). https://doi.org/10.3390/fermentation6020059
75. Chebli, Y., Kaneda, M., Zerzour, R., Geitmann, A.: The cell wall of the *Arabidopsis* pollen tube-spatial distribution, recycling, and network formation of polysaccharides. Plant Physiol. **160**, 1940–1955 (2012). https://doi.org/10.1104/pp.112.199729
76. Shen, Z., Reese, J.C., Reeck, G.R.: Purification and characterization of polygalacturonase from the rice weevil, *Sitophilus oryzae* (Coleoptera: Curculionidae). Insect Biochem. Mol. Biol. **26**, 427–433 (1996)
77. Bhardwaj, V., Garg, N.: Exploitation of micro-organisms for isolation and screening of pectinase from environment. In: Globelics 2010 8th International Conference, University of Malaya, Kuala Lumpur, Malaysia (2010)
78. Gummadi, S.N., Panda, T.: Purification and biochemical properties of microbial pectinases—a review. Process Biochem. **38**, 987–996 (2003). https://doi.org/10.1016/S0032-9592(02)00203-0
79. Satapathy, S., Rout, J.R., Kerry, R.G., Thatoi, H., Sahoo, S.L.: Biochemical prospects of various microbial pectinase and pectin: an approachable concept in pharmaceutical bioprocessing. Front. Nutr. **7**, 117 (2020). https://doi.org/10.3389/fnut.2020.00117
80. Shet, A.R., Desai, S.V., Achappa, S.: Pectinolytic enzymes: classification, production, purification and applications. Res. J. Life Sci. Bioinform. Pharm. Chem. Sci. **4**, 337–348 (2018). https://doi.org/10.26479/2018.0403.30
81. Suykerbuyk, M.E.G., Schaap, P.J., Stam, H., Musters, W., Visser, J.: Cloning, sequence and expression of the gene coding for rhamnogalacturonase of *Aspergillus aculeatus*; a novel pectinolytic enzyme. Appl. Microbiol. Biotechnol. **43**, 861–870 (1995). https://doi.org/10.1007/BF02431920
82. Chaudhri, A., Suneetha, V.: Microbially derived pectinases: a review. IOSR J Pharm Biol Sci. **2**, 1–5 (2012)
83. Singh, R., Kumar, M., Mittal, A., Mehta, P.K.: Microbial enzymes: industrial progress in 21st century. 3 Biotech. **6**, 174 (2016). https://doi.org/10.1007/s13205-016-0485-8
84. Anisa, S.K., Girish, K.: Pectinolytic activity of *Rhizopus* sp. and *Trichoderma viride*. Int. J. Res. Pure Appl. Microbiol. **4**, 28–31 (2014)
85. Underkofler, L.A., Barton, R.R., Rennert, S.S.: Production of microbial enzymes and their applications. Appl. Microbiol. **6**, 212 (1958)

86. Bharadwaj, P.S., Udupa, P.M.: Isolation, purification and characterization of pectinase enzyme from *Streptomyces thermocarboxydus*. J. Clin. Microbiol. Biochem. Technol. **5**, 1–6 (2019)
87. Raveendran, S., Parameswaran, B., Ummalyma, S.B., Abraham, A., Mathew, A.K., Madhavan, A., Rebello, S., Pandey, A.: Applications of microbial enzymes in food industry. Food Technol. Biotechnol. **56**, 16–30 (2018). https://doi.org/10.17113/ftb.56.01.18.5491
88. Boddy, L., Hiscox, J.: Fungal Ecology: Principles and mechanisms of colonization and competition by saprotrophic fungi. In: The Fungal Kingdom, pp. 293–308. Wiley (2017)
89. Mamma, D., Kourtoglou, E., Christakopoulos, P.: Fungal multienzyme production on industrial by-products of the citrus-processing industry. Bioresour. Technol. **99**, 2373–2383 (2008). https://doi.org/10.1016/j.biortech.2007.05.018
90. Doughari, J.H., Onyebarachi, G.C.: Production, purification and characterization of polygalacturonase from *Aspergillus flavus* grown on orange peel. Appl. Microbiol. Open Access. **4**, 1–7 (2019). https://doi.org/10.4172/2471-9315.1000155
91. Jacob, N.: Pectinolytic enzymes. In: Biotechnology for Agro-Industrial Residues Utilisation, pp. 383–396. Springer, Berlin (2009)
92. Schuster, E., Dunn-Coleman, N., Frisvad, J.C., Van Dijck, P.W.: On the safety of *Aspergillus niger*—a review. Appl. Microbiol. Biotechnol. **59**, 426–435 (2002). https://doi.org/10.1007/s00253-002-1032-6
93. Hölker, U., Höfer, M., Lenz, J.: Biotechnological advantages of laboratory-scale solid-state fermentation with fungi. Appl. Microbiol. Biotechnol. **64**, 175–186 (2004). https://doi.org/10.1007/s00253-003-1504-3
94. Bayoumi, R.A., Yassin, H.M., Swelim, M.A., Abdel-All, E.Z.: Production of bacterial pectinase (s) from agro-industrial wastes under solid state fermentation conditions. J. Appl. Sci. Res. **4**, 1708–1721 (2008)
95. Jayani, R.S., Shukla, S.K., Gupta, R.: Screening of bacterial strains for polygalacturonase activity: Its production by *Bacillus sphaericus* (MTCC 7542). Enzyme Res. **2010** (2010). https://doi.org/10.4061/2010/306785
96. Daskaya-Dikmen, C., Karbancioglu-Guler, F., Ozcelik, B.: Cold active pectinase, amylase and protease production by yeast isolates obtained from environmental samples. Extremophiles **22**, 599–606 (2018). https://doi.org/10.1007/s00792-018-1020-0
97. Moharib, S.A., El-Sayed, S.T., Jwanny, E.W.: Evaluation of enzymes produced from yeast. Food Nahrung **44**, 47–51 (2000). https://doi.org/10.1002/(SICI)1521-3803(20000101)44:1%3c47::AID-FOOD47%3e3.0.CO;2-K
98. Blanco, P., Sieiro, C., Villa, T.G.: Production of pectic enzymes in yeasts. FEMS Microbiol. Lett. **175**, 1–9 (1999). https://doi.org/10.1111/j.1574-6968.1999.tb13595.x
99. Karaoğlan, M., Erden-Karaoğlan, F.: Effect of codon optimization and promoter choice on recombinant endo-polygalacturonase production in *Pichia pastoris*. Enzyme Microb. Technol. **139**, 109589 (2020). https://doi.org/10.1016/j.enzmictec.2020.109589
100. Teixeira, J.A., Gonçalves, D.B., De Queiroz, M.V., De Araújo, E.F.: Improved pectinase production in *Penicillium griseoroseum* recombinant strains. J. Appl. Microbiol. **111**, 818–825 (2011). https://doi.org/10.1111/j.1365-2672.2011.05099.x
101. Esquivel, J.C.C., Voget, C.E.: Purification and partial characterization of an acidic polygalacturonase from *Aspergillus kawachii*. J. Biotechnol. **110**, 21–28 (2004). https://doi.org/10.1016/j.jbiotec.2004.01.010
102. Schwan, R.F., Cooper, R.M., Wheals, A.E.: Endopolygalacturonase secretion by *Kluyveromyces marxianus* and other cocoa pulp-degrading yeasts. Enzyme Microb. Technol. **21**, 234–244 (1997). https://doi.org/10.1016/S0141-0229(96)00261-X
103. Whitehead, M.P., Shieh, M.T., Cleveland, T.E., Cary, J.W., Dean, R.A.: Isolation and characterization of polygalacturonase genes (*pecA* and *pecB*) from *Aspergillus flavus*. Appl. Environ. Microbiol. **61**, 3316 (1995). https://doi.org/10.1128/aem.61.9.3316-3322.1995
104. Bussink, H.J.D., Buxton, F.P., Fraaye, B.A., de Graaff, L.H., Visser, J.: The polygalacturonases of *Aspergillus niger* are encoded by a family of diverged genes. Eur. J. Biochem. **208**, 83–90 (1992). https://doi.org/10.1111/j.1432-1033.1992.tb17161.x

105. Mohamed, S.A., Farid, N.M., Hossiny, E.N., Bassuiny, R.I.: Biochemical characterization of an extracellular polygalacturonase from *Trichoderma harzianum*. J. Biotechnol. **127**, 54–64 (2006). https://doi.org/10.1016/j.jbiotec.2006.06.009
106. Gainvors, A., Frezier, V., Lemaresquier, H., Lequart, C., Aigle, M., Belarbi, A.: Detection of polygalacturonase, pectin-lyase and pectin-esterase activities in a *Saccharomyces cerevisiae* strain. Yeast **10**, 1311–1319 (1994). https://doi.org/10.1002/yea.320101008
107. Martos, M.A., Zubreski, E.R., Garro, O.A., Hours, R.A.: Production of Pectinolytic enzymes by the yeast *Wickerhanomyces anomalus* isolated from citrus fruits peels. Biotechnol. Res. Int. **2013** (2013). https://doi.org/10.1155/2013/435154
108. Maisuria, V.B., Patel, V.A., Nerurkar, A.S.: Biochemical and thermal stabilization parameters of polygalacturonase from *Erwinia carotovora* subsp. carotovora BR1. J. Microbiol. Biotechnol. **20**, 1077–1085 (2010). https://doi.org/10.4014/jmb.0908.08008
109. Alaña, A., Gabilondo, A., Hernando, F., Moragues, M.D., Dominguez, J.B., Llama, M.J., Serra, J.L.: Pectin lyase production by a *Penicillium italicum* strain. Appl. Environ. Microbiol. **55**, 1612 (1989). https://doi.org/10.1128/aem.55.6.1612-1616.1989
110. Zhang, Z., Dong, J., Zhang, D., Wang, J., Qin, X., Liu, B., Xu, X., Zhang, W., Zhang, Y.: Expression and characterization of a pectin methylesterase from *Aspergillus niger* ZJ5 and its application in fruit processing. J. Biosci. Bioeng. **126**, 690–696 (2018). https://doi.org/10.1016/j.jbiosc.2018.05.022
111. Lemaire, A., Garzon, C.D., Perrin, A., Habrylo, O., Trezel, P., Bassard, S., Lefebvre, V., Van Wuytswinkel, O., Guillaume, A., Pau-Roblot, C.: Three novel rhamnogalacturonan I-pectins degrading enzymes from *Aspergillus aculeatinus*: Biochemical characterization and application potential. Carbohydr. Polym. **248**, 116752 (2020). https://doi.org/10.1016/j.carbpol.2020.116752
112. Rahman, M., Choi, Y.S., Kim, Y.K., Park, C., Yoo, J.C.: Production of novel polygalacturonase from *Bacillus paralicheniformis* CBS32 and application to depolymerization of ramie fiber. Polymers (Basel). **11**, 1525 (2019). https://doi.org/10.3390/polym11091525
113. Andrade, M.V.V. de, Delatorre, A.B., Ladeira, S.A., Martins, M.L.L.: Production and partial characterization of alkaline polygalacturonase secreted by thermophilic *Bacillus* sp. SMIA-2 under submerged culture using pectin and corn steep liquor. Food Sci. Technol. **31**, 204–208 (2011). https://doi.org/10.1590/S0101-20612011000100031
114. Zhou, C., Xue, Y., Ma, Y.: Cloning, evaluation, and high-level expression of a thermo-alkaline pectate lyase from alkaliphilic *Bacillus clausii* with potential in ramie degumming. Appl. Microbiol. Biotechnol. **101**, 3663–3676 (2017). https://doi.org/10.1007/s00253-017-8110-2
115. Bekli, S., Aktas, B., Gencer, D., Aslim, B.: Biochemical and molecular characterizations of a novel pH-and temperature-stable pectate lyase from *Bacillus amyloliquefaciens* S6 for industrial application. Mol. Biotechnol. **61**, 681–693 (2019). https://doi.org/10.1007/s12033-019-00194-2
116. Yuan, P., Meng, K., Wang, Y., Luo, H., Shi, P., Huang, H., Tu, T., Yang, P., Yao, B.: A low-temperature-active alkaline pectate lyase from *Xanthomonas campestris* ACCC 10048 with high activity over a wide pH range. Appl. Biochem. Biotechnol. **168**, 1489–1500 (2012). https://doi.org/10.1007/s12010-012-9872-8
117. Cheng, L., Duan, S., Zheng, K., Feng, X., Yang, Q., Liu, Z., Liu, Z., Peng, Y.: An alkaline pectate lyase D from *Dickeya dadantii* DCE-01: clone, expression, characterization, and potential application in ramie bio-degumming. Text. Res. J. **89**, 2075–2083 (2019). https://doi.org/10.1177/0040517518790971
118. Liao, C.H., McCallus, D.E., Wells, J.M.: Calcium-dependent pectate lyase production in the soft-rotting bacterium *Pseudomonas fluorescens*. Phytopathology **83**, 813–824 (1993)
119. Arbige, M.V., Shetty, J.K., Chotani, G.K.: Industrial enzymology: the next chapter. Trends Biotechnol. **37**, 1355–1366 (2019). https://doi.org/10.1016/j.tibtech.2019.09.010
120. Tarafdar, A., Sirohi, R., Gaur, V.K., Kumar, S., Sharma, P., Varjani, S., Pandey, H.O., Sindhu, R., Madhavan, A., Rajasekharan, R.: Engineering interventions in enzyme production: Lab to industrial scale. Bioresour. Technol. **124771** (2021). https://doi.org/10.1016/j.biortech.2021.124771

121. Ramesh, A., Devi, P.H., Chattopadhyay, S., Kavitha, M.: Commercial applications of microbial enzymes. In: Microbial Enzymes: Roles and Applications in Industries, pp. 137–184. Springer, Berlin (2020)
122. Taragano, V.M., Pilosof, A.M.R.: Application of Doehlert designs for water activity, pH, and fermentation time optimization for *Aspergillus niger* pectinolytic activities production in solid-state and submerged fermentation. Enzyme Microb. Technol. **25**, 411–419 (1999)
123. Mandalari, G., Bennett, R.N., Bisignano, G., Trombetta, D., Saija, A., Faulds, C.B., Gasson, M.J., Narbad, A.: Antimicrobial activity of flavonoids extracted from bergamot (*Citrus bergamia* Risso) peel, a byproduct of the essential oil industry. J. Appl. Microbiol. **103**, 2056–2064 (2007). https://doi.org/10.1111/j.1365-2672.2007.03456.x
124. Zema, D.A., Calabrò, P.S., Folino, A., Tamburino, V., Zappia, G., Zimbone, S.M.: Valorisation of citrus processing waste: a review. Waste Manag. **80**, 252–273 (2018). https://doi.org/10.1016/j.wasman.2018.09.024
125. Kozłowski, R.M., Różańska, W.: Enzymatic treatment of natural fibres. In: Handbook of Natural Fibres, pp. 227–244. Elsevier, Amsterdam (2020)
126. Pasha, K.M., Anuradha, P., Subbarao, D.: Applications of pectinases in industrial sector. Int. J. pure Appl. Sci. Technol. **16**, 89 (2013)
127. Brando, C.H.J., Brando, M.F.: Methods of coffee fermentation and drying. Cocoa coffee Ferment, pp. 367–396 (2014)
128. Costa, J.A. V, Treichel, H., Kumar, V., Pandey, A.: Advances in solid-state fermentation (Chap. 1). In: Pandey, A., Larroche, C., Soccol, C.R. (eds.) Current Developments in Biotechnology and Bioengineering, pp. 1–17. Elsevier, Amsterdam (2018)
129. Srivastava, N., Srivastava, M., Ramteke, P.W., Mishra, P.K.: Chapter 23 - Solid-state fermentation strategy for microbial metabolites production: an overview. In: Gupta, V.K., Pandey, A. (eds.) New and Future Developments in Microbial Biotechnology and Bioengineering, pp. 345–354. Elsevier, Amsterdam (2019)
130. Pandey, A.: Solid-state fermentation. Biochem. Eng. J. **13**, 81–84 (2003). https://doi.org/10.1016/S1369-703X(02)00121-3
131. López-Gómez, J.P., Manan, M.A., Webb, C.: Solid-state fermentation of food industry wastes (Chap. 7). In: Kosseva, M.R., Webb, C. (eds.) Food Industry Wastes, 2nd edn, pp. 135–161. Academic Press, Cambridge (2020)
132. de Castro, R.J.S., Sato, H.H.: Enzyme production by solid state fermentation: general aspects and an analysis of the physicochemical characteristics of substrates for agro-industrial wastes valorization. Waste Biomass Valorizat. **6**, 1085–1093 (2015). https://doi.org/10.1007/s12649-015-9396-x
133. Couto, S.R., Sanromán, M.A.: Application of solid-state fermentation to food industry—a review. J. Food Eng. **76**, 291–302 (2006). https://doi.org/10.1016/j.jfoodeng.2005.05.022
134. Irfan, M., Nadeem, M., Syed, Q.: One-factor-at-a-time (OFAT) optimization of xylanase production from *Trichoderma viride*-IR05 in solid-state fermentation. J. Radiat. Res. Appl. Sci. **7**, 317–326 (2014). https://doi.org/10.1016/j.jrras.2014.04.004
135. Huerta, S., Favela, E., Lopez-Ulibarri, R., Fonseca, A., Viniegra-Gonzalez, G., Gutierrez-Rojas, M.: Absorbed substrate fermentation for pectinase production with *Aspergillus niger*. Biotechnol. Tech. **8**, 837–842 (1994)
136. Panchami, P.S., Gunasekaran, S.: Extraction and characterization of pectin from fruit waste. Int. J. Curr. Microbiol. Appl. Sci. **6**, 943–948 (2017). https://doi.org/10.20546/ijcmas.2017.602.116
137. Waites, M.J., Morgan, N.L., Rockey, J.S., Higton, G.: Industrial Microbiology: An Introduction. Wiley, Hoboken (2009)
138. Thomas, L., Larroche, C., Pandey, A.: Current developments in solid-state fermentation. Biochem. Eng. J. **81**, 146–161 (2013). https://doi.org/10.1016/j.bej.2013.10.013
139. Oumer, O.J., Abate, D.: Screening and molecular identification of pectinase producing microbes from coffee pulp. Biomed. Res. Int. **2018**, 2961767 (2018). https://doi.org/10.1155/2018/2961767

140. Karthik, L., Kumar, G., Rao, K.V.B.: Screening of pectinase producing microorganisms from agricultural waste dump soil. Asian J. Biochem. Pharm. Res. **1**(2), 329–337 (2011)
141. Sandhya, R., Kurup, G.: Screening and isolation of pectinase from fruit and vegetable wastes and the use of orange waste as a substrate for pectinase production. Int. Res. J. Biol. Sci. **2**, 34–39 (2013)
142. Manan, M.A., Webb, C.: Design aspects of solid state fermentation as applied to microbial bioprocessing. J. Appl. Biotechnol. Bioeng. **4**, 91 (2017). https://doi.org/10.15406/jabb.2017.04.00094
143. Ali, H.K.Q., Zulkali, M.M.D.: Design aspects of bioreactors for solid-state fermentation: a review. Chem. Biochem. Eng. Q. **25**, 255–266 (2011)
144. Manpreet, S., Sawraj, S., Sachin, D., Pankaj, S., Banerjee, U.C.: Influence of process parameters on the production of metabolites in solid-state fermentation. Malays. J. Microbiol. **2**, 1–9 (2005)
145. Siddiqui, M., Pande, V., Arif, M.: Production, purification, and characterization of polygalacturonase from *Rhizomucor pusillus* isolated from decomposting orange peels. Enzyme Res. **2012** (2012). https://doi.org/10.1155/2012/138634
146. Roopesh, K., Ramachandran, S., Nampoothiri, K.M., Szakacs, G., Pandey, A.: Comparison of phytase production on wheat bran and oilcakes in solid-state fermentation by *Mucor racemosus*. Bioresour. Technol. **97**, 506–511 (2006). https://doi.org/10.1016/j.biortech.2005.02.046
147. Erdal, S., Taskin, M.: Production of alpha-amylase by *Penicillium expansum* MT-1 in solid-state fermentation using waste Loquat (*Eriobotrya japonica* Lindley) kernels as substrate. Rom. Biotechnol. Lett. **15**, 5342–5350 (2010)
148. Mitchell, D.A., Krieger, N., Berovič, M.: Group I bioreactors: unaerated and unmixed. In: Solid-State Fermentation Bioreactors, pp. 65–76. Springer, Berlin (2006)
149. Nava, I., Favela-Torres, E., Saucedo-Castañeda, G.: Effect of mixing on the solid-state fermentation of coffee pulp with *Aspergillus tamarii*. Food Technol. Biotechnol. **49**, 391 (2011)
150. Valdez, A.L., Babot, J.D., Schmid, J., Delgado, O.D., Fariña, J.I.: Scleroglucan production by *Sclerotium rolfsii* ATCC 201126 from amylaceous and sugarcane molasses-based media: promising insights for sustainable and ecofriendly scaling-up. J. Polym. Environ. **27**, 2804–2818 (2019). https://doi.org/10.1007/s10924-019-01546-4
151. Mamo, J., Kangwa, M., Fernandez-Lahore, H.M., Assefa, F.: Optimization of media composition and growth conditions for production of milk-clotting protease (MCP) from *Aspergillus oryzae* DRDFS13 under solid-state fermentation. Brazilian J. Microbiol. 1–14 (2020). https://doi.org/10.1007/s42770-020-00243-y
152. Mazutti, M., Ceni, G., Di Luccio, M., Treichel, H.: Production of inulinase by solid-state fermentation: effect of process parameters on production and preliminary characterization of enzyme preparations. Bioprocess Biosyst. Eng. **30**, 297–304 (2007). https://doi.org/10.1007/s00449-006-0096-6

# Zero-Waste Concept in the Seafood Industry: Enzymatic Hydrolysis Perspective

**Siti Balqis Zulfigar and Anis Najiha Ahmad**

**Abstract** The Zero-waste concept in the fish and seafood industry is consistent with the Sustainable Development Goal (SDG) 14, which aims to conserve and sustainably use the oceans, seas and marine resources for sustainable development. However, approximately 70% of raw materials in the fish and seafood industries are discarded and go underutilised. The increasing demand for processed fish products is expected to induce larger amounts of by-products in the near future. Fish by-products harbour valuable components like protein and essential amino acids which are potentially converted into high-value products achieved through enzymatic treatment of the by-products.

**Keywords** Seafood · Zero waste · Enzymatic · Amino acids

## 1 Introduction

### *1.1 Zero-Waste Policy*

Consumerism and a consumption culture have resulted in a demand for mass production, which necessitates significant energy and material inputs. This in turn generates a large amount of unwanted material, substance, or by-product—waste [1, 2]. There is, unfortunately, a disproportionate amount of waste generated daily compared to the waste treated or recycled. For example, the World Bank estimates that about 2.01 billion tonnes of municipal solid waste (MSW) are generated annually worldwide. However, only about a quarter of this waste is recycled or composted. Up to 40%

S. B. Zulfigar (✉)
Bioprocess Technology Division, School of Industrial Technology, Universiti Sains Malaysia, 11800 Gelugor, Penang, Malaysia
e-mail: balqiszulfigar@usm.my

A. N. Ahmad
International Institute for Halal Research and Training (INHART), International Islamic University Malaysia (IIUM), Level3, KICT Building, Gombak, Selangor, Malaysia
e-mail: anisnajiha@iium.edu.my



A. Z. Yaser et al. (eds.), *Waste Management, Processing and Valorisation*,
https://doi.org/10.1007/978-981-16-7653-6_11

of the waste generated is either dumped or openly burned. Waste continues to be generated in an exponential way. By 2050, about 3.4 billion tonnes of MSW are expected to be generated annually. This amount is about a 70% increment from the current percentage of waste disposed.

Besides municipal waste, a significant portion of waste is also generated by industrial, agricultural, commercial, construction activities [3]. In China alone, 3.3 billion tons of industrial waste are produced each year [4]. Industrial waste encompasses a wide range of materials produced from composite (and possibly hazardous) materials with varying environmental toxicity [5]. The complexity of different waste materials and streams (e.g. municipal and industrial) makes appropriate and sustainable waste management challenging. Inappropriate waste disposal could pollute the environment, causing a huge waste of land resources and disturbing the regional ecosystem [4]. The massive amount of waste that is accumulating in nature has the potential to deplete natural resources such as drinking water sources, agricultural soil and even the air. [3, 6, 7]. This puts huge pressure on the authorities to manage waste in a more holistic and sustainable manner [8, 9] to prevent catastrophic social and environmental consequences.

Nowadays, the zero-waste concept has been implemented worldwide, from cities (Los Angeles, San Francisco, Austin), states (e.g. Nova Scotia, California, Adelaide and Vancouver) to countries (e.g. South Africa, New Zealand, China, India) and provinces or states [10, 11] and cities. These different entities have also devised different zero-waste definitions to suit their waste management goals and strategies. For example, in the city of Los Angeles, zero waste is defined as "maximising diversion from landfills and reducing waste at the source, with the ultimate goal of striving for more sustainable solid waste management practices". The city aims to achieve zero waste via radical changes in three areas: (1) product creation (manufacturing and packaging), (2) product use (use of sustainable, recycled and recyclable products) and (3) product disposal (resource recovery or landfilling) [12]. Different definitions of zero waste by different entities are compiled in Table 1.

The zero-waste movement has emerged as a response to the escalating waste problems worldwide [18]. This term was first coined in the 1970s by Paul Palmer, a chemist from Yale University [5]. The concept of zero waste, in simple words, refers to "sending nothing to landfill or incineration". The Zero-Waste International Alliance [14] defines zero waste as "the conservation of all resources through responsible production, consumption, reuse and recovery of products, packaging and materials without burning and with no discharges to land, water, or air that threaten the environment or human health." This definition reflects a few fundamentals of zero waste, which are about waste prevention through sustainable design and consumption practices and optimum resource recovery from waste [19]. Zero waste firmly rejects managing waste by incineration or landfills [20]. It represents a paradigm shift, away from the traditional model that regards waste as a norm. It integrates ideas, principles and systems in which everything has its own use. In contrast to the concept of burning waste, the concept not only saves money, protects the environment and public health, but it may also be used to create jobs [21].

**Table 1** Different definitions of zero waste adopted by different entities

| Definition | Adopted by | Source |
|---|---|---|
| Zero waste is a holistic approach to addressing the problem of unsustainable resource flows. Zero waste encompasses waste eliminated at the source through product design and producer responsibility and waste reduction strategies further down the supply chain, such as recycling, reuse and composting | Guam | [13] |
| Efforts to reduce Solid Waste generation waste to nothing, or as close to nothing as possible, by minimising excess consumption and maximising the recovery of Solid Wastes through Recycling and Composting | Solid Waste Association of North America (SWANA) | [13] |
| Zero waste is a philosophy and a design principle for the 21st Century. It includes 'recycling' but goes beyond recycling by taking a 'whole system' approach to the vast flow of resources and waste through human society | Connecticut | [13] |
| "Zero waste is the conservation of all resources by means of responsible production, consumption, reuse and recovery of Journal Pre-proof three products, packaging and materials without burning and with no discharges to land, water, or air that threaten the environment or human health" | Arkadelphia, AR; Austin, TX; Burbank, CA; Glendale, CA; Oakland, CA; Oceanside, CA; Palo Alto, CA and Telluride, CO | [14] |
| "… a new goal that seeks to redesign the way that resources and materials flow through society, taking a 'whole system' approach. It is both an 'end of the pipe' solution that maximises recycling and waste minimisation and a design principle that ensures that products are made to be reused, repaired or recycled back into nature or the marketplace. Zero-waste envisions the complete redesign of the industrial system so that we no longer view nature as an endless supply of materials" | Zero-Waste New Zealand Trust | [15] |
| "Sending nothing to landfill or incineration" | San Francisco | [16] |
| "A simple way of encapsulating the aim to go as far as possible in reducing the environmental impact of waste" | United Kingdom | [17] |

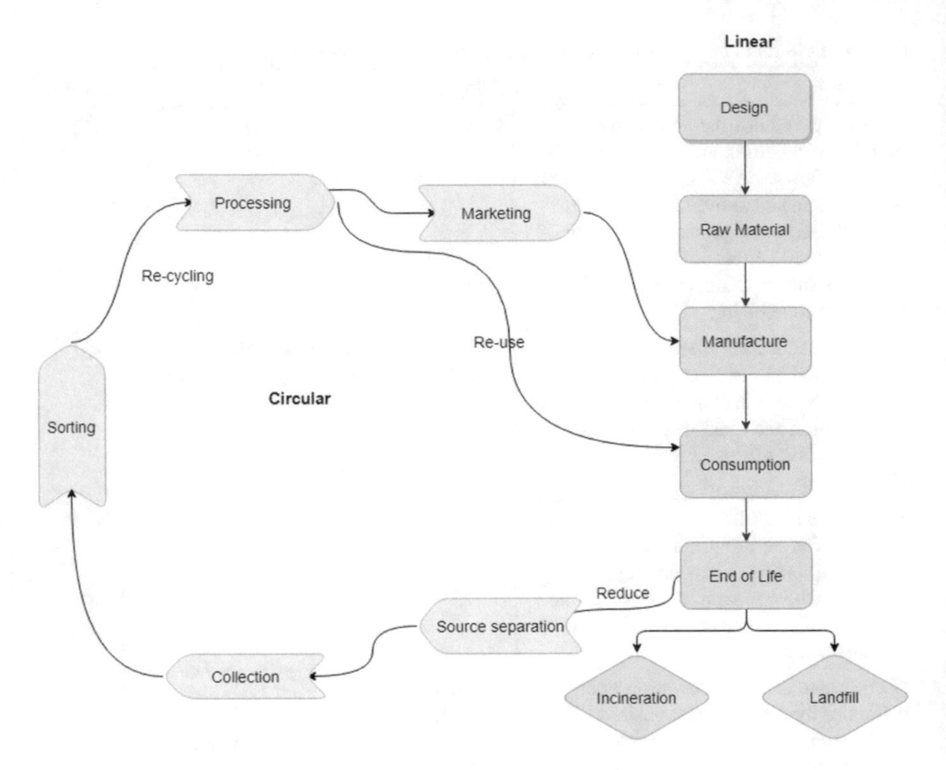


**Fig. 1** Linear and circular resource flow [3]

The zero-waste concept is often linked with the circular economy. Conventionally, waste management is shaped to serve a linear economy in which products are merely made, used and later disposed [21]. To achieve zero waste, waste management must move from this linear system towards a cyclical system. In a circular economy, material resources will be recirculated and reused for new product development, thus eliminating production waste to the extent possible. Figure 1 shows the different approaches to linear and cyclical resource flows.

The integration of the zero-waste concept into the fish and seafood industry is coherent with the Sustainable Development Goal (SDG) 14, which aims to conserve and sustainably use the oceans, seas and marine resources for sustainable development [22]. The value of fish and seafood produced by marine and aquaculture production is estimated to be around USD 401 billion in 2018 and this figure is expected to rise further as global fish consumption rises [23]. Out of this value, up to 35% of the global harvest is lost throughout the chain before reaching the final consumer [23]. This huge percentage of biomass is either discarded or incinerated. Seafood processing industries generate high percentages of by-products that harbour valuable components potentially reaped in the sector of the bioeconomy.

### 1.2 Policies on Seafood Waste Management

Marine animal processing facilities generate effluents loaded with organic compounds with high biological oxygen demand (BOD), chemical oxygen demand (COD), total suspended solids (TSS) and various microflora which pose a serious threat to the environment [24]. Due to the high organic content of animal by-products, their disposal is governed by tight regulations by the authorities to minimise the impact on the environment and on human health. For example, seafood waste treatment and by-product utilization in the EU are strictly regulated under the Regulations of European Communities Disposal, Processing and Placing on the Market and Regulation EC No 1069/2009, which sets the guidelines for the disposal and utilization of animal by-products which are not intended for human consumption [25]. This regulation lays out the categories of animal by-products and derivatives according to the level of risk to human health as depicted in Table 2. The methods of disposal and the utilization of the materials must strictly adhere to the guidelines of the respective groups. According to the classifications, aquatic animals and their by-products are low risk materials and are classified as Category 3. These materials have human applications (cosmetic and medical devices), animal applications (fishmeal, animal feed, pet food) and fertiliser applications.

## 2 Seafood Processing Industry

### 2.1 Global Outlook on the Seafood Industry

Fisheries and seafood commodities account for approximately 17% of global animal protein consumption, with total fish and seafood supply expected to reach 200 million tonnes by 2030 [26]. In 2018, this sector contributed 11% of total agricultural export value, totalling USD 164 billion [23]. Out of this proportion, 88 percent of the supplies were utilised for human consumption while the remaining were used for non-food purposes [26] as depicted in Fig. 2. Although fresh or chilled fish represents the largest portion of seafood commodities (44%), the current trend has witnessed a surge in the trade of processed fish products, driven by demand from consumers, modern technological capabilities with enhanced supplier–consumer linkages [26]. The fish processing industry has also diversified into the production of high-value products, primarily in developed countries.

The processing of fish and seafood is mainly aimed to preserve the quality of this highly perishable food. Fish processing refers to mechanical or chemical activities to transform the raw materials into desired products or as a means of preservation method (Food waste and loss website). Primary fish processing involves activities like deheading, fin trimming, scaling, degutting, filleting and skinning. The edible portion of the primary processing will either be distributed as fillets, steak, loins or undergo secondary processing such as canning, curing, prepared as ready-to-eat

**Table 2** Categories and methods of disposal of animal by-products under the Regulation of the European Council (EC) No 1069/2009

| Category | Materials | Disposal recommendation |
|---|---|---|
| 1 (high risk) | • Carcasses and all body parts of animals are suspected of being infected with transmissible spongiform encephalopathy (TSE)<br>• Carcasses of wild animals are suspected of being infected with a disease that humans or animals could contract<br>• Carcasses of animals are used in experiments<br>• Parts of animals that are contaminated due to illegal treatments<br>• International catering waste<br>• Carcasses and body parts from zoo and circus animals or pets<br>• Specified risk materials (body parts that pose a particular disease risk, e. g, cows' spinal cords) | • Incineration or co-incineration at an approved plant<br>• Pressure sterilisation (apart from possible TSE cases or animals killed under TSE eradication laws) followed by permanent marking, then landfill<br>• Using them as fuel for combustion at an approved combustion plant<br>• Sending them for burial at an authorised landfill, if they are international catering waste |

(continued)

**Table 2** (continued)

| Category | Materials | Disposal recommendation |
| --- | --- | --- |
| 2 (high risk) | • Animals rejected from abattoirs due to having infectious diseases<br>• Carcasses containing residues from authorised treatments<br>• Unhatched poultry that has died in its shell<br>• Carcasses of animals killed for disease control purposes<br>• Carcasses of dead livestock<br>• Manure<br>• Digestive tract content | • Incinerating or co-incinerating without processing or with prior processing, when the resulting material has to be marked with glyceroltriheptanoate (GTH)<br>• Sending them to an authorised landfill after processing by pressure sterilisation and marking with GTH<br>• Making them into organic fertilisers/soil improvers, after processing and marking with GTH<br>• Composting or anaerobic digestion after processing by pressure sterilisation and marking with GTH (milk, milk products, eggs, egg products, digestive tract content, manure do not need processing, providing any risk of spreading serious transmissible disease)<br>• In the case of manure, digestive tract content, milk, milk products and colostrum, they can be applied to land without being processed<br>• Using them in composting or anaerobic digestion, if they are materials coming from aquatic animals ensiled<br>• Using them as fuel for combustion<br>• Using them for the manufacture of certain cosmetic products, medical devices and for safe industrial or technical uses |

(continued)

**Table 2** (continued)

| Category | Materials | Disposal recommendation |
|---|---|---|
| 3 (low risk) | • Carcasses or body parts are not fit for humans to eat, at a slaughterhouse<br>• Products or foods of animal origin are originally meant for human consumption but are withdrawn for commercial reasons, not because they are unfit to eat<br>• Domestic catering waste<br>• Shells from shellfish with soft tissue<br>• Eggs, egg by-products, hatchery by-products and eggshells<br>• *Aquatic animals, aquatic and terrestrial invertebrates*<br>• *Aquatic animal by-products are derived from establishments or plants that produce goods for human consumption*<br>• Hides and skins from slaughterhouses<br>• Animal hides, skins, hooves, feathers, wool, horns and hair that had no signs of infectious disease at death<br>• Processed animal proteins (PAP) | • Incineration to co-incineration<br>• Sending them to a landfill after they have been processed<br>• Processing them, if they're not decomposed or spoiled and using them to make feed for farm animals (where allowed by the TSE/animal by-product regulations)<br>• Processing them and using them to make pet food<br>• Processing them and using them to make organic fertilisers and soil improvers<br>• Using them in composting or anaerobic digestion<br>• Ensiling (turning them into silage) if they come from aquatic animals<br>• Applying them to land as a fertiliser, in some cases<br>• Using them as fuel for combustion<br>• Using them to make cosmetic products or medical devices |

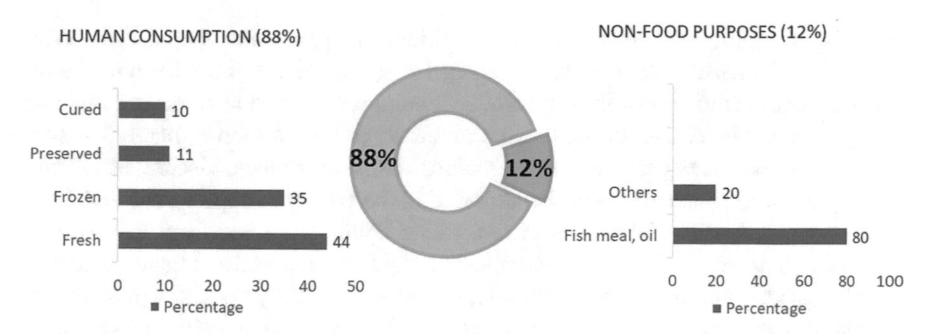


**Fig. 2** Global fish product utilization (%). *Source* The State of World Fisheries and Aquaculture, Sustainability in Action [26]

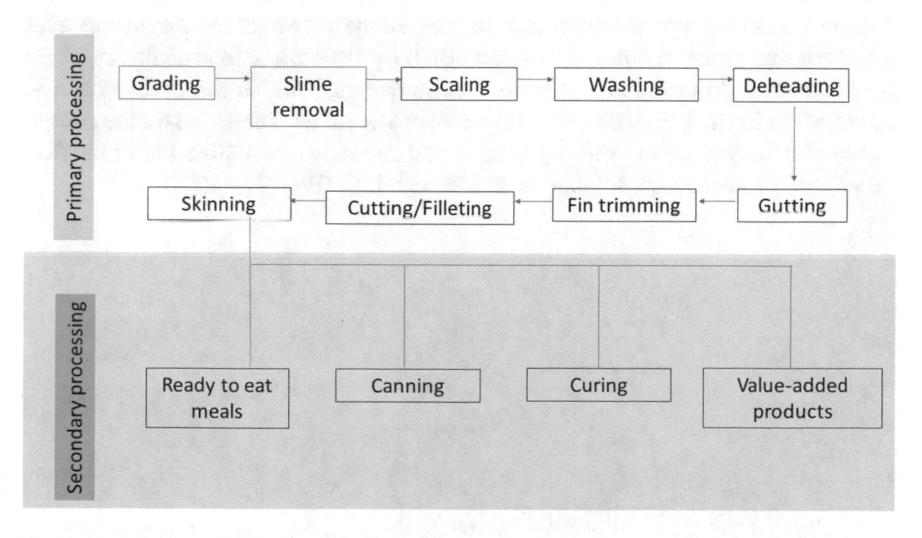


**Fig. 3** Flow process of fish processing activities

meals or turned into value-added products. Based on the report of World Fisheries and Aquaculture (FAO, 2020), the most popular fish caught are anchoveta (*Engraulis ringens*), allaska pollock (*Theraga chalcogramma*) and skipjack tuna (*Katsuwonus pelamis*). Figure 3 depicts the process flow in a typical fish processing facility.

## 3 Processing of Fish and Seafood By-Products

Activities like gutting, washing, scaling, de-boning and filleting in the fish processing industry, however, generate a high portion of by-products composed of inedible components like fish heads, bones, scales, viscera that makeup to 70% of the total wet weight, commonly discarded and considered as processing waste [26]. With the trend of increasing demand for processed fish, especially in the form of ready-to-eat meals, the number of by-products generated is expected to surge and must be tackled effectively. The Australian seafood industry produces approximately 100,000 tonnes of by-products per year, which include fish skin, frames, viscera, heads and roe [27]. A similar scenario was also reported in the UK, with approximately 12.7% or approximately 133,000 tonnes of fish waste being generated from the seafood processing industry [23]. Being perishable in nature, the disposal and management of these by-products comes with a cost and is governed by tightly regulated procedures. In Australia, the removal of the by-products from the processing facilities to landfills could cost up to AUD 150/t and increase the financial burden of processing facilities [27].

Apart from being reduced into low-valued products such as fish meal, fertilisers or animal feed, recent developments have witnessed the growth of interest in the development of high-valued commodities from fish by-products like food additives, bioactive peptides and medicinally important components. This, in turn, will maximise profit generation and lead to better utilization of resources. The approximate compositions of fish by-products vary by species and can range from 80% protein to 20% fat, with some species containing up to 25% ash [28] (Fig. 4).

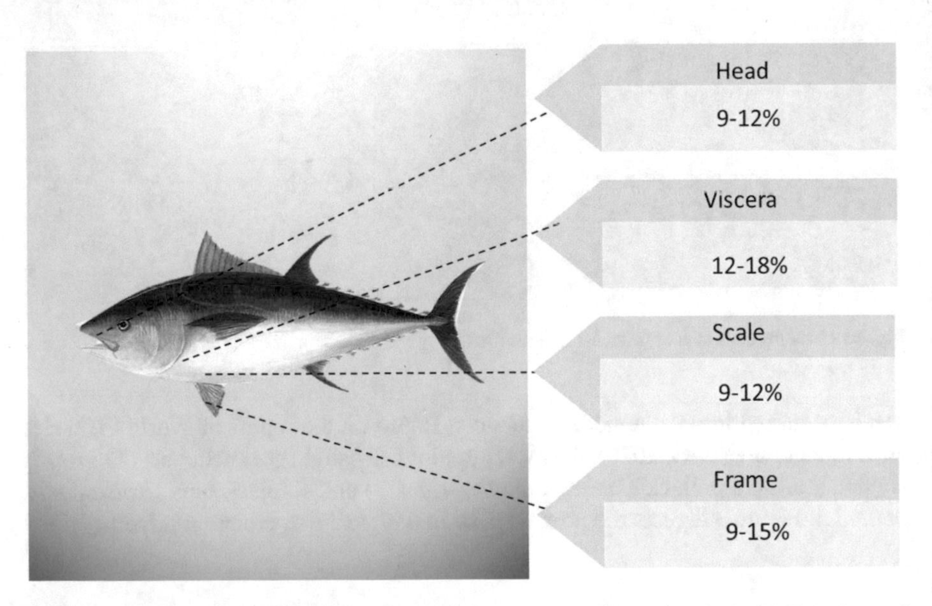


**Fig. 4** Average wet weight (%) of fish by-products. *Source* The State of World Fisheries and Aquaculture, Sustainability in action (2020) [26]

### 3.1 Fish By-Product Hydrolysate

Recent advances have accelerated research into the bioconversion of fish by-products via the hydrolysis method. Protein hydrolysate can be produced through several means, which include chemical (acid or alkaline) hydrolysis, autolysis, which is regulated by endogenous enzymes, thermal hydrolysis or enzymatic hydrolysis [29]. Although the cost of chemical hydrolysis is relatively lower, amino acid profiles generated from this type of hydrolysis are not consistent and the extent of the degradation is difficult to control [28]. Similarly, autolysis, which is fueled by endogenous enzymes, can cause changes in product profiles. Enzymatic hydrolysis is a preferred method due to its specificities, ease of control and being more environmentally friendly as it does not utilise hazardous or toxic reagents.

### 3.2 Bioactive Peptides

Bioactive peptides are protein fragments of 2–20 amino acids derived from various sources and can exert various biological activities when liberated from their parent protein [30]. Apart from the amino acid composition, the effectiveness of a peptide in delivering its physiological effects is highly dependent on its bioavailability [31]. Peptides ingested by humans must stay intact throughout the digestive system until they reach the targeted organ [32]. According to [31], peptides with three and more amino acids tend to be more susceptible and may be hydrolyzed extracellularly by the brush border membrane of the intestinal epithelium. Di- and tripeptides, on the other hand, are more resistant and can retain their activity through the intestinal tract, although they might still be hydrolyzed after passing the absorption stage.

Fish skin is characterised by the presence of collagen components in its structure. Materials containing collagen (skin, fish frame and scales) account for up to 30% of the total by-products after filleting activity [33]. Initial characterizations of fish skin may vary according to the species and the processing methods involved [34]. The hydroxyproline content of a material is used to quantify the collagen content of a tissue sample. Peptides with proline and hydroxyproline amino acids are of interest, due to their resistance to digestive enzymes and their capability to remain intact through the intestinal tract. This allows effective delivery of the bioactive compound into the bloodstream and successfully reaches the targeted organ [31, 35]. It was reported that the collagen content is affected by the fish species and processing activities like treatment in brine or freezing, which may result in protein denaturation, higher crosslinking, lower solubility and low extraction yield [34]. Previous reports have also indicated that proteolytic hydrolysis of fish skin collagen has resulted in hydrolysate with better functional characteristics such as solubility, foaming, emulsifying and gelation properties compared to acid-solubilised fish collagen [34]. Meanwhile, fish skin gelatin hydrolysate from shortfin scad (*Decapterus macrosoma*) have the bioactive properties of angiotensin-converting enzyme (ACE) inhibitor and antioxidant activity [36]. ACE inhibitor peptides were reported to be commonly dominated by glycine, proline, phenylalanine, tyrosine and leucine [37].

Antioxidant capacity of the fish skin was hypothesized to be influenced by the amino acid composition and the size of the peptides. Cysteine, which contains sulfhydryl group, can interact with free radicals. The antioxidative property of a peptide was also reported to be influenced by the position of the amino acid in a peptide [38]. Non-polar hydrophobic amino acids such as proline, alanine and phenylalanine at the N-terminus of a peptide were suggested to contribute to their antioxidative properties. Other hydrophobic amino acids like valine, histidine and tyrosine equally contribute to this bioactivity. Aside from that, hydrophobic amino acids have been reported to contribute to peptide solubility [39]. A report by [40] mentioned that short, low molecular weight peptides have the capacity to act as electron donors and attack free radicals. This is supported by the findings of [38], which discovered that oligopeptides with a molecular weight of less than 1000 Da were more efficient as an antioxidant.

# 4 Conclusion

The Zero-waste concept aims to reduce waste disposal while encouraging responsible societies at all levels, including industrial key players, to engage in responsible production and consumption. Waste and by-products are diverted from reaching 'landfills' by fully optimizing the resources available. It is evident that fish and seafood by-products are high protein materials which harbour important amino acids, potentially developed into high-value commodities through the efficient method of enzymatic hydrolysis. Future studies, however, may emphasise the aspect of the bioavailability of the peptides to ascertain the successful delivery and effectiveness of the substance.

# References

1. Lopez, R.: The environment as a factor of production: The effects of economic growth and trade liberalization. J. Environ. Econ. Manage. **27**, 163–184 (1994). https://doi.org/10.1006/jeem.1994.1032
2. Orecchia, C., Zoppoli, P.: Consumerism and environment: does consumption behaviour affect environmental quality? SSRN Electron. J. 1–17 (2007). https://doi.org/10.2139/ssrn.1719507
3. Song, Q., Li, J., Zeng, X.: Minimizing the increasing solid waste through zero waste strategy. J. Clean. Prod. **104**, 199–210 (2015). https://doi.org/10.1016/j.jclepro.2014.08.027
4. Liao, B., Wang, T.: Research on industrial waste recovery network optimization: opportunities brought by artificial intelligence. Math. Probl. Eng. **2020**, 1–10 (2020). https://doi.org/10.1155/2020/3618424
5. Zaman, A.U.: A comprehensive review of the development of zero waste management: lessons learned and guidelines. J. Clean. Prod. **91**, 12–25 (2015). https://doi.org/10.1016/j.jclepro.2014.12.013

6. Troschinetz, A.M., Mihelcic, J.R.: Sustainable recycling of municipal solid waste in developing countries. Waste Manag. **29**, 915–923 (2009). https://doi.org/10.1016/j.wasman.2008.04.016
7. Menikpura, S.N.M., Gheewala, S.H., Bonnet, S.: Framework for life cycle sustainability assessment of municipal solid waste management systems with an application to a case study in Thailand. Waste Manag. Res. **30**, 708–719 (2012). https://doi.org/10.1177/0734242X12444896
8. Shekdar, A.V.: Sustainable solid waste management: an integrated approach for Asian countries. Waste Manag. **29**, 1438–1448 (2009). https://doi.org/10.1016/j.wasman.2008.08.025
9. Cheng, H., Hu, Y.: Municipal solid waste (MSW) as a renewable source of energy: current and future practices in China. Bioresour. Technol. **101**, 3816–3824 (2010). https://doi.org/10.1016/j.biortech.2010.01.040
10. Greyson, J.: An economic instrument for zero waste, economic growth and sustainability. J. Clean. Prod. **15**, 1382–1390 (2007). https://doi.org/10.1016/j.jclepro.2006.07.019
11. Matete, N., Trois, C.: Towards zero waste in emerging countries—a South African experience. Waste Manag. **28**, 1480–1492 (2008). https://doi.org/10.1016/j.wasman.2007.06.006
12. Zero Waste: Zero-waste policy and legislation—zero waste. https://www.zerowaste.com/blog/zero-waste-policy-and-legislation/
13. US EPA, R. 09: How communities have defined zero waste. https://www.epa.gov/transforming-waste-tool/how-communities-have-defined-zero-waste
14. ZWIA: Zero waste definition. https://zwia.org/zero-waste-definition/
15. Tennant-Wood, R.: Going for zero: a comparative critical analysis of zero waste events in southern New South Wales. Australas. J. Environ. Manag. **10**, 46–55 (2003). https://doi.org/10.1080/14486563.2003.10648572
16. SF Environment: Zero Waste. https://sfenvironment.org/zero-waste-in-SF-is-recycling-composting-and-reuse
17. Phillips, P.S., Tudor, T., Bird, H., Bates, M.: A critical review of a key waste strategy initiative in England: zero waste places projects 2008–2009. Resour. Conserv. Recycl. **55**, 335–343 (2011). https://doi.org/10.1016/j.resconrec.2010.10.006
18. Connett, P.: The zero waste solution: untrashing the planet one community at a time. Chelsea Green Publishing (2013)
19. Zaman, A.U., Lehmann, S.: Challenges and opportunities in transforming a city into a "zero waste city." Challenges **2**, 73–93 (2011)
20. Zaman, A.U.: A strategic framework for working toward zero waste societies based on perceptions surveys. Recycling. **2**, 1 (2017)
21. Wautelet, T.: Exploring the role of independent retailers in the circular economy: a case study approach (2018)
22. RES/70/1: Transforming our world: the 2030 agenda for sustainable development. Seventieth United Nations General Assembly New York 25 (2015). https://doi.org/10.1163/157180910X12665776638740
23. Anon.: Food Loss and Waste. http://www3.cec.org/islandora/en/item
24. Islam, M.S., Khan, S., Tanaka, M.: Waste loading in shrimp and fish processing effluents: potential source of hazards to the coastal and nearshore environments. Mar. Pollut. Bull. **49**, 103–110 (2004). https://doi.org/10.1016/j.marpolbul.2004.01.018
25. European Parliament and Council: Regulation (EC) No 1069/2009. Off. J. Eur. Union. **300**, 1–33 (2009)
26. Food and Agriculture Organization: The State of World Fisheries and Aquaculture 2020. Sustainability in action (2020)
27. Knuckey, I., Sinclair, C., Surapaneni, A., Ashcroft, W.: Utilisation of seafood processing waste—challenges and opportunities. Supersoil (2002)
28. Siddik, M.A.B., Howieson, J., Fotedar, R., Partridge, G.J.: Enzymatic fish protein hydrolysates in finfish aquaculture: a review. Rev. Aquac. **13**, 406–430 (2021). https://doi.org/10.1111/raq.12481
29. Zamora-Sillero, J., Gharsallaoui, A., Prentice, C.: Peptides from fish by-product protein hydrolysates and its functional properties: an overview. Mar. Biotechnol. **20**, 118–130 (2018). https://doi.org/10.1007/s10126-018-9799-3

30. Daliri, E.B.M., Oh, D.H., Lee, B.H.: Bioactive peptides. Foods **6**, 1–21 (2017). https://doi.org/10.3390/foods6050032
31. Segura-Campos, M., Chel-Guerrero, L., Betancur-Ancona, D., Hernandez-Escalante, V.M.: Bioavailability of bioactive peptides. Food Rev. Int. **27**, 213–226 (2011). https://doi.org/10.1080/87559129.2011.563395
32. Bhandari, D., Rafiq, S., Gat, Y., Gat, P., Waghmare, R., Kumar, V.: A review on bioactive peptides: physiological functions, bioavailability and safety. Int. J. Pept. Res. Ther. **26**, 139–150 (2020). https://doi.org/10.1007/s10989-019-09823-5
33. Gomez-Guillen, M.C., Turnay, J., Fernandez-Diaz, M.D., Ulmo, N., Lizarbe, M.A., Montero, P.: Structural and physical properties of gelatin extracted from different marine species: a comparative study. Food Hydrocoll. 25–34 (2002)
34. Blanco, M., Vázquez, J.A., Pérez-Martín, R.I., Sotelo, C.G.: Hydrolysates of fish skin collagen: an opportunity for valorizing fish industry byproducts. Mar. Drugs. **15**, 1–15 (2017). https://doi.org/10.3390/md15050131
35. Roberts, P.R., Zaloga, G.P.: Dietary bioactive peptides. New Horiz. **2**, 237–243 (1994)
36. Rasli, H.I., Sarbon, N.M.: Preparation and physicochemical characterization of fish skin gelatine hydrolysate from shortfin scad (Decapterus macrosoma). Int. Food Res. J. **26**, 287–294 (2019)
37. Hong, F., Ming, L., Yi, S., Zhanxia, L., Yongquan, W., Chi, L.: The antihypertensive effect of peptides: a novel alternative to drugs? Peptides **29**, 1062–1071 (2008). https://doi.org/10.1016/j.peptides.2008.02.005
38. Zheng, L., Wei, H., Yu, H., Xing, Q., Zou, Y., Zhou, Y., Peng, J.: Fish skin gelatin hydrolysate production by ginger powder induces glutathione synthesis to prevent hydrogen peroxide induced intestinal oxidative stress via the Pept1-p62-Nrf2 cascade. J. Agric. Food Chem. **66**, 11601–11611 (2018). https://doi.org/10.1021/acs.jafc.8b02840
39. Cody Ellis: World Bank: Global waste generation could increase 70% by 2050 | Waste Dive
40. Chi, C.F., Cao, Z.H., Wang, B., Hu, F.Y., Li, Z.R., Zhang, B.: Antioxidant and functional properties of collagen hydrolysates from Spanish mackerel skin as influenced by average molecular weight. Molecules **19**, 11211–11230 (2014). https://doi.org/10.3390/molecules190811211

# Utilization of Lignocellulosic Agro-Waste as an Alternative Carbon Source for Industrial Enzyme Production

**Ja'afar Nuhu Ja'afar and Awwal Shitu**

**Abstract** Waste, as a by-product of most human activities, especially of agricultural origin, contains an equivalent useful product substance that is available within it. Synthetic media used for enzyme production in biotechnology industries is becoming costlier and has become a subject of concern for many food-processing industries. Hence, manufacturers are now in search of alternative approaches to cut down the cost of production. Therefore, this review focuses on how agro-industrial wastes have been utilized as the sole carbon source for enzyme production in a solid-state fermentation culture. Similarly, the review limits its scope to enzymes produced from fungal sources only. Conclusively, the review suggests that the ability to use agro-waste as an alternative carbon source in the production of enzymes calls for its wider utilization in the food-processing industries, thus, alleviating the cost concern.

**Keywords** Agro-waste · Solid-state fermentation · Enzyme

Currently, there is rapid growth in the food and agricultural industry. Simultaneously, the increase in population *vis-a-vis* improved economic growth has attracted significant investments in this industry [1, 2]. Expectedly, excess waste generated from this industry will pose a significant environmental challenge. For example, millions of metric tons of biomass are reported to be produced annually from agriculture to which contains high amounts of nutrients that can form breeding grounds for disease-causing pathogens if left unprocessed or inadequately treated [1]. Agro-industrial wastes are highly nutritious in nature and can facilitate microbial growth. Most agricultural wastes are lignocellulosic in nature (Table 1) [3, 4], thus, can be used for the

J. N. Ja'afar (✉)
Chevron Biotechnology Center, Modibbo Adama University, Yola P.M.B. 2076, Adamawa State, Nigeria
e-mail: jnjaafar@mautech.edu.ng

Center for Renewable Energy, Modibbo Adama University, Yola P.M.B. 2076, Adamawa State, Nigeria

J. N. Ja'afar · A. Shitu
Department of Biotechnology, Faculty of Life Sciences, Modibbo Adama University, Yola P.M.B. 2076, Adamawa State, Nigeria



A. Z. Yaser et al. (eds.), *Waste Management, Processing and Valorisation*,
https://doi.org/10.1007/978-981-16-7653-6_12

**Table 1** Composition of lignocellulosic components in agro-industrial lignocellulosic waste [4]

| Lignocellulosic materials | Cellulose (% ww) | Hemicellulose (% ww) | Lignin |
|---|---|---|---|
| Banana waste | 13.2 | 14.8 | 14 |
| Corn cobs | 45.0 | 35.0 | 15 |
| Grasses | 25–40 | 25–50 | 10–30 |
| Sugar cane bagasse | 42.0 | 25.0 | 20 |
| Wheat straw | 32.9 | 24.0 | 8.9 |
| Rice straw | 38.0 | 26.0 | 15.0 |
| Soft wood | 45–50 | 25–35 | 25–35 |
| Hard wood | 40–55 | 24–35 | 25–35 |

production of various value-added products, such as industrially important enzymes [5].

Agro-industrial wastes, such as sugar cane bagasse, corncob and rice bran, have been widely investigated via different fermentation strategies for the production of value-added products such as enzymes, biogas and bioethanol. Solid-state fermentation (SSF) holds much potential compared to submerged fermentation (SmF) methods for the utilization of agro-based wastes for enzyme production [2]. This is due to the simplicity of the technique, cost effectiveness, semblance of microbe's natural habitat, ease of product recovery with stable physical–chemical properties and high quality of the product [6, 7]. More so, SSF permits the use of agricultural and agro-industrial residues as substrates, which are converted into products with high commercial value like secondary metabolites. In the SSF process, the solid substrate not only supplies nutrients to the culture but also serves as an anchorage for the microbial cells [8, 9]. In addition, studies have shown that pre-treatment of the wastes can greatly enhance enzyme yields by several folds [2]. Pre-treatments are techniques widely used in production processes involving lignocellulose-based raw materials where the complex plant structure is disrupted by physical, chemical or a combination of methods to improve digestibility [2–5, 10–12]. Chemical pre-treatments were shown to be superior to other techniques [4, 5, 11].

Commonly used substrates in SSF are cereal grains (rice, wheat, barley and corn), legume seeds, wheat bran, lignocellulose materials such as straws, sawdust or wood shavings (Table 2) and a wide range of plant and animal materials. These substrates' components are polymeric and insoluble or sparingly soluble in water, yet they are often inexpensive and readily available and they provide a rich source of nutrients for microbial growth [8].

Various studies have demonstrated that using little or no quantity of water in SSF has several benefits, including simple product recovery, lower overall production costs, smaller fermenter sizes, minimal downstream processing and lower energy requirements for stirring and sterilizing [7–9]. Before beginning any SSF fermentation process, several parameters such as microorganisms, solid support utilized, water activity, temperature, aeration and the kind of fermenter used should be examined [3,

**Table 2** Common microbes and substrates used in solid-state fermentation [4, 13]

| Microbe | Substrate |
|---|---|
| Bacteria | |
| *Xanthomonas campestries* MTCC 2286 | Potato peel |
| *Bacillus licheniformis* MTCC 1483 | Wheat straw, sugarcane bagasse, maize straw and paddy straw |
| Fungi | |
| *Aspergillus niger* | Rice bran, wheat bran, black gram bran and soybean |
| *Aspergillus oryzae* | Soybean meal (waste) |
| *Rhizopus arrhizus* and *Mucor subtillissimus* | Corncob, cassava peel, soybean, wheat bran and citrus pulp |
| *Aspergillus terreus* | Palm oil cake |

4]. Single pure cultures, mixed identifiable cultures, or a consortium of mixed indigenous microorganisms can all be employed in SSF [9]. Some SSF processes necessitate the development of moulds that require low moisture levels in order to carry out fermentation using extracellular enzymes produced by fermenting microorganisms [9].

Moulds are commonly employed in SSF to increase the production of value-added products since they naturally grow on solid substrates including wood, seeds, stems and roots [8, 9]. Bacteria and yeasts, which require a greater moisture level for effective fermentation, can be utilized for SSF as well, although the output will be reduced [11, 13]. SSF is a multistep process involving the following steps: (1) Selection of substrate. (2) Pre-treatment of substrate by either mechanical, chemical or biochemical processing to improve the availability of the bound nutrients and to reduce the size of the components [4, 10–12]. (3) Hydrolysis. (4) Fermentation and (5) Downstream processing for purification and quantification of products.

Enzymes are biological molecules responsible for metabolic activities of plants, animals and all other living microorganisms. They are mainly proteins though some contain non-protein components such as prosthetic groups, co-factors and co-enzymes. When excreted or extracted from the producing organisms, enzymes are capable of acting independently, hence, the property that attracts attention to their industrial utilization [13]. Microbes are considered the most convenient sources of commercial enzymes due to efficient production processes [14]. Specifically, microbial sources of enzymes are preferred due to their economic feasibility, high production yield, consistency, ease of genetic modification and optimization, regular supply due to absence of seasonal and weather fluctuations, rapid proliferation of microbes on inexpensive substrate, greater stability and catalytic activity [15]. The advantage of using microbes for the production of enzymes is that bulk production is economical and easy to manipulate to obtain enzymes with desired characteristics. Recovery,

**Table 3** Agro-industrial wastes used for enzyme production [5, 11, 20, 23, 26]

| Agro-industrial waste(s) | Microbe | Enzyme |
|---|---|---|
| Papaya waste | *A. niger* | α-Amylase |
| Wheat bran and orange peel | *P. notatum* | Pectin methyl esterase |
| Orange peel | *A. niger* | α-Amylase |
| Coconut oil cake | *A. oryzae* | α-Amylase |
| Rice bran | *Bacillus* sp. | α-Amylase |
| Corn bran | *Bacillus* sp. | α-Amylase |
| Rice bran, wheat bran and soybean | *A. niger* | α-Amylase |
| Fruits peel waste | *A. niger* | Invertase |

isolation and purification processes of microbial enzymes are easier than those from other sources. Fungal enzymes are preferred over other microbial sources owing to their widely accepted "generally regarded as safe" (GRAS) status [16] (Table 3). While media formulations based on agricultural wastes are heterogeneous by nature, the production of enzymes of microbial source is influenced by optimal process parameters such as temperature, pH, ideal substrate selection and microorganism, size of the substrate, concentration of the inoculum and the moisture content of the substrate [14]. Several studies have been conducted on alternative media optimization to improve enzyme yield [17–19].

## 1 α-Amylase (EC 3.2.1.1)

Amylases are one of the most widely used enzymes in biotechnological industries as they catalyse the breakdown of starch into simple sugars. The enzyme accounts for approximately 25% of the enzyme market covering most biotechnological industrial processes such as sugar, textile, paper, food, pharmaceuticals, cosmetics, fermentation baking, brewing, detergent, distilling and pharmaceutical industries [20–24]. Similarly, they have gained spectrum of applications in many sectors such as clinical, medical and analytical chemistry. Although they can be derived from different sources, including microorganisms, plants and animals [25], microbial enzymes generally meet industrial demands because of their inherent features including specificity, stability, optimum temperature and pH performance [5, 10, 11, 26]

## 2 Amyloglucosidases (EC 3.1.2.3)

Amyloglucosidases (AMG), also known as glucoamylases, cleave starch to release glucose molecules. As an exoamylase, it releases β-D glucose from non-reducing ends of amylose, amylopectin and glycogen [4, 27]. The enzymes have optimal activities within the pH range of 4.5–5.0 and temperature range of 40–60 °C [31]. Like amylases, lignocellulose residues such as wheat bran have been used as alternative carbon sources for AMG production in solid-state fermentation conditions with high yields [28, 29]. Other residues used include rice bran [30, 31], cassava, yam banana including its peels and plantain [28, 31, 32]. AMG finds applications in the food, brewing and bakery industry [28].

## 3 Cellulase (EC 3.2.1.4)

Cellulases also known as endoglucanases are enzymes of great biotechnological and commercial importance owing to their role in bioethanol production [33]. The enzyme is shown to be produced by a range of both fungal and bacterial organisms. Fungal organisms producing cellulases include strains of *Trichoderma reesei* [34, 35], *Schizophyllum commune, Melanocarpus* sp., *Aspergillus* sp., *Penicillium* sp. and *Fusarium* sp. [36]. In addition to biofuels, cellulases find application in bread, brewing, textiles, detergents, paper and pulp industries [37]. Key researches over the past two (2) decades on the utilization of lignocellulosic food waste as a suitable carbon source for cellulase production have been documented including banana peel [38], mango peel [39] and apple pomace [33, 37].

## 4 Xylanases (EC 3.2.1.8)

Xylanases are enzymes that break down the plant polysaccharide, xylan. As xylan is a complex polysaccharide [37], a consortium of enzymes is required for its complete hydrolysis including endoxylanases, β-xylosidases, ferulic acid esterase, p-coumaric acid esterase, acetylxylan esterase and α-glucuronidase [40]. Compared to bacteria, fungi synthesize the enzyme in large quantities [41–43]. Lignocellulosic food wastes investigated as potential carbon sources for xylanases commercial production include wheat straw [42, 43], wheat bran [37], grape pomace [38], apple pomace, melon peel and hazelnut shell [44]. Xylanases have wide applications in the food, biomedical, animal feed and bioethanol sectors of the industries [37, 41–46].

## 5 Inulinase (EC3.2.1.7)

Inulinases can be sub-classified into exo-inulinases and endo-inulinases depending upon their modes of activity [47, 48]. Inulinases gained importance in the food and pharmaceutical industry due to their preferred appeal by humans as sweeteners [49]. Inulinase acts upon inulin, which is a polyfructose (fructan) chain terminated by a glucose molecule [49]. The fructose units in inulin are bonded together by a β-2,1-linkage [49]. The activity of inulinase results in the complete conversion of the substrate to fructose [47, 48]. Inulinase also finds application in the production of bioethanol, citric acid, butanediol and lactic acid. The production of inulinase using lignocellulosic substrates such roots of *Cichorium intybus* [49, 50], banana peel, wheat bran, rice bran, orange peel and bagasse [47, 51], coconut oil cake [50] have been investigated. Fungal species known to synthesize the enzyme include *Streptococcus salivarius*, *Actinomyces viscosus*, *Kluyveromyces fragilis*, *Chrysosporium pannorum*, *Penicillium* sp. and *Aspergillus niger* [52]. Others are *Fusarium oxysporum* [49, 53], *Saccharomyces* spp. [49] and *Pencillium rugulosum* [53].

## 6 Mannanase (EC 3.2.1.25)

Mannanases are a group of enzymes that degrade mannan, which is an integral part of the plant cell wall. Because it removes hemicellulose effectively, it has gained interest within the paper and pulp industry [54]. It also has applications in the food, oil, feed and textile industries [55–57]. Lignocellulose and agro-food industry wastes studied as substrates in mannanase production include apple pomace and cottonseed powder [58], lime, grape, tangerine and sweet orange peels [59], palm kernel cake [60, 61], sugar cane pulp, soybean meal, locust bean gum and peels from plantain, mango, potato and passion fruit [55, 62, 63]. Some fungal organisms studied to synthesize mannanase include *Aspergillus* spp. [55, 58, 60, 64], *Pencillium* spp. and *Sterptomyces* spp. [50, 58, 62].

## 7 Invertase (EC 3.2.1.26)

Also known as β-fructofuranosidase, invertase is a glycoprotein that catalyses the hydrolysis of sucrose to glucose and fructose. The enzyme is optimally active at a pH of 4.5 and a temperature of 55 °C. Because of its industrial importance and cost, alternative agro-wastes have been utilized as substrates including apple pomace, carrot residue, wheat bran and peel waste from orange, pineapple and pomegranate [65–68]. *Saccharomyces* spp, *Cladosporium* spp and *Aspergillus* spp were the common fungal species involved in the enzyme's synthesis [65–68].

## 8 Pectinase (EC 3.2.1.15)

As pectin is an integral part of the plant cell wall, pectinases catalyse its disintegration [69]. Pectinases have been widely utilized in the fruit and wine industries for clarification and removal of turbidity in finished product [70]. In addition, the enzyme stabilizes and intensifies the colour of the fruit extract [70]. Agricultural residues used for the production of pectinases include apple and grape pomace [71, 72], seedless sunflower heads and barley spent grain [73, 74]. Fungal organisms for the production of pectinases include *Penicillium* spp and *Aspergillus* spp [70, 72–75].

## 9 Protease (EC 3.4.21.62)

Proteases (EC 3.4.21.62) are enzymes that hydrolyse the peptide bonds linking amino acids together in polypeptide chains. They dominate the enzyme industry with a market share of almost 60% [76], to which detergent proteases constitute the major component, as they remain stable and active at the alkaline pH of the wash environment [77]. Genetically modified microbes [78, 79] that are thermostable and can withstand pH variations are employed in producing most of the proteases utilized in industries. Lignocellulosic substrates used as raw materials in proteases production include brewer's spent grain and corn steep liquor using *Streptomyces malaysiensis* [80], soybeans and tomato pomace using *Aspergillus oryzae* [81, 82] and tomato pomace and Jatropha seed cake using *Aspergillus versicolor* [82, 83].

## 10 Transglutaminase (EC 2.3.2.13)

Transglutaminases are a class of transferase enzymes, which catalyse the formation of isopeptide bonds between the $\gamma$-carboxyamide groups of glutamine residues and the $\gamma$-amino group of lysine residues [84]. Transglutaminases are used in products such as flour, baked goods, processed cheese/milk/meat/and fish, cosmetics, gelled food products and leather finishing [84, 85]. When wool is treated with transglutaminase post protease treatment, the wool's strength is increased, thus, improving the longevity of the wool fibres [86]. Agro-wastes investigated as potential media alternatives in transglutaminase production include sorghum straw [87], untreated corn grits, milled brewer's rice, industrial fibrous soy residue, soy hull and malt bagasse [85, 88].

## 11 Lipases (EC 3.1.1.3)

Lipases facilitate the breakdown and mobilization of lipids within the cells. In the industries, lipases are involved with hydrolysis, transesterification, alcoholysis, acidolysis, aminolysis and esterification [89]. They have applications in the detergents, oil, dairy, pharmaceutical and bakery industries [89–91]. Similarly, they are involved in biopolymer synthesis, biodiesel production and the treatment of fat-containing waste effluents [92, 93]. Lipases have been produced using lignocellulose and wastes from peanut cake [94], sheanut cake [95], coconut oil cake [96, 97] and palm oil mill effluent [17].

## 12 Summary

Agro-wastes, as well as municipal solid wastes, are subject to natural complex aerobic and anaerobic processes by consortia of microorganisms, which are responsible for the production of enzyme mixtures necessary for the effective breakdown of the biomass [98]. Successful utilization of agro-waste and other lignocellulosic materials as an alternative carbon source for enzyme production depends on careful selection of the waste and optimizing the process parameters. Agro-waste utilization would go a long way in reducing the cost of enzyme production thus, making it economically viable for commercial and industrial applications. Enzymes have played numerous roles in products and processes that are commonly encountered in food and beverages, animal feeds, cleaning supplies, clothing, paper products, transportation fuels, pharmaceuticals and in the diagnosis of some diseases [99, 100]. The demand for industrial enzymes especially those of microbial origin in a variety of industrial processes continues to increase and the contribution of enzyme costs to the economy of biochemical production continues to improve as new technological innovations and other breakthroughs are overcoming major production barriers [101–104]. For this reason, enzyme mediated processes are now rapidly gaining grounds in modern industrial processes than it was many years ago; due to the numerous advantages, enzymes offer over the conventional chemical transformation processes [102, 105, 106].

## References

1. Bharathiraja, S., Suriya, J., Krishnan, M., Manivasagan, P., Kim, S.-K.: Production of enzymes from agricultural wastes and their potential industrial applications. Adv. Food Nutr. Res. **80**, 125–148 (2017)
2. Goh, K.M., Kahar, U.M., Chai, Y.Y., Chong, C.S., Chai, K.P., et al.: Recent discoveries and applications of *Anoxybacillus*. Appl. Microbiol. Biotechnol. **97**, 1475–1488 (2013)

3. Aguilar, C.N., Gutierrez-Sánchez, G., rado-Barragán, P.A., Rodríguez-Herrera, R., Martínez-Hernandez, J.L., Contreras-Esquivel, J.C.: Perspectives of solid state fermentation for production of food enzymes. Am. J. Biochem. Biotechnol. **4**(4), 354–366 (2008)
4. Ravindran, R., Jaiswal, A.: Microbial enzyme production using lignocellulosic food industry wastes as feedstock: a review. Bioengineering **3**, 30 (2016)
5. Rose's, P.R., Guerra, N.P.: Optimization of amylase production by *Aspergillus niger* in solid-state fermentation using sugarcane bagasse as solid support material. World J. Microbiol. Biotechnol. **25**, 1929–1939 (2009)
6. Ajay, P., Farhath, K.: Production and extraction optimization of xylanase from *Aspergillus niger* DFR-5 through solid-state fermentation. Biores. Technol. **101**, 7563–7569 (2010)
7. Baysal, Z., Uyar, F., Aytekin, C.: Solid-state fermentation for production of alpha amylase by a thermotolerant *Bacillus subtilis* from hot spring water. Process Biochem. **93**, 1020–1025 (2003)
8. Admassu, H., Wei, Z., Ruijin, Y., Mohammed, A.A.G., Wenbin, Z.: Recent Advances on Efficient Methods for α-Amylase Production by Solid State Fermentation (SSF). Int. J. Adv. Res. **3**(9), 1485–1493 (2015)
9. Lopes, A.S., Rodrigues, C., Jefferson, D.L.C., Marília, P.M., Rafaela, D.O.P., Luiz, A.B., Luciana, P.D.S.V., Carlos, R.S.: Gibberellic acid fermented extract obtained by solid-state fermentation using citric pulp by *Fusarium moniliforme*: Influence on *Lavandula Angustifolia* Mill., Cultivated *In Vitro*. Pak. J. Bot. **45**(6), 2057 (2013)
10. Hölker, U.: Lenz J Solid-state fermentation–are there any biotechnological advantages? Curr. Opin. Microbiol. **8**, 301–306 (2005)
11. Gupta, R., Gigras, P., Mohapatra, H., Goswami, V.K., Chauhan, B.: Microbial alpha-amylases: a biotechnological perspective. Process Biochem. **38**, 1599–1616 (2003)
12. Gangadharan, D., Sivaramakrishnan, S., Nampoothiri, K.M., Pandey, A.: Solid culturing of *Bacillus amyloliquefaciens* for alpha amylase production. Food Technol. Biotechnol. **44**, 269–274 (2006)
13. Babu, K.R., Satyanarayana, T.: Production of bacterial enzymes by solid state fermentation. J. Sci. Ind. Res. **55**, 464–467 (1996)
14. Sodhi, H.K., Sharma, K., Gupta, J.K., Soni, S.K.: Production of a thermostable alpha amylase from *Bacillus* sp. PS-7 by solid-state fermentation and its synergistic use in the hydrolysis of malt starch for alcohol production. Process Biochem. **40**, 525–534 (2005)
15. Neelam, G., Samanta, R., Subtapa, B., Vivek, R.: Microbial enzymes and their relevance in industries, medicine and beyond. BioMed. Res. Int. 329121 (2013)
16. Sindhu, R., Suprabha, G.N., Shashidhar, S.: Optimization of process parameters for the production of alpha-amylase from *Penicillum janthinellum* (NCIM 4960) under solid-state fermentation. Afr. J. Microb. Res. **3**(9), 498–503 (2009)
17. Aliyu, S., Alam, M.Z., Ismail, A.M., Hamzah, S.M.: Optimization of lipase production by Candida cylindracea in palm oil mil effluent based medium using statistical experimental design. J. Mol. Catal. B Enzym. **69**, 66–73 (2011)
18. Solange, M.I., Jose, T.A.: Increase in the fructo oligosaccharides yield and productivity by solid-state fermentation with Aspergillus japonicas using agro-industrial residues as support and nutrient source. Biochem. Eng. J. **53**, 154–157 (2010)
19. Francis, F., Sabu, A., Nampoothiri, K.M., Ramachandran, S., Ghosh, S., Szakacs, G., Pandey, A.: Use of response surface methodology for optimizing process parameters for the production of α-amylase by *Aspergillus oryzae*. Biochem. Eng. J. **15**, 107–115 (2003)
20. Mamo, G., Gashe, B.A., Gessesse, A.: A highly thermostable amylase from a newly isolated thermophilic Bacillus sp. WN11. J. Appl. Microbiol. **86**, 557–560 (1999)
21. Pandey, R., Nigam, P., Soccol, C.R., Singh, D., Soccol, V.T., Mohan, R.: Advances in microbial amylases. Biotechnol. Appl. Biochem. **31**, 135–152 (2000)
22. Oudjeriouat, N., Moreau, Y., Santimone, M., Svensson, B., Marchis-Mouren, G., Desseaux, V.: On the mechanism of amylase. Eur. J. Biochem. FEBS. **270**, 3879 (2003)
23. Ramachandran, S., Patel, A.K., Nampoothiri, K.M., Francis, F., Nagy, V., Szakacs, G., Pandey, A.: Coconut oil cake, a potential raw material for the production of alpha amylase. Bioresour. Techno. **93**, 167–174 (2004)

24. Kathiresan, K., Manivannan, S.: Amylase production by *Penicillium fellutanum* isolated from mangrove rhizosphere soil. Afri. J. Biotech. **5**, 829–832 (2006)
25. Banks, K.R., Satyanarayana, T.: Production of bacterial enzymes by solid-state fermentation. J. Sci. Ind. Res. **55**, 464–467 (1975)
26. Liu, X.D., Xu, Y.: A novel raw starch digesting alpha-amylase from a newly isolated Bacillus sp. YX-1: purification and characterization. Bioresour. Technol. **99**, 4315–4320 (2008)
27. Veana, F., Martínez-Hernández, J.L., Aguilar, C.N., Rodríguez-Herrera, R., Michelena, G.: Utilization of molasses and sugar cane bagasse for production of fungal invertase in solid state fermentation using Aspergillus niger GH1. Braz. J. Microbiol. **45**, 373–377 (2014)
28. Kumar, P., Satyanarayana, T.: Microbial glucoamylases: Characteristics and applications. Crit. Rev. Biotechnol. **29**, 225–255 (2009)
29. Diler, G., Chevallier, S., Pöhlmann, I., Guyon, C., Guilloux, M., Le-Bail, A.: Assessment of amyloglucosidase activity during production and storage of laminated pie dough. Impact on raw dough properties and sweetness after baking. J. Cereal Sci. **61**, 63–70 (2015)
30. Singh, H., Soni, S.K.: Production of starch-gel digesting amyloglucosidase by *Aspergillus oryzae* HS-3 in solid state fermentation. Process Biochem. **37**, 453–459 (2001)
31. Shin, H.K., Kong, J.Y., Lee, J.D., Lee, T.H.: Syntheses of hydroxybenzyl-α-glucosides by amyloglucosidase-catalyzed transglycosylation. Biotechnol. Lett. **22**, 321–325 (2000)
32. Pandey, A.: Improvements in solid-state fermentation for glucoamylase production. Biol. Wastes **34**, 11–19 (1990)
33. Dhillon, G.S., Kaur, S., Brar, S.K., Verma, M.: Potential of apple pomace as a solid substrate for fungal cellulase and hemicellulase bioproduction through solid-state fermentation. Ind. Crops Prod. **38**, 6–13 (2012)
34. Hai-Yan Sun, H., Li, J., Zhao, P., Peng, M.: Banana peel: A novel substrate for cellulase production under solid-state fermentation. Afr. J. Biotechnol. **10**, 17887–17890 (2011)
35. Saravanan, P., Muthuvelayudham, R., Viruthagiri, T.: Application of Statistical Design for the Production of Cellulase by *Trichoderma reesei* Using Mango Peel. Enzyme Res. **2012**, 157643 (2012)
36. Adeniran, H.A., Abiose, S.H., Ogunsua, A.O.: Production of fungal β-amylase and Amyloglucosidase on some Nigerian agricultural residues. Food Bioprocess Technol. **3**, 693–698 (2010)
37. Leite, P., Salgado, J.M., Venâncio, A., Domínguez, J.M., Belo, I.: Ultrasounds pretreatment of olive pomace to improve xylanase and cellulase production by solid-state fermentation. Bioresour. Technol. **214**, 737–746 (2016)
38. Hassan, S.S., Williams, G.A., Jaiswal, A.K.: Emerging technologies for the pretreatment of lignocellulosic biomass. Bioresour. Technol. **262**, 310–318 (2018)
39. Salim, A.A., Grbavˇci´c, S., Šekuljica, N., Stefanovi´c, A., Jakoveti´c Tanaskovi´c, S., Lukovi´c, N., Kneževi´c-Jugovi´c, Z.: Production of enzymes by a newly isolated Bacillus sp. TMF-1 in solid state fermentation on agricultural by-products: The evaluation of substrate pretreatment methods. Bioresour. Technol. **228**, 193–200 (2017)
40. Prakash, B., Vidyasagar, M., Madhukumar, M.S., Muralikrishna, G., Sreeramulu, K.: Production, purification and characterization of two extremely halotolerant, thermostable and alkali-stable α-amylases from Chromohalobacter sp. TVSP 101. Process Biochem. **44**, 210–215 (2009)
41. Harris, A.D., Ramalingam, C.: Xylanases and its application in food industry: a review. J. Exp. Sci. **1**, 1–11 (2010)
42. Knob, A., Beitel, S.M., Fortkamp, D., Terrasan, C.R.F., de Almeida, A.F.: Production, purification and characterization of a major *Penicillium glabrum* Xylanase using brewer's spent grain as substrate. Biomed. Res. Int. **2013**, 1–8 (2013)
43. Goswami, G., Pathak, R.: Microbial xylanases and their biomedical applications: a review. Int. J. Basic Clin. Pharmacol. **2**, 237 (2013)
44. Polizeli, M.L.T.M., Rizzatti, A.C.S., Monti, R., Terenzi, H.F., Jorge, J.A., Amorim, D.S.: Xylanases from fungi: properties and industrial applications. Appl. Microbiol. Biotechnol. **67**, 577–591 (2005)

45. Lowe, S.E., Theodorou, M.K., Trinci, A.P.: Cellulases and xylanase of an anaerobic rumen fungus grown on wheat straw, wheat straw holocellulose, cellulose and xylan. Appl. Environ. Microbiol. **53**, 1216–1223 (1987)
46. Gawande, P.V., Kamat, M.Y.: Production of *Aspergillus xylanase* by lignocellulosic waste fermentation and its application. J. Appl. Microbiol. **87**, 511–519 (1999)
47. Vandamme, E.J., Derycke, D.G.: Microbial inulinases: Fermentation process, properties and applications. Adv. Appl. Microbiol. **29**, 139–176 (1983)
48. Vijayaraghavan, K., Yamini, D., Ambika, V., Sravya Sowdamini, N.: Trends in inulinase production—a review. Crit. Rev. Biotechnol. **29**, 67–77 (2009)
49. Chesini, M., Neila, L.P., Fratebianchi de la Parra, D., Rojas, N.L., Contreras Esquivel, J.C., Cavalitto, S.F., Ghiringhelli, P.D., Hours, R.A.: Aspergillus kawachii produces an inulinase in cultures with yacon (*Smallanthus sonchifolius*) as substrate. Electron. J. Biotechnol. **16** (2013)
50. Chi, Z., Chi, Z., Zhang, T., Liu, G., Yue, L.: Inulinase-expressing microorganisms and applications of inulinases. Appl. Microbiol. Biotechnol. **82**, 211–220 (2009)
51. Mazutti, M., Bender, P., Treichel, H., Di, L.M.: Optimization of inulinase production by solid-state fermentation using sugarcane bagasse as substrate. Enzyme Microb. Technol. **39**, 56–59 (2006)
52. Dilipkumar, M., Rajasimman, M., Rajamohan, N.: Utilization of copra waste for the solid-state fermentatative production of inulinase in batch and packed bed reactors. Carbohydr. Polym. **102**, 662–668 (2014)
53. Gupta, A.K., Kaur, N., Singh, R.: Fructose and inulinase production from waste *Cichorium intybus* roots. Biol. Wastes **29**, 73–77 (1989)
54. Clarke, J.H., Davidson, K., Rixon, J.E., Halstead, J.R., Fransen, M.P., Gilbert, H.J., Hazlewood, G.P.: A comparison of enzyme-aided bleaching of softwood paper pulp using combinations of xylanase, mannanase and alpha-galactosidase. Appl. Microbiol. Biotechnol. **53**, 661–667 (2000)
55. Naganagouda, K., Salimath, P.V., Mulimani, V.H.: Purification and characterization of endo-beta-1,4 mannanase from *Aspergillus niger* gr for application in food processing industry. J. Microbiol. Biotechnol. **19**, 1184–1190 (2009)
56. Mamma, D., Hatzinikolaou, D.G., Christakopoulos, P.: Biochemical and catalytic properties of two intracellular β-glucosidases from the fungus *Penicillium decumbens* active on flavonoid glucosides. J. Mol. Catal. B Enzym. **27**, 183–190 (2004)
57. Dhawan, S., Kaur, J.: Microbial mannanases: an overview of production and applications. Crit. Rev. Biotechnol. **27**, 197–216 (2007)
58. Yin, J.-S., Liang, Q.-L., Li, D.-M., Sun, Z.-T.: Optimization of production conditions for β-mannanase using apple pomace as raw material in solid-state fermentation. Ann. Microbiol. **63**, 101–108 (2013)
59. Olaniyi, O.O., Osunla, C.A., Olaleye, O.O.: Exploration of different species of orange peels for mannanase production. E3 J. Biotechnol. Pharm. Res. 5, 12–17 (2014)
60. Rashid, J.I.A., Samat, N., Yusoff, W.M.W.: Studies on extraction of mannanase enzyme by *Aspergillus terreus* SUK-1 from fermented Palm Kernel Cake. Pak. J. Biol. Sci. **16**, 933–938 (2013)
61. Onilude, A.A., Festus Fadahunsi, I., Antia, E., Garuba, E.O., Inuwa, M., Afaru, J.: Characterization of Crude Alkaline β-mannosidase produced by *Bacillus* sp. 3A Isolated from Degraded Palm Kernel Cake. AU J. Technol. 15, 152–158 (2012)
62. Almeida, J.M., Lima, V.A., Giloni-Lima, P.C., Knob, A.: Passion fruit peel as novel substrate for enhanced β-glucosidases production by *Penicillium verruculosum*: Potential of the crude extract for biomass hydrolysis. Biomass Bioenergy **72**, 216–226 (2015)
63. McCleary, B.V., Matheson, N.K.: Action patterns and substrate-binding requirements of β-d-mannanase with mannosaccharides and mannan-type polysaccharides. Carbohydr. Res. **119**, 191–219 (1983)
64. Chauhan, P.S., Puri, N., Sharma, P., Gupta, N.: Mannanases: Microbial sources, production, properties and potential biotechnological applications. Appl. Microbiol. Biotechnol. **93**, 1817–1830 (2012)

65. Hang, Y.D., Woodams, E.E.: β-Fructofuranosidase production by *Aspergillus species* from apple pomace. LWT Food Sci. Technol. **28**, 340–342 (1995)
66. Rashad, M.M., Nooman, M.U.: Production, purification and characterization of extracellular invertase from *Saccharomyses Cerevisiae* NRRL Y-12632 by solid-state fermentation of red carrot residue. Aust. J. Basic Appl. Sci. **3**, 1910–1919 (2009)
67. Alegre, A.C.P., Polizeli, M.d.L.T.d.M., Terenzi, H.F., Jorge, J.A., Guimarães, L.H.S.: Production of thermostable invertases by *Aspergillus caespitosus* under submerged or solid state fermentation using agroindustrial residues as carbon source. Braz. J. Microbiol. 40, 612–622 (2009)
68. Uma, C., Gomathi, D., Ravikumar, G., Kalaiselvi, M., Palaniswamy, M.: Production and properties of invertase from a *Cladosporium cladosporioides* in SmF using pomegranate peel waste as substrate. Asian Pac. J. Trop. Biomed. **2**, S605–S611 (2012)
69. Sakai, T., Sakamoto, T., Hallaert, J., Vandamme, E.J.: Pectin, pectinase and protopectinase: production, properties and applications. Adv. Appl. Microbiol. **39**, 213–294 (1993)
70. Servili, M., Begliomini, A.L., Montedoro, G., Petruccioli, M., Federici, F.: Utilisation of a yeast pectinase in olive oil extraction and red wine making processes. J. Sci. Food Agric. **58**, 253–260 (1992)
71. Hours, R.A., Voget, C.E., Ertola, R.J.: Some factors affecting pectinase production from apple pomace in solid-state cultures. Biol. Wastes **24**, 147–157 (1988)
72. Botella, C., Diaz, A., de Ory, I., Webb, C., Blandino, A.: Xylanase and pectinase production by *Aspergillus awamori* on grape pomace in solid state fermentation. Process Biochem. **42**, 98–101 (2007)
73. Almeida, C., Brányik, T., Moradas-Ferreira, P., Teixeira, J.: Continuous production of pectinase by immobilized yeast cells on spent grains. J. Biosci. Bioeng. **96**, 513–518 (2003)
74. Patil, S.R., Dayanand, A.: Production of pectinase from deseeded sunflower head by *Aspergillus niger* in submerged and solid-state conditions. Bioresour. Technol. **97**, 2054–2058 (2006)
75. Silva, D., Tokuioshi, K., da Silva Martins, E., Da Silva, R., Gomes, E.: Production of pectinase by solid-state fermentation with *Penicillium viridicatum* RFC3. Process Biochem. **40**, 2885–2889 (2005)
76. Sawant, R., Nagendran, S.: Protease: an enzyme with multiple Industrial Applications. World J. Pharm. Pharm. Sci. **3**, 568–579 (2014)
77. Gupta, R., Beg, Q., Lorenz, P.: Bacterial alkaline proteases: Molecular approaches and industrial applications. Appl. Microbiol. Biotechnol. **59**, 15–32 (2002)
78. Pillai, P., Mandge, S., Archana, G.: Statistical optimization of production and tannery applications of a keratinolytic serine protease from Bacillus subtilis P13. Process Biochem. **46**, 1110–1117 (2011)
79. Radha, S., Nithya, V.J., Himakiran Babu, R., Sridevi, A., Prasad, N., Narasimha, G.: Production and optimization of acid protease by *Aspergillus* spp under submerged fermentation. Arch. Appl. Sci. Res. **3**, 155–163 (2011)
80. Nascimento, R.P., Junior, N.A., Coelho, R.R.R.: Brewer's spent grain and corn steep liquor as alternative culture medium substrates for proteinase production by Streptomyces malaysiensis AMT-3. Braz. J. Microbiol. **42**, 1384–1389 (2011)
81. Chancharoonpong, C., Hsieh, P.-C., Sheu, S.-C.: Enzyme production and growth of *Aspergillus oryzae* S. on Soybean Koji Fermentation. APCBEE Procedia **2**, 57–61 (2012)
82. Belmessikh, A., Boukhalfa, H., Mechakra-Maza, A., Gheribi-Aoulmi, Z., Amrane, A.: Statistical optimization of culture medium for neutral protease production by *Aspergillus oryzae.* Comparative study between solid and submerged fermentations on tomato pomace. J. Taiwan Inst. Chem. Eng. **44**, 377–385 (2013)
83. Veerabhadrappa, M.B., Shivakumar, S.B., Devappa, S.: Solid-state fermentation of Jatropha seed cake for optimization of lipase, protease and detoxification of anti-nutrients in Jatropha seed cake using *Aspergillus versicolor* CJS-98. J. Biosci. Bioeng. **117**, 208–214 (2014)
84. Kieliszek, M., Misiewicz, A.: Microbial transglutaminase and its application in the food industry. A review. Folia Microbiol. **59**, 241–250 (2014)

85. Motoki, M., Seguro, K.: Transglutaminase and its use for food processing. Trends Food Sci. Technol. **9**, 204–210 (1998)
86. Cortez, J., Bonner, P.L., Griffin, M.: Application of transglutaminases in the modification of wool textiles. Enzym. Microb. Technol. **34**, 64–72 (2004)
87. Téllez-Luis, S.J., González-Cabriales, J.J., Ramírez, J.A., Vázquez, M.: Production of Transglutaminase by *Streptoverticillium ladakanum* NRRL-3191 using glycerol as carbon source. Food Technol. Biotechnol. **42**, 75–81 (2004)
88. de Souza, C.F.V., Rodrigues, R.C., Heck, J.X., Ayub, M.A.Z.: Optimization of transglutaminase extraction produced by *Bacillus circulans* BL32 on solid-state cultivation. J. Chem. Technol. Biotechnol. **83**, 1306–1313 (2008)
89. Aravindan, R., Anbumathi, P., Viruthagiri, T.: Lipase applications in food industry. Indian J. Biotechnol. **6**, 141–158 (2007)
90. Fernandez-Lafuente, R.: Lipase from Thermomyces lanuginosus: uses and prospects as an industrial biocatalyst. J. Mol. Catal. B Enzym. **62**, 197–212 (2010)
91. Mohammadi, M., Habibi, Z., Dezvarei, S., Yousefi, M., Samadi, S., Ashjari, M.: Improvement of the stability and selectivity of *Rhizomucor miehei* lipase immobilized on silica nanoparticles: Selective hydrolysis of fish oil using immobilized preparations. Process Biochem. **49**, 1314–1323 (2014)
92. Prasad, M.P., Manjunath, K.: Comparative study on biodegradation of lipid-rich wastewater using lipase producing bacterial species. Indian J. Biotechnol. **10**, 121–124 (2011)
93. Sabat, S., et al.: Study of enhanced lipase production using agrowaste product by bacillus stearothermophilus Mtcc 37. IJPCBS **2**(3), 266–274 (2012)
94. Rohit, S., Chisti. Y., Banerjee. U.C.: Production, purification, characterization and applications of lipases. Biotechnol. Adv. (2001)
95. Salihu, A., Bala, M., Alam, M.Z.: Lipase production by *Aspergillus niger* using sheanut cake: an optimization study. J. Taibah Univ. Sci. **10**, 850–859 (2016)
96. Kanmani, P., Kumaresan, K., Aravind, J.: Utilization of coconut oil mill waste as a substrate for optimized lipase production, oil biodegradation and enzyme purification studies in *Staphylococcus pasteuri*. Electron. J. Biotechnol. **18**, 20–28 (2015)
97. Toscano, L., Gochev, V., Montero, G., Stoytcheva, M.: Enhanced production of extracellular lipase by novel mutant strain of aspergillus niger. Biotechnol. Biotechnol. Equip. 2243–2247 (2011)
98. Prasad, S., Singh, A., Joshi, H.C.: Ethanol as an alternative fuel from agricultural, industrial and urban residues. Resour. Conserv. Recycl. **50**, 1–39 (2007)
99. Gavrilescu, M., Chisti, Y.: Biotechnology—a sustainable alternative for chemical industry. Biotechnol. Adv. **23**, 471–499 (2005). https://doi.org/10.1016/j.biotechadv.2005.03.004
100. Gurung, N., Ray, S., Bose, S., Rai, V.: A broader view: microbial enzymes and their relevance in industries, medicine and beyond. BioMed. Res. Int. **2013**, 329121 (2013)
101. Li, S., Yang, X., Yang, S., Zhu, M., Wang, X.: Technology prospecting on enzymes: application, marketing and engineering. Comput. Struct. Biotechnol. J. **2**, e201209017 (2012)
102. Pandey, A., Binod, P., Palkhiwala, P., Gaikaiwari, R., Madhavan, N., Duggal, A., Dey, K.: Industrial enzymes—present status and future perspectives for India. J. Sci. Ind. Res. **72**, 271–286 (2013)
103. Adrio, J.L., Demain, A.L.: Microbial enzymes: tools for biotechnological processes. Biomolecules **4**(1), 117–139 (2014)
104. Kalim, B., Böhringer, N., Ali, N., Schäberle, T.F.: Xylanases—from microbial origin to industrial application. Br. Biotechnol. J. **7**(1), 1–20 (2015)
105. Heipieper, H.J., Neumann, G., Cornelissen, S., Meinhardt, F.: Solvent-tolerant bacteria for biotransformations in two-phase fermentation systems. Appl. Microbiol. Biotechnol. **74**(5), 961–973 (2007)
106. Qingchun, J., Sujing, X., Bingfang, H., Xiaoning, L.: Purification and characterization of an organic solvent-tolerant lipase from Pseudomonas aeruginosa LX1 and its application for biodiesel production. J. Mol. Catal. B Enzym. **66**(3–4), 264–269 (2010)

# The Future Promising Alternative Renewable Energy from Microbial Fuel Cell

**Nurul Atiqah Shamsudin, Muhammad Najib Ikmal Mohd Sabri, Husnul Azan Tajarudin, Ana Masara Ahmad Mokhtar, and Muaz Mohd Zaini Makhtar**

**Abstract** The concerns about pollution, resource depletion, and climate change caused by prolonged use of traditional fossil and nuclear fuels have generated renewed interest in renewable energy alternatives. The energy policies were put in place to encourage the usage of renewable energy. Renewable energy is expected to meet roughly 15–20% of the world's energy needs by 2030. As a result, novel technologies that could make renewable resources more accessible are urgently needed to meet this expanding need. These technologies must be developed in order to conserve non-renewable energy reserves while also developing more sustainable technology. Microbial fuel cells (MFCs) are bioelectrochemical devices that use the power of respiring organisms to directly transform organic substrates into electrical energy. The device is made up of an MFC reactor, an external circuit, and a voltage monitoring system. Organic materials such as domestic, poultry, agricultural, and wastewater can be used as carbon sources for MFCs. Protons, electrons, carbon dioxide, and electricity were produced by electrogenic bacteria of various sorts breaking down the carbons in the organic fuel. This paper describes the theory and current developments in the field of MFCs, as well as some potential future uses, such as a bioremediation process for a sustainable environment that also generates electricity. Significant associations for a great potential through implementation of MFC technology were identified based on the use of diverse substrates and a wide spectrum of contaminants as a future promising renewable energy.

**Keywords** Membraneless microbial fuel cell · Bioremediation · Bioconversion · Electricity · Renewable energy

N. A. Shamsudin · M. N. I. M. Sabri · H. A. Tajarudin · A. M. A. Mokhtar · M. M. Z. Makhtar (✉)
School of Industrial Technology, Bioprocess Technology Division, Universiti Sains Malaysia, 11800 Penang, Malaysia
e-mail: muazzaini@usm.my

M. M. Z. Makhtar
Fellow of Centre for Global Sustainability Studies, Universiti Sains Malaysia, 11800 Penang, Malaysia



A. Z. Yaser et al. (eds.), *Waste Management, Processing and Valorisation*,
https://doi.org/10.1007/978-981-16-7653-6_13

## 1 Introduction

Statistically, over 1.06 billion people (14% averagely of the global population) live without electricity [1]. This statement was supported by Fig. 1; hence, both Africa and Malaysia have reported that over 600 million, 603 village in Sabah did not have access on electricity, respectively [2, 3]. There problems arise from non-renewable energy such as oil, coal, and natural gas as shown in Fig. 2. The two major problems

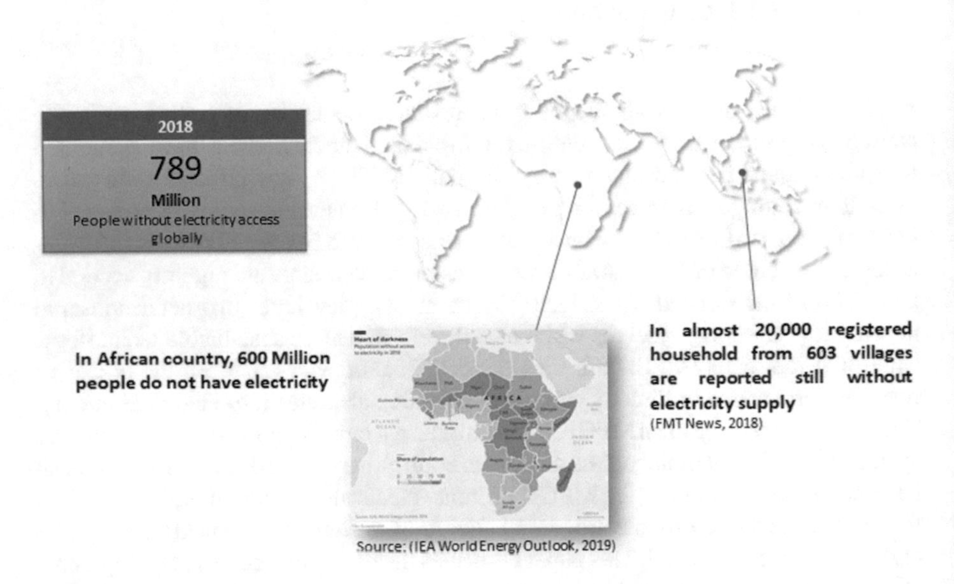


**Fig. 1** Energy statistic [2, 3]

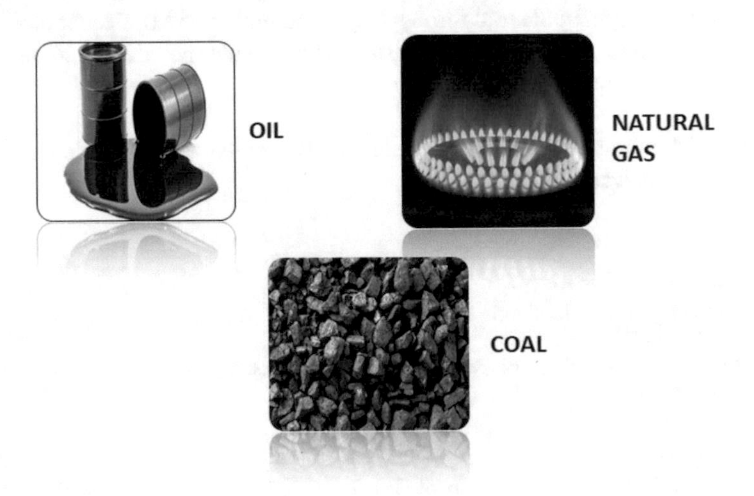


**Fig. 2** Non-renewable sources [2, 3]

**Fig. 3** Renewable sources [2, 3]

from these conventional energies which first is ready to be exhausted and cannot be replaced. Additionally increased coal usage in Southeast Asia contributes for one-third of all $CO_2$ emissions [4]. Data from [5] reported that a deficit on crude oil reservation from 2015 to 2017 about 1181 billion barrel in Malaysia [5]. With all this circumstances, renewable energy is one of the solutions which can be implemented due to embedded hugely. As stated in Fig. 3, there various types of renewable energy are solar, wind, geothermal, hydropower, and biomass. As shown in a report by Martha [6], rising the share of renewable energy in final energy consumption from 19% in 2017 to 65% by 2050 is critical to satisfy the 2 °C climate target [6].

## 2 World Current Problem

Recent decades, an exponentially rise on usage of energy worldwide ranging from 8588.9 million tonnes (Mtoe) in 1995 to 13, 147.3 Mtoe in 2015 may cause a depletion of natural resource which driven an impact on deficit on amount of natural resource, crude oil mainly. Crude oil affected hugely as stated by [7] that reservation of crude oil may last for another 19 years [7]. Since the early 1970s, emissions have doubled, hastening environmental change and climate depletion due to $CO_2$ emission data [8, 9].

Much of the literature on price tag of non-renewable energy such as coal and natural gas acknowledges that it may get more expensive and fluctuations. Perhaps, escalation of price for coal been nearly up to USD45 per tonne by 2030 due to tighter demands and limited stocks additionally natural gas also getting more expensive in the world's market [7].

Rapid population growth leads to uncontrolled solid waste as statistically production rate of solid waste is expected to exceed 9 Mt/year by 2020 [10, 11]. It has been inclusively shown that food waste makes up the majority of Malaysian garbage, which is estimated to raise to 9, 820, 000 tonnes by 2020 in Peninsular Malaysia alone [12]. While carried out on waste management by Malaysia government, there is a report that 70% and 50% for both collection of waste budget and municipal operating budget, respectively [10, 13]. In order to handle substantial quantities of solid waste, respected authorities must construct adequate sewage treatment plants and a particular land filling space [14, 15].

All above problems can be carter well with a green technology named as microbial fuel cell (MFC). MFC is a device which converts chemical energy to electrical energy by catalytic reaction of microorganism as shown by Fig. 4. Advantages of this device are (1) can use organic matter as fuel, (2) can produce energy in remote areas, (3) can produce energy and simultaneously treat wastewater when used wastewater as fuel, and (4) avoiding the toxicity problems of methanol [16]. With this technology, Sustainable Development Goal (SDG) can be achieved on clean water and sanitation; pillar 6, affordable and clean energy; pillar 7, industry, innovation, and infrastructure; pillar 9 and climate action; pillar 13 as determined by Fig. 5. Malaysia has begun to move forward in enforcing the SDG since 11th Malaysia Plan in which highlighted the

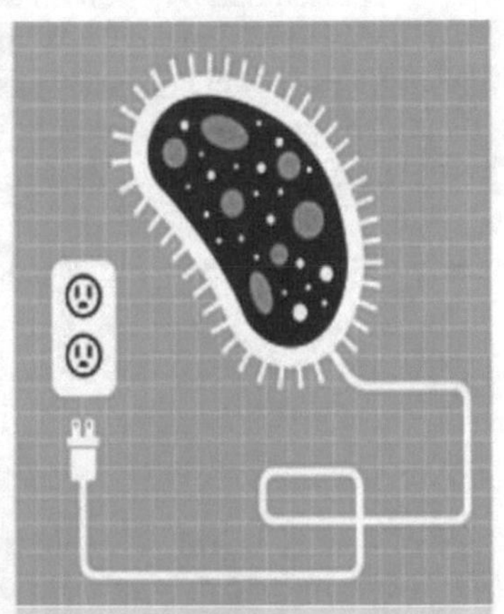

**MICROBIAL FUEL CELL**

"A microbial fuel cell is a device that converts chemical energy to electrical energy by the catalytic reaction of microorganisms."

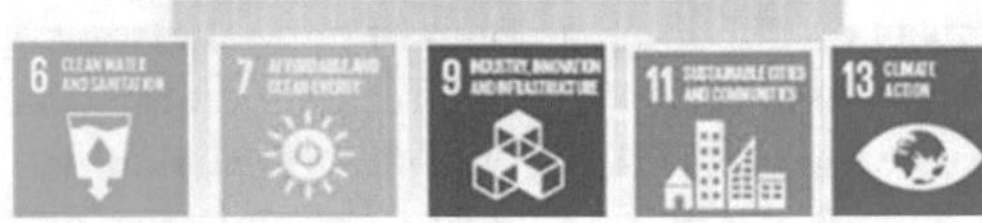

**Fig. 4** Fundamental definition of MFC

**Fig. 5** SDG opportunity on MFC perspective

significance of implementing green growth for long-term sustainability and resilience [17].

As stated by Table 1, differentiation between conventional electricity generation and MFC can be stated through raw material, availability of raw material, impact to environment and cost.

**Table 1** Comparison between conventional electricity generation and MFC

| Technical description | Conventional electricity generation | MFC |
|---|---|---|
| Raw material | Fossil fuel such as natural gas and diesel | Carbon wastes from the landfill |
| Availability of raw material | Fossil fuel resources around the world will be depleted soon | Carbon waste was generated continuously from the anaerobic digester and industry which are plenty and keep on increasing and free of charge |
| Impact to environment | Would create greenhouse gas effect which will cause great global warming due to pollution | Solve environmental pollution by utilizing the carbon waste from the treatment plants/landfill |
| Cost | The cost of fossil fuel keeps increasing over year | The cost is basically free due to carbon waste is an unwanted waste |

## 3 Microbial Fuel Cell at Emerging Stage

Microbial fuel cells (MFCs) technology has emerged in recent years as a promising yet challenging technology. The first knowledge report on bacteria that can generate electric current was by [18]. Systematically, based on Fig. 6a, 'Scopus' search with keyword 'microbial fuel cell' indicates almost 60-fold increase in the number of articles published over the last one decade (1998–2008) perpendicularly over the recent years it had been reported an electric current output from the MFCs. Perhaps Fig. 6b also shows the countrywise distribution of MFC researchers; the data also drawn from Scopus Web site [19]. Figure 7 shows a support on how obviously huge amount of citation and publication through another journal logger, Web of Science (WOS), related to MFC are increasing annually (ISI Web of Science, January 2017).

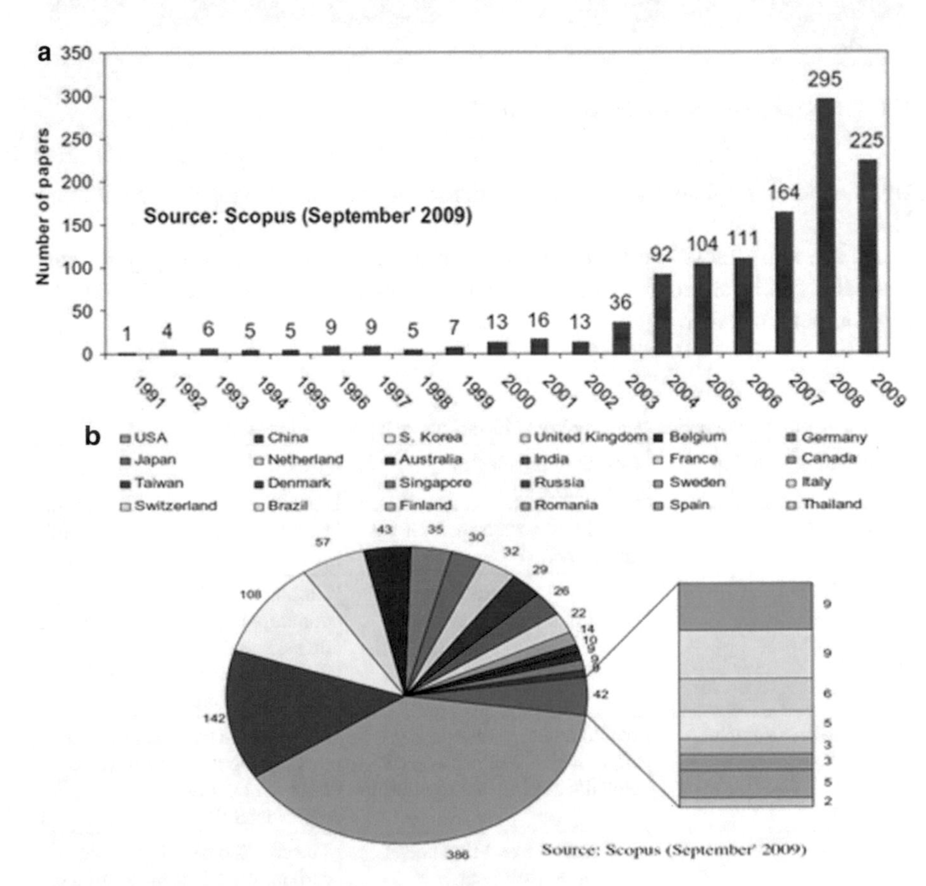


**Fig. 6** **a** Number of articles on MFCs. The data is based on the number of articles mentioning MFC in the citation database Scopus in September' 2009. **b** The countrywide distribution in MFC research [19]. Reprinted from Pant et al. [19], with permission from Elsevier

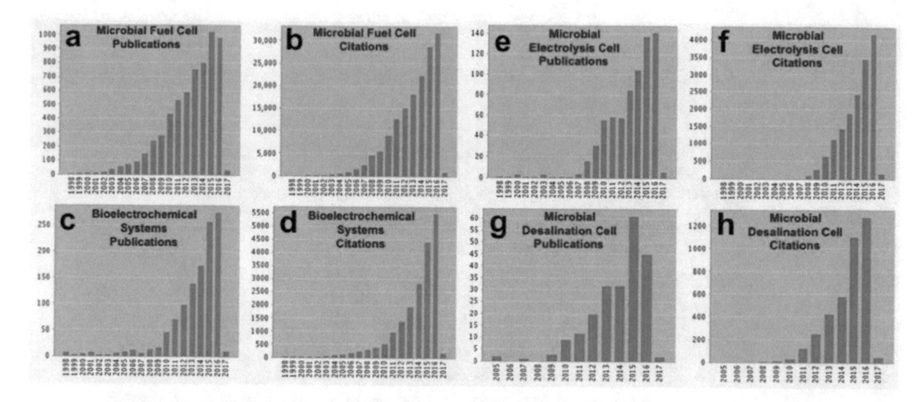


**Fig. 7** Quantitative analysis of the scientific literature on microbial fuel cells and bioelectrochemical systems (*Source* ISI WEB OF SCIENCE, January 2017)

Figure 8 concludes that how big and varieties of scope that researcher may looking at and engage with MFC; correlation between bioremediation (activity of microorganism) simultaneously generation of electricity is the most frequent topic engagement. Almost half of million results appear on 'Google Scholar' search engine if keyword of MFC been searched based on Fig. 9.

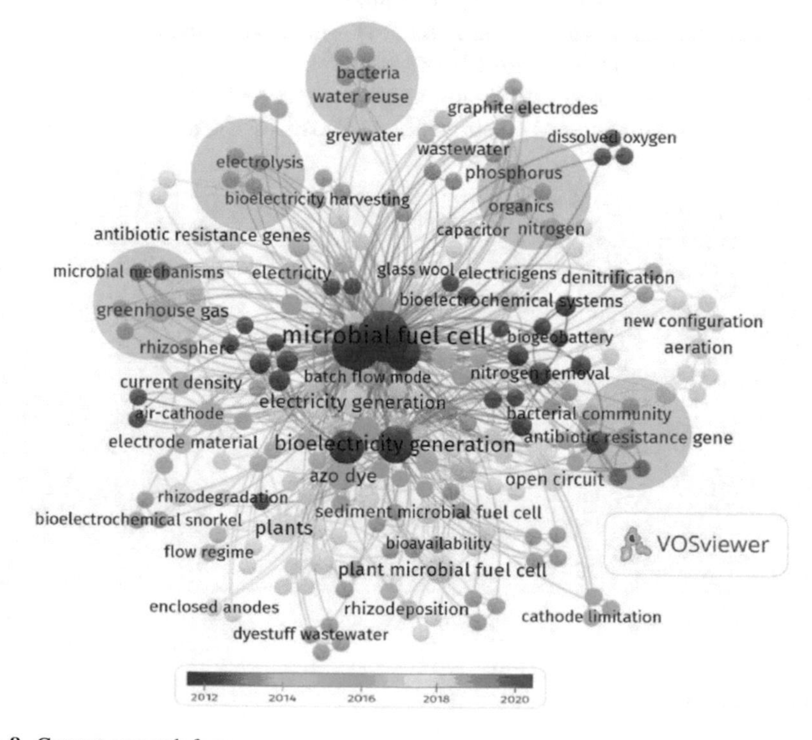


**Fig. 8** Current research focus

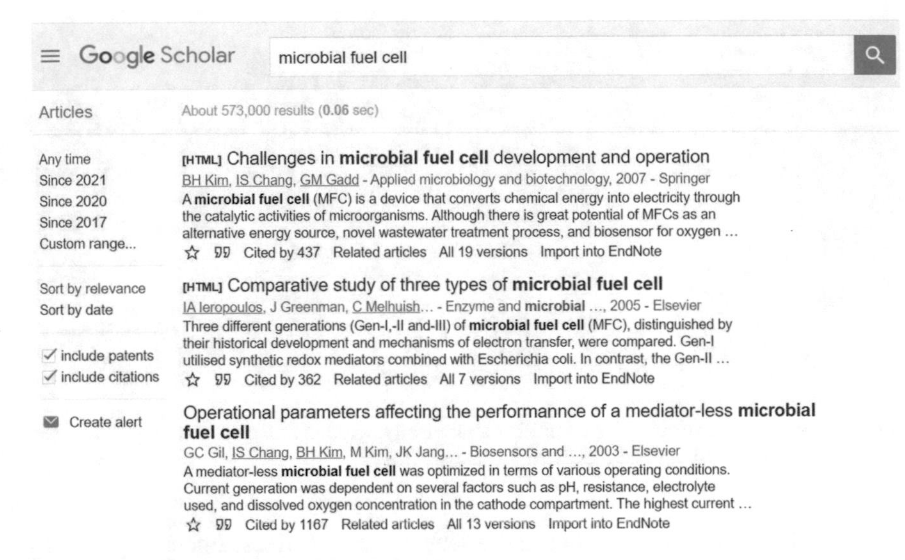


**Fig. 9** MFC visibility through 'Google Scholar'

Figure 10 illustrated three major disciplines of knowledge that had been covered by MFCs which are biological, chemical, and electrical. Biological methodology covers on the mechanism of microorganism attachment on anode (biofilm), biologically current production (electron transfer) and biological treatment efficiency (COD removal) [20, 21]. It has been similarly demonstrated by biological fuel cell which distinguished feature use of the living organism to convert chemical energy such as carbohydrates, directly into electric energy. Those MFCs, in reality, incorporate

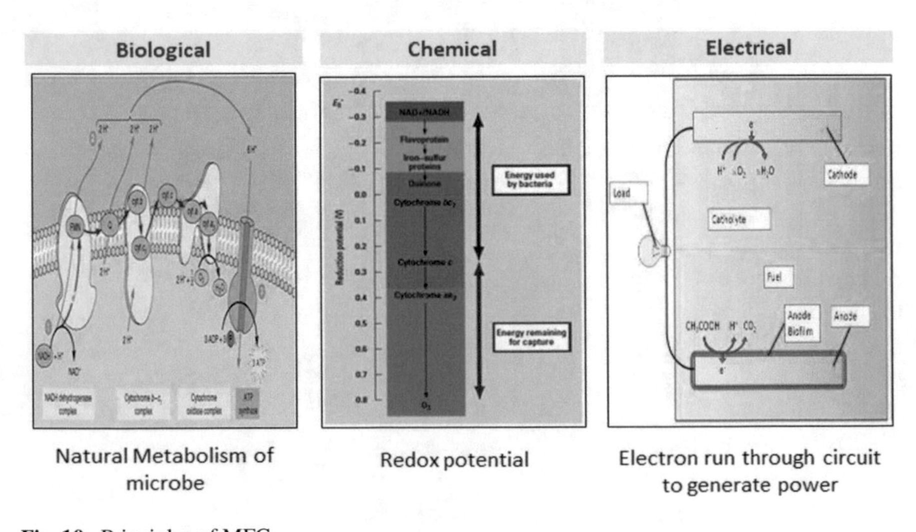


**Fig. 10** Principles of MFC

microorganisms or enzymes in the anode, cathode, or both compartments. As non-biological catalysts in chemical fuel cells, bacteria or enzymes are functioning as 'electrocatalysts' to significantly enable the electron transfer [22]. The final discipline is chemical, namely as electrochemical term in fuel cell technology. This methodology takes place at the electrodes to produce an electric current through an electrolyte. Two equations below are the overall and fundamental of reactions that occur during reduction and oxidation. Two equations below are the overall and fundamental of reactions that occur during reduction and oxidation (redox reaction):

$$C_6H_{12}O_6 + 6H_2O = 6CO_2 + 24H^+ + 24e^- \quad (1)$$

Anode: Oxidation of glucose, releasing electrons and creating $H^+$ ion; energy released

$$O_2 + 4H^+ + 4e^- \rightarrow 2H_2O \quad (2)$$

Cathode: Reduction reaction of oxygen with protons and electrons to form water.

Electrical part comes in order to evaluated how much power generation and time required for bioclectrochcmical reaction based on [20]:

$$V = I \times R \quad (3)$$

$$P = V \times I \quad (4)$$

where V is voltage, I is current, R is resistance, and P is power.

Based on Fig. 11, three types of bacterial are named as (1) electron donating microorganisms, (2) electron accepting microorganisms, and (3) electroactive microorganisms. Firstly, electron donating microorganism can easily divided into anode-respiring bacteria, anode-reducing bacteria and exoelectrogenic. Anode-respiring bacteria can catalyze electron transfer in organic substrates onto the anode as a replacement for natural extracellular electron acceptors by a variety of mechanisms [23]. This is also supported by Scott and Yu [22], Torres et al. [24] that this bacteria is able to conserve energy by respiration with an anode as electron acceptor. Meanwhile anode-reducing bacteria is microorganism that is able to donate electrons to an anode [22, 25]. Exoelectrogens came from 'exo-' from exocellular and 'electrogens' for the ability to transfer electrons to insoluble electron acceptors. This ability is suitable for mediatorless microbial fuel cell which is formation of biofilm on the anode surface. It is supposed to be motivated by the utilization of higher quantities of energy by the electrode materials [20, 26]. Iron reduction feature by *S.algae* known to be exoelectrogens whose function is conversion of glucose to current [26]

Secondly is electron accepting microorganisms divided into two categories which are electrotroph and cathodophile. Electrotroph is the ability by microorganism to

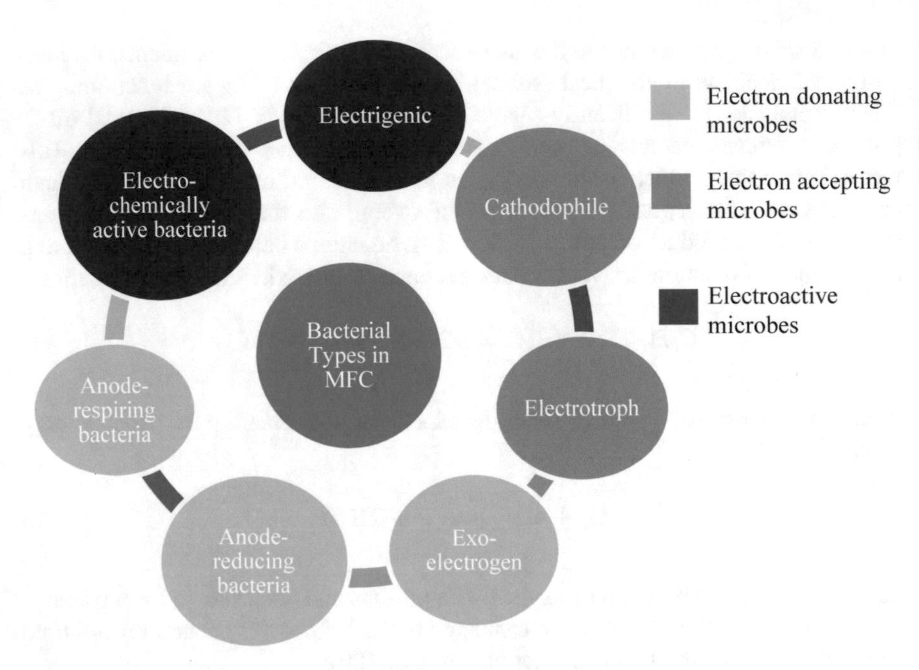


**Fig. 11** Type of electrogenic bacteria

draw electrons from a cathode via a direct contact mechanism or self-secreted mediators to conserve energy for growth [27, 28]. This type of microbes is not well established in the case of biocathode as stated by [29, 30]. Cathodophile favorably colonized on surface of the cathode in which inorganic carbon has been used as the sole carbon source for the growth of cathodophile bacteria [31]. This bacteria can grow in most of the aerobic and anaerobic biocathode MFCs [32–34]. New term is accelerated as ano-cathodophilic when the poised potential of an electroactive biofilm is to oxidize organic matter and reduce nitrogen contaminants alternatively [35, 36].

Thirdly is electroactive microorganism classified into two classes which are electrigenic and electrochemically active bacteria. Electrigens is a microorganism capable of fully oxidizing organic molecules to $CO_2$ using an electrode as the only electron acceptor while conserving energy for development [37]. Some of the electrigens are *Enterococcus faecalis* (isolated from faeces, facultative anaerobic, and gram positive bacteria), *Shewanella,* and *Escherichia coli* [38]. Electrochemically active bacteria (EAB) is an microorganism that is widely used in mediatorless MFC, gain electron by oxidizing the organic matter and transfer electrons to the surface of the anode [20]. EAB has been emerged as ability to donate electrons to or accept electrons from an electrode via a direct contact mechanism or self-secreted mediators [22, 39]. It has been assumed that understanding of EAB is still in early stage due to several unnoticed electrochemical capabilities of various microbes, which can be exploited for different MFC applications [40, 41].

The electron transportation can be divided into three types which are (1) outer membrane bound cytochrome, (2) nanowires, and (3) redox mediators [22] as shown Fig. 12. Firstly cytochromes came from bacteria that capable of donating electrons out of their cells toward a solid-state electron acceptors [42]. For example, thermophile *Thermincolapotens* likely uses cytochrome to directly transfer electrons to Fe (III) and an electrode in its genome and many located in the outer membrane [43]. So far, nanowire is a technique that consists of electrically conductive pili and considered as new way of transferring electron to electrode [20]. This technique enables electron transfer over micrometer—long distance. For example, *Geobactersulfurreducens* is capable of transfer electrons beyond cell surfaces to electrodes through membrane proteins or nanowires; meanwhile, *Shewanellaoneidensis* MR1 is capable of producing both nanowires and soluble redox mediators [22]. Since 1980s, redox-active mediators were widely used in MFCs to improve power performance [44]. Chemical modification on perforated pores and channels on cell membranes can accelerate electron transfer which facilitated redox mediator that resulted in improvement of power output for an MFC powered by *P.aeruginosa* [45].

Figure 13 shown varieties types of MFC which can be classified onto (1) based on mediator, (2) based on nutrition, (3) based on temperature, (4) based on light dependency, and (5) based on configuration. MFC based on mediator is mediatorless and mediator. Next is MFC based on nutrition classified onto phototrophic MFC, heterotrophic MFC, and mixotrophic MFC. Meanwhile, MFC based on temperature is only divided into thermophilic and mesophilic MFC. Moreover, light dependency MFC also can be classified into light-mediated cathodic MFC and light-mediated anodic MFC. Typically, worldwide MFCs focused based on configurations are (1) single-chambered MFC, (2) dual-chambered MFC, (3) up flow MFC, (4) tubular MFC, and (5) stacked MFC.

Firstly, single-chambered MFC possessed aeration on anodic chamber without including cathodic chamber. MFCs' internal resistance is reduced, resulting in increased electricity output [46]. Double-chambered MFC that was built by one

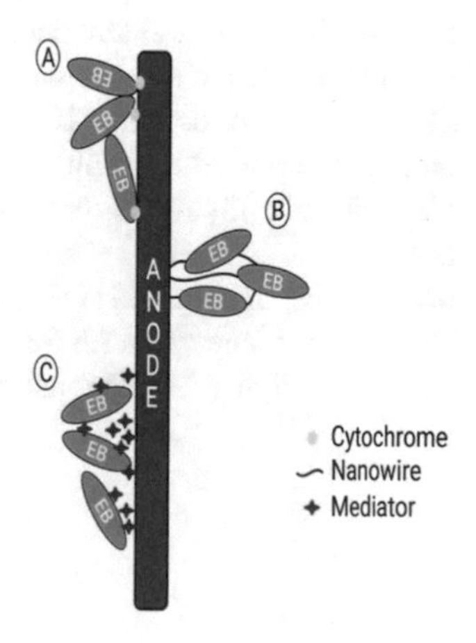


**Fig. 12** Electron transport in microbial fuel cells A outer membrane bound cytochrome, B solid conductive appendages, called 'nanowires,' and C redox mediators. Created with BioRender.com

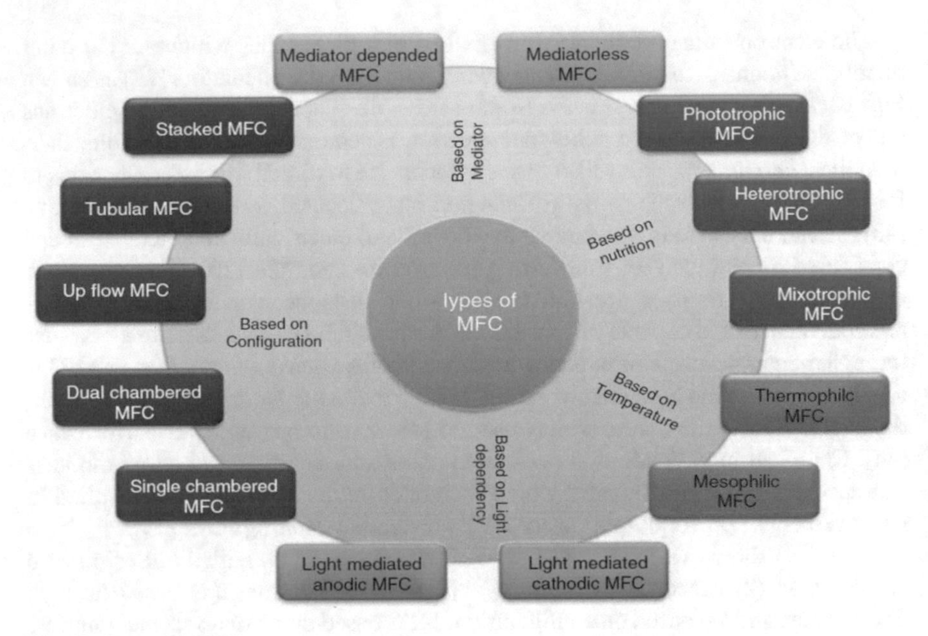


**Fig. 13** Type of MFCs

cathode chamber and one anode chamber, which was joined by a bridge and divided by a proton or cation exchange membrane, which allowed protons to pass to the cathode while blocking oxygen passage into the anode. Chemically the plain carbon cathode was catalyst and coated in ferryanide due to platinum expensively [46, 47]. Both single- and double-chambered MFC constitute the most common types of MFC. There is few study that studies on nutrient recovery using conventional double-chamber MFC [48]. Next is flat-plate MFC that can achieve a maximum power density for domestic wastewater obtained about 72 mW/m$^2$ increment 2.8 times compared to single-chambered MFC [47, 49]. The major highlighten on falt-plate design feature is the decreased distance between electrodes, which can reduce internal resistance and increases ionic diffusion rates compared to other designs [22, 50]. High power density (20.2 W/m$^2$) is achieved and the highest COD removal efficiency is reported by Mahdi Mardanpour [51] by using tubular MFC. Research by He et al. [52, 53] stated that both flat plate and tubular are categorized as configurations with low internal resistance and high power density [20, 52, 53]. Stack MFC operated by Li et al. [54] is capable of generating a maximum electrical power about 133 mW/m$^2$ linearly with 88% chemical oxygen demand (COD) removal in about 91 h [54]. Figure 14 demonstrated the configurations of typical MFCs worldwide.

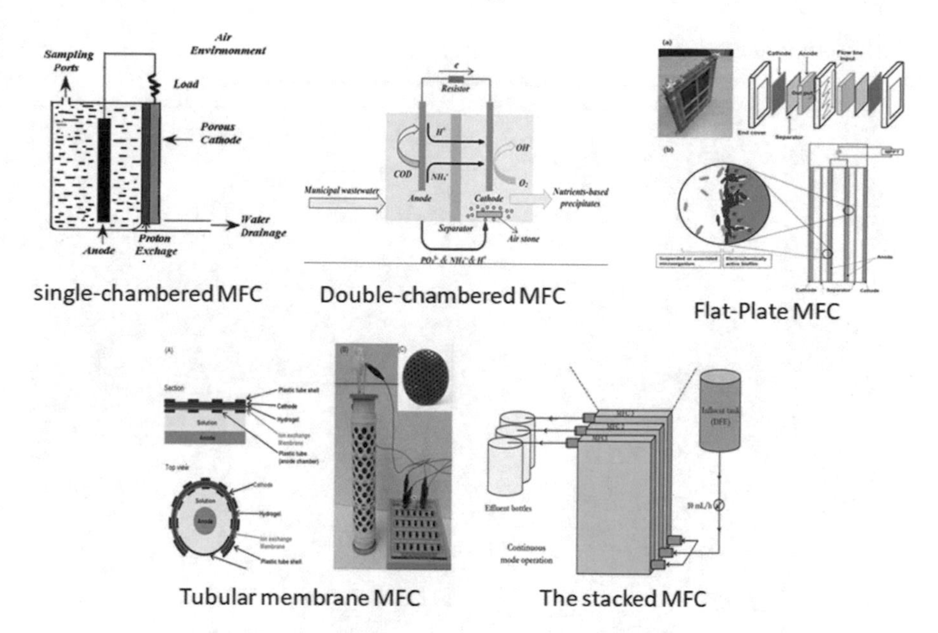


**Fig. 14** Typical MFCs [48, 55, 56]. Reprinted from [48, 55, 56] with permission from Elsevier

# 4 Basic Design of MFC

Common MFCs system is composed of anode chamber, cathode chamber, the separator, and external circuit. In basic MFC, the anodic chamber contains organic substrates that are metabolized by microbes for growth and energy production while generating electrons and proton while the cathode chamber contains a high potential electron acceptor to complete the circuit [57]. The configuration of MFCs systems can be developed into several types depending on the application, such as a single chamber or double chamber [58]. The MFC system process includes anodic processes, in which electron donors such as organic compounds and sulfide are oxidized, and cathodic reactions, in which electron acceptors such as oxygen, nitrate, nitrite, or perchlorate are reduced, that are examples of the process that happen in this technology [59].

Anode chamber is considered as the most important part of the MFC system because at this part various reactions happen such as electricity driven by the microbes, reduction of chemical oxygen demand (COD), reduction of the organic compound, wastewater treatment, biofilm formation, and hydrogen production occurred in this chamber [60]. The anode chamber is made up of a microorganism (catalyst) and an electrode (anode), and it can be fed with anolyte and redox mediators such as growth media or wastewater Debabrata [61]. The electrochemically active microbe in the anode chamber converts the organic material that obtains from the anolyte to produce carbon dioxide, proton, and electrons Debabrata [61]. The electrons released by microorganism metabolic activity are transported to and

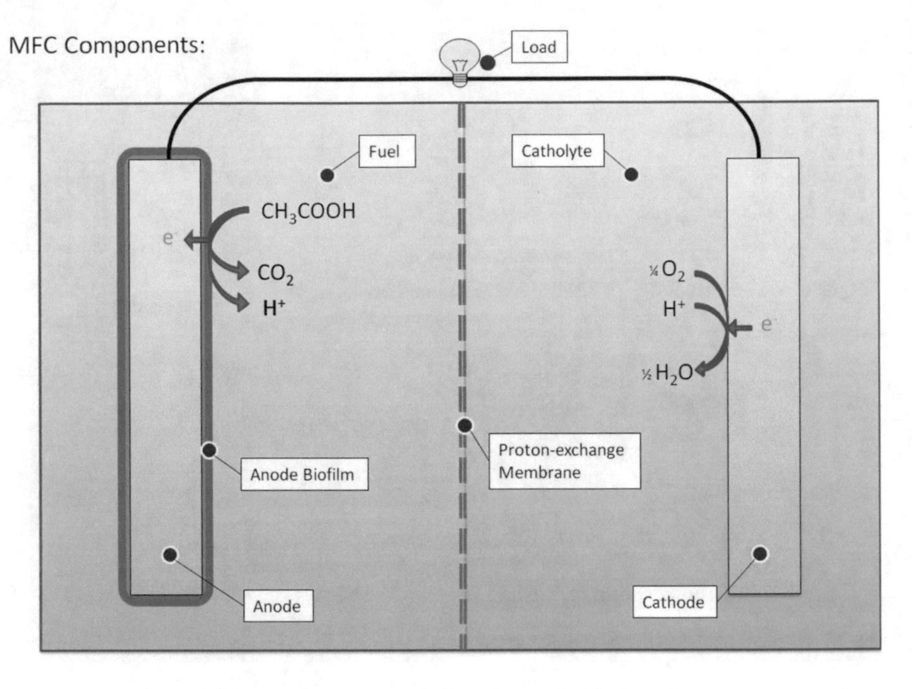


**Fig. 15** Basic design of the MFC component [48] with permission from Elsevier

collected on the anode's electrode surface transferred to the cathode via cytochromes or redox-active proteins [62].

In MFC, the cathode chamber is the electron and proton acceptor [60]. Electrons usually flow to the cathode through a conductive substance with external resistance. Accepting these electrons reduces the photons that travel through the membrane, and water is formed on the cathode when the requisite protons and electrons are removed during bacterial substrate catabolism and combined with oxygen Debabrata [61]. A salt bridge (or electrolytes) or an ion or proton exchange membrane (PEM) must connect the two chambers and complete the circuit. The produced protons can only pass in one direction: from the anode to the cathode chamber [57]. Biocatalyst is divided from oxygen by putting a membrane between two different chambers that allows charge to be passed between the electrodes, the anode chamber, where the bacteria grow, and the cathode chamber, where the electrons react with the oxygen [63] (Fig. 15).

## 5 Sources of MFC Substrate

The substrate is one of the most critical biological factors determining energy generation in MFCs [19]. The substrate is one of the most important factors in determining electricity generation since it gives sustenance and energy to the microorganisms employed in MFCs [59]. Various substrates can be used in MFC systems to generate electricity (Table 2). The qualities and components of the waste material, such as the chemical makeup of the organic compound that may be transformed into a product or fuel, determine the efficiency and economic viability of converting organic wastes to bioenergy [19]. Pure substrates such as glucose, acetate, butyrate, lactate, proteins, cellulose, cysteine, glycine, and glycerol were utilized in several studies [64]. As an example, the performance of MFC using acetate as substrate and is known that acetate is a non-fermentable substrate that can be used as an electron donor by dissimilatory iron-reducing bacteria, allowing them to generate up to 66% more power than butyrate [64].

Several studies have been published on the generation of energy directly from complex organic wastewater, such as municipal wastewater, ocean sediments, and industrial wastewater, such as starch processing wastewater and brewery wastewater [64]. Because many exoelectrogenic bacteria can only use a limited range of substrates, the complex mixture of organics found in most wastewater streams suggests that diverse microbial communities are required to oxidize the organic matter [59]. Because microorganisms catalyze power generation, much attention has been paid to characterize the microbial communities in MFCs powered by various fuels, and the type of substrate fed to an MFC was discovered to affect the structure and composition of the microbial community, which in turn influenced the efficiency of the MFCs [59].

The only goal of the various treatment processes so far has been to eliminate pollutants from waste streams before they can be safely discharged into the environment [19]. Due to differences in operating circumstances, surface area and kind of electrodes, and microorganisms involved, comparing MFC performances from the literature is difficult [19]. As a result, comprehensive tables for both defined and undefined substrates have been included to provide comparative information on various aspects of MFC performance, even though tabulating all parameters is often difficult due to the use of different MFC designs, operational conditions, various measurement units, and representation styles, to name a few [72]. Based on the table and figure below, this is the common substrate that commonly being used in MFC technology.

## *5.1 Acetate*

Acetate has been the substrate of choice for electricity generation in the majority of MFC studies thus far. When compared to acetate, the recalcitrance of many types of wastewater makes them more difficult to utilize [19]. Acetate is a straightforward substrate that is widely used as a carbon source to induce electroactive bacteria, and because of its inertness to alternative microbial conversions (fermentation and methanogenesis) at room temperature, it is commonly used as a substrate [19]. In other terms, acetate is a non-fermentable substrate and a very suitable electron donor

**Table 2** Type of substrate used in current research using various types of culture as the source of inoculum in a different type of MFC configurations [65] Debabrata [61]

| Type of substrate | Concentration | Source of inoculum | Type of MFC (with electrode surface area and/or cell volume) | Current density (mA/cm$^2$) | References |
|---|---|---|---|---|---|
| Acetate | 1 g/L | Pre-acclimated bacteria from MFC | Cube-shaped one-chamber MFC with graphite fiber brush anode (7170 m$^2$/m$^3$ brush volume) | 0.8 | Logan et al. [66] |
| Cellulose | 4 g/L | Pure culture of *Enterobacter cloacae* | U-tube MFC with carbon cloth anode (1.13 cm$^2$) and carbon fibers as cathode | 0.02 | Rezaei et al. [67] |
| Ethanol | 10 mM | Anaerobic sludge from wastewater plant | Two-chambered aqueous cathode MFC with carbon paper electrodes (22.5 cm$^2$) | 0.025 | Kim et al. [68] |
| Leachate | 6000 mg/L | Leachate and sludge | Two-chambered MFC with carbon veil electrode (30 cm$^2$) | 0.0004 | Gálvez et al. [69] |
| Starch | 10 g/L | Pure culture of *Clostridium butyricum* | Two-chambered MFC with woven graphite anode (7 cm$^2$) and ferricyanide catholyte | 1.3 | Niessen et al. [70] |
| Domestic wastewater | 600 mg/L | Anaerobic sludge | Two-chambered mediatorless MFC with plain graphite electrode (50 cm$^2$) | 0.06 | Wang et al. [71] |

for the dissimilatory iron-reducing bacterium that could generate power up to 66% power generation much higher compared to butyrate [64].

## 5.2 *Carbohydrate*

Various organics can be used to generate electricity in MFCs, and the monomers of all complex and high molecular weight wastewaters are the major metabolic fuels—carbohydrates, fatty acids, and amino acids. Carbohydrates are by far the most abundant group of these organics [72]. Because of their easy availability and abundance, lignocellulosic biomass, such as agricultural residues and woody biomass, has received a lot of attention for fuel and energy generation [72]. Agricultural residue-derived lignocellulosic materials are viable feedstock for cost-effective energy production due to their abundance and renewability [19].

## 5.3 *Glucose*

Glucose is another common substrate in MFCs. The performance of an MFC containing Proteus vulgaris has also been observed to be reliant on the carbon source in the microorganism's beginning medium, with glucose beginning cells in MFC running for a shorter period than galactose beginning cells [19]. It is reported a glucose fed-batch MFC with 100 mM ferric cyanide as cathode oxidant produced a maximum power density of 216 W/m$^3$ [19].

## 5.4 *Wastewater*

Wastewater can be used to provide the carbon source for bacterial growth and thus the end products of the oxidation process, that is, electrons and protons, for long-term bioelectricity generation [73]. Several bacterial growth mediums, such as cysteine, include considerable levels of redox mediators, while high-strength wastewater contains reduced sulfur species, which can operate as an abiotic electron donor and temporarily boost power output [19]. Inoculum or biocatalysts for substrate oxidation can be obtained from primary wastewaters from an industry, such as chocolate industry wastewater or palm oil mill effluent (POME) [73]. COD levels in brewery wastewater can exceed 5000 mg/L, which has been widely explored in several MFCs for treatment and bioenergy production [73]. Furthermore, it contains a high concentration of carbohydrates or sugars, which can be used as electron donors in MFCs [73]. Compared to industrial wastewater, municipal wastewater has the potential to be a promising substrate for MFC-based bioenergy production. This method can be used to create eco-friendly public toilets that generate electricity and help to keep the surrounding environment neat and clean [73] (Fig. 16).

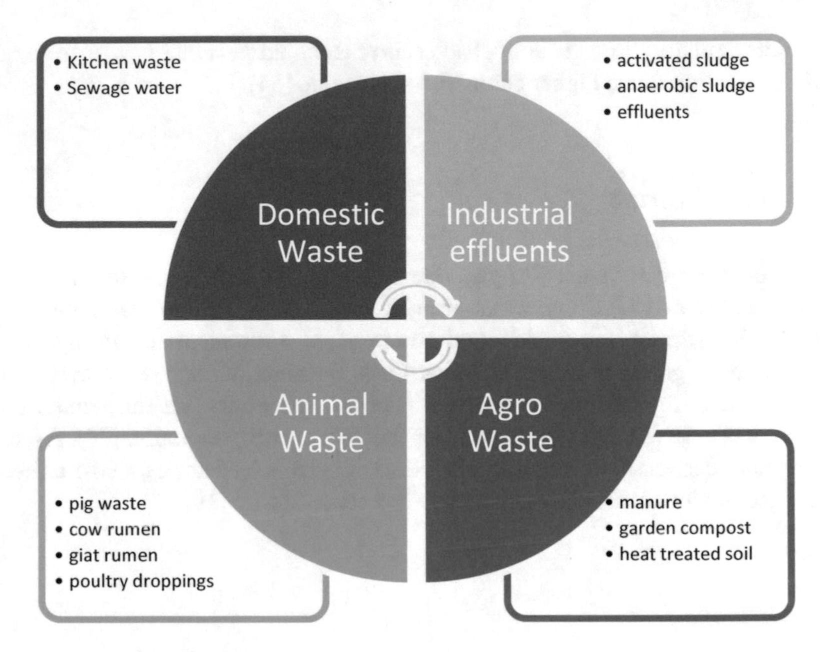


**Fig. 16** Source of MFC substrate

# 6 Bioremediation MFC

In a microbial fuel cell, a wide range of microorganisms can contribute to the generation of power. Researchers recently found a novel metabolic type of energy-producing microbes, indicating that a wide range of organic molecules can be converted to power in self-sustaining microbial fuel cells that are known as electrogens that can totally oxidize organic compounds to carbon dioxide with an electrode acting as the sole electron acceptor while conserving energy to support growth from electron transfer [74]. Another fascinating aspect of MFC is the development of large-scale MFC for the conversion of sewage and other organic waste to power, as well as the bioremediation of pollutants in the environment [74].

An MFC can be adjusted in a variety of fascinating and useful ways, leading to the development of novel fuel-cell-based technologies that the modification of the basic two-electrode setup for bioremediation is one such application [75]. MFC technology is not only utilized to generate electricity; instead, it might be utilized to drive reactions that remove or decompose waste [75]. Sustainable water management and soil bioremediation should strive to recover and reuse nutrients as well as degrade organic contaminants. MFCs can be employed as self-powered bioelectrochemical devices in natural situations for bioremediation of inorganic-contaminated water [76].

Biomass is an abundant and energy-rich resource, and it is projected to be an ideal fuel for MFC, and MFC would be an appealing energy conversion technology for the biomass business [77]. Biomass includes wood from trees and plants, as well as charcoal, crop residues, and animal waste. However, it is projected that it might meet 7–15% of the world's energy needs [60]. Wood, sugarcane bagasse, rice husk, rice straw, maize cob, and other lignocellulosic biomass (cellulose, hemicellulose, and lignin) and grass biomass (cellulose, hemicellulose, and lignin) are among the biomass types used [77]. The use of wood-based materials biomass for renewable energy production has sparked a lot of attention. For the synthesis of soluble sugars from solid lignocellulosic materials such as agricultural waste and hardwoods, the steam explosion is currently the most cost-effective treatment method (Logan & Regan, 2006). The use of a neutral hydrolysate created by the steam explosion of corn stover in an MFC has recently been demonstrated to be possible, with MFC testing yielding as much as 933 $mW/m^2$ [25, 78].

In some cases, the potential of MFC microbial communities to digest a wide range of environmental pollutants may be more beneficial than the production of energy itself, particularly when MFC technology can be employed for in situ environmental clean-up [79]. Thus, the abundance and renewability of cellulosic and lignocellulosic materials make them cheap renewable energy and carbon resources. The abundance of biomass substance has the advantage of being able to be cultivated and may be used to meet energy demands. It might also be easily kept and used for a longer period of time. It is possible to match the usage with the rate of growth to ensure long-term viability (Table 3).

## 7 Simultaneous of Bioremediation and Electricity

Rapid industrialization has resulted in an increase in the use of fossil fuels, restricting their availability and making future extraction difficult, and as a result, research has been focused on the development of clean and green alternative energy generation technologies to meet the rising population's energy demands Debabrata [61]. On numerous levels around the world, there have been ongoing debates and conversations about developing technologies that can help with the burning issue of climate change, both in terms of being based on renewable and sustainable energy and maybe assisting in the conversion of trash to valuable items Debabrata [61]. Microbial fuel cells (MFCs) use microbes to create electricity from a range of reduced materials, such as organic matter [60]. MFCs can use a wide range of organic and inorganic materials as substrates, including organic wastes, lipids, carbohydrates, and proteins.

In comparison to other technologies, MFCs have a high energy transformation efficiency because they transfer the chemical energy held in the substrates into electrical power energy [83]. MFCs have the primary function of harvesting bioelectricity via microbial redox processes, and they offer a wide range of uses that are environmentally friendly [60]. MFC has a low environmental effect since it can remove and reuse a wide range of toxins and wastes, including home, agricultural, and industrial wastewater [84]. Furthermore, when organic material degrades, it only produces $CO_2$, which has a lower influence on climate change than methane and in comparison to other treatment approaches, and the risk of secondary contamination is considerably

**Table 3** Performance of MFC technologies using variety types of biomass substrate

| Substrate | MFC type | Feeding mode | Volume (NAC) (mL) | Anode material | Cathode material | Substrate concentration | COD removal (ηCOD) | Power density (max) | CE | References |
|---|---|---|---|---|---|---|---|---|---|---|
| Powdered rice straw | Two-chambered MFC | Batch | 160 | Carbon paper (10 $cm^2$) | Carbon paper (10 $cm^2$) | 1000 mg COD/L | | 190 $mW/m^2$ | 37% | [80] |
| Powdered rice straw | H-type MFC | Batch | 200 | Carbon paper (10 $cm^2$) | Carbon paper (10 $cm^2$) | 1000 mg COD/L | | 145 $mW/m^2$ | 54.3–45.3% | [75] |
| Steam exploded corn stover | Single-chamber air–cathode MFC | Batch | 28 | Carbon paper (7.1 $cm^2$) | Carbon cloth + Pt catalyst (0.5 $mg/cm^2$) (7.1 $cm^2$) | 1 g COD/L | 60–70% | 861 ± 37 $mW/m^2$ | 20–30% | [77] |
| Raw corn stover | Bottle-type air–cathode MFC | Batch | 250 | Carbon paper | Carbon cloth + Pt catalyst (0.35 $mg/cm^2$) (4.9 $cm^2$) | | 42% ± 8% (cellulose) 17% ± 7% (hemicellulose) | 331 $mW/m^2$ | 3.6% | Wang et al. (2009) |
| Rice milling | Earthen pot MFC | Batch | 400 | Stainless–steel mesh (190 $cm^2$) | Graphite plate (231 $cm^2$) | 2250 mg COD/L | 96.5% | 2.3 $W/m^3$ | 21% | [72] |
| Rice straw hydrolysate | Air–cathode single-chamber MFC | Batch | 220 | Carbon brush | Gas diffusion electrode + Pt catalyst (0.5 mg/ $cm^2$) | 400 mg COD/L | 49–72% | 137.6 ± 15.5 $mW/m^2$ | 8.5–17% | Wang et al. (2009) |
| Steam exploded corn stover residue | Bottle-type air–cathode MFC | Batch | 250 | Carbon paper | Carbon cloth + Pt catalyst (0.35 $mg/cm^2$) (4.9 $cm^2$) | | 60% ± 4% (cellulose) 15% ± 4% (hemicellulose) | 406 $mW/m^2$ | 1.6% | [19] |

(continued)

**Table 3** (continued)

| Substrate | MFC type | Feeding mode | Volume (NAC) (mL) | Anode material | Cathode material | Substrate concentration | COD removal (ηCOD) | Power density (max) | CE | References |
|---|---|---|---|---|---|---|---|---|---|---|
| Wheat straw hydrolysate | H-type dual-chambered MFC | Batch | 300 | Carbon paper (42 $cm^2$) | Carbon paper (42 $cm^2$) | 250 - 2000 mg COD/L | | 123 $mW/m^2$ | 15.5–37.1% | [65] |
| Beet sugar wastewater | Upflow anaerobic sludge blanket reactor MFC (air–cathode MFC | Continuous | 650 mL | Granular graphite (100 g) + graphite rod | Carbon paper + PTFE + Pt catalyst (0.5 $mg/cm^2$) | 127, 500 mg COD/L (diluted 20–5-0 times) | 53.2% | 1410 $mW/m^2$ | 1% | [65] |
| Beet sugar wastewater | ABSMFC MFC (4 units) | Continuous | 690 mL | Carbon fiber felt (56 $cm^2$) | Carbon cloth (wet proofed) + PTFE + Pt catalyst (0.5 $mg/cm^2$) | 127, 500 mg COD/L (diluted 40–20-10 times) | 70% | 115.5 $mW/m^2$ | | [72] |
| Cassava mill wastewater | Two-chambered MFC | Continuous | 15 L | Graphite plate (0.16 $m^2$) | Graphite plate (0.16 $m^2$) | 16, 000 mg COD/L | 72% | 1771 $mW/m^2$ | 20% | [65] |
| Cereal processing wastewater | Dual-chambered | Batch | 310 mL | Carbon paper | Carbon paper + Pt catalyst (0.5 $mg/cm^2$) | 595 mg COD/L | 95% | 81 $mW/m^2$ | 40.5% | [81] |
| Distillery wastewater (molasses based) | Single-chambered open air–cathode MFC | Batch | 500 mL | Graphite plate (70 $cm^2$) | Graphite plate (70 $cm^2$) | 15.2 kg COD/ ($m^3$ day) | 72.8% | 124.35 $mW/m^2$ | | [75] |

(continued)

**Table 3** (continued)

| Substrate | MFC type | Feeding mode | Volume (NAC) (mL) | Anode material | Cathode material | Substrate concentration | COD removal (ηCOD) | Power density (max) | CE | References |
|---|---|---|---|---|---|---|---|---|---|---|
| Molasses wastewater mixed with sewage | Single-chambered MFC | Continuous | 25 mL | Graphite fiber brush (10 $cm^2$) + stainless steel mesh current collector | Pressed AC + PTFE + Pt catalyst + stainless steel mesh current collector | 9958 mg COD/L | 59% | 382 $mW/m^2$ | | [75] |
| Mustard tuber wastewater | Dual-chambered MFC | Batch | 150 mL | Carbon cloth (40.5 $cm^2$) | Carbon cloth (40.5 $cm^2$) | 550 mg COD/L | 57.1% | 246 $mW/m^2$ | | Guo et al. (2013) |
| Olive mill wastewater diluted (1:10) | Single-chambered air–cathode MFC | Batch | 16 mL | Carbon cloth (7 $cm^2$) | Carbon cloth (7 $cm^2$) | 20.9 g COD/L | 65% | | | [82] |

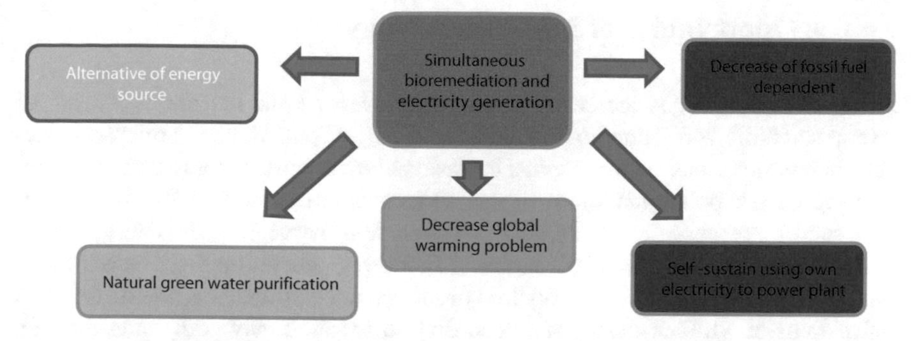


**Fig. 17** Impact of bioremediation and electricity generation using MFC technology

decreased in MFCs [84]. Thus, the capability of MFC technology for simultaneous electricity generation and bioremediation has large potential in new promising energy sources in near future (Fig. 17).

## 8 Commercialization Potential of MFC

The MFC has a number of advantages that distinguish it as one of the most promising cleaner technologies [84]. The development of large-scale MFC for the conversion of sewage and other organic waste to power, as well as the bioremediation of contaminated areas, is another fascinating field of MFC research [74]. MFC technology has been employed in a variety of applications, but there are significant problems that must be overcome in order for the technology to be economically feasible [73]. As a result, any pilot study, field trial, or prototype installation add to a valuable body of knowledge that helps to prepare technologies for real-world deployment and a wider market and several critical performance aspects are explored for optimization, as with any prototype development, setting the agenda for field testing a particular MFC technology [85]. Thus, MFC system technology can be scalable to produce electricity of higher power and voltage output. Electricity generation, wastewater treatment, heavy metal/toxic chemical bioremediation, and other niche uses are all possible using MFC technology [73]. Thus, it can be said that MFC technology is promising alternative technology that can utilize to generate bioelectricity using various types of biomass substrate simultaneously with further treatment.

# 9 The Opportunity of MFC Technology

MFC power output has increased significantly over the last decade as a result of several scientific and technological advances [72]. Applications for microbe–electrode interactions have also expanded to waste removal, wastewater treatment, bioremediation, toxic pollutants removal, and the recovery of commercially viable products, such as resource recovery, $CO_2$ sequestration, harvesting the energy stored in marine sediments, and desalination [72]. Power generation from wastewaters combined with the oxidation of organic or inorganic compounds is one of the most active areas of MFC research, and according to research, any compound that can be degraded by bacteria can be converted into electricity [79]. The online water-monitoring system is essential for maintaining proper wastewater usage from industries or municipalities in order to conserve the aquatic environment and public health [73].

Conventional biosensors typically require a transducer, whereas MFC acts as a transducer in and of itself, making MFC a cost-effective biosensor and the MFC has been demonstrated to be a successful biosensor for detecting organic compounds and contaminants in wastewater [73]. Data from the natural environment can be useful in understanding and modeling ecosystem responses, but sensors scattered throughout the environment require power to operate [78]. MFCs could be used to power such devices, particularly in river and deep-water environments where access to the system to replace batteries is difficult [78]. An unusual opportunity application of MFC technology is the use of glucose and oxygen from the blood to power-implanted medical devices and the implanted MFC could provide power indefinitely, eliminating the need for battery replacement surgery [79]. Interlinking MFCs with other technologies should also be considered in order to improve the chances of their integration into existing wastewater treatment plants and thus advance MFCs' prospects for practical applications [72].

# 10 Conclusion

The numerous substrates that have been employed in MFCs for current production, as well as waste treatment, are summarized in this paper. However, the list is far from complete, since new substrates are introduced into these systems with improved outputs in terms of both power generation and waste treatment. Simple substrates like acetate and glucose were often utilized in the early years, but in recent years, researchers have been experimenting with more unusual substrates with the goal of reusing waste biomass or treating wastewater on the one hand, while enhancing MFC production on the other. Bioenergy in the form of electricity generated from renewable and waste biomass using MFCs has a lot of potential in terms of energy self-sufficiency as well as minimizing competition with food production, which is a concern with traditional biofuels. Bioenergy in the form of electricity generated from renewable and waste biomass using MFCs has tremendous development potential, both in terms of energy self-sufficiency and lowering competition with food production, which is a concern with traditional biofuels.

**Acknowledgements** The authors would like to express their gratitude to the Ministry of Higher Education Malaysia for funding this study through the Fundamental Research Grant Scheme (FRGS) (203/PTEKIND/6711823) and the Universiti Sains Malaysia for support through the Research University Grant Short Term Grant (304/PTEKIND/6315353). The authors have stated that the work does not contain any conflicts of interest.

## References

1. Hales, D.: Renewables 2018, Global Status report (2018)
2. (IEA) I. E. A.: World Energy Outlook 2019 (2019). https://www.iea.org/reports/world-energy-outlook-2019
3. Patrick, T.: 20,000 left in The Dark in Sabah, says Minister (2018). https://www.freemalaysiatoday.com/category/nation/2018/08/22/20000-left-in-the-dark-in-sabah-says-minister/
4. UN ESCAP: Energy and Development in the ASEAN Region (2019). https://www.unescap.org/resources/energy-and-development-asean-region
5. Statistics, E.: Handbook Malaysia Energy Statistics (2019)
6. Martha, E.: Global energy transformation: a roadmap to 2050. In: Global Energy Transformation. A Roadmap to 2050 (2018)
7. Ali, R., Daut, I., Taib, S.: A review on existing and future energy sources for electrical power generation in Malaysia. Renew. Sustain. Energy Rev. **16**(6), 4047–4055 (2012). https://doi.org/10.1016/j.rser.2012.03.003
8. IPCC: Climate change. Synthesis report. Contribution of working groups I, II and III to the fifth assessment report of the intergovernmental panel on climate change (2014a).
9. IPCC: Climate change 2014, mitigation of climate change, Contribution of Working Group III to the Fifth Assessment Report of the Intergovernmental Panel on Climate Change (2014b).
10. Phun, C., Bong, C., Shin, W., Hashim, H., Shiun, J.: Review on the renewable energy and solid waste management policies towards biogas development in Malaysia. Renew. Sustain. Energy Rev. **70**(July 2015), 988–998 (2017). https://doi.org/10.1016/j.rser.2016.12.004
11. Saeed, M.O., Hassan, M.N., Mujeebu, M.A.: Assessment of municipal solid waste generation and recyclable materials potential in Kuala Lumpur, Malaysia. Waste Manag. **29**(7), 2209–2213 (2009). https://doi.org/10.1016/j.wasman.2009.02.017
12. Hanum, F., Yuan, L.C., Kamahara, H., Aziz, H.A., Atsuta, Y., Yamada, T., Daimon, H.: Treatment of sewage sludge using anaerobic digestion in Malaysia: Current state and challenges. Frontiers in Energy Research, **7**(MAR), 1–7 (2019). https://doi.org/10.3389/fenrg.2019.00019
13. United Nation Environmental Programme: Waste Generation—How Many Million Tons? (2002) http://www.vitalgraphics.net/waste/html_file/08-9_waste_generation.html
14. Bhattacharyya, J.K., Shekdar, A.V.: Treatment and disposal of refinery sludges: Indian scenario. Waste Manag. Res. J. Int. Solid Wastes Public Cleansing Assoc. ISWA **21**(3), 249–261 (2003). https://doi.org/10.1177/0734242X0302100309
15. Indah Water Konsortium: Indah Water Cleaning the Unseen (2013). https://www.iwk.com.my/cms/upload_files/resource/sustainabilityreport/Sustainability_Report2012_2013.pdf
16. Oliveira, V., Carvalho, T., Melo, L., Pinto, A., Simoes, M.: Effects of hydrodynamic stress and feed rate on the performance of a microbial fuel cell. Environ. Eng. Manag. J. **15** (2016). https://doi.org/10.30638/eemj.2016.273

17. Ministry of Economic Affairs: Mid-term review of the Eleventh Malaysia Plan, 2016–2020: new priorities and emphases (2018). http://www.epu.gov.my/sites/default/files/2020-08/Mid-TermReviewof11thMalaysiaPlan.pdf
18. Potter, M.C., Waller, A.D.: Electrical effects accompanying the decomposition of organic compounds. Proc. Royal Soc. London. Ser. B, Contain. Papers Biol. Char. **84**(571), 260–276 (1911). https://doi.org/10.1098/rspb.1911.0073
19. Pant, D., Van Bogaert, G., Diels, L., Vanbroekhoven, K.: A review of the substrates used in microbial fuel cells (MFCs) for sustainable energy production. Biores. Technol. **101**(6), 1533–1543 (2010). https://doi.org/10.1016/j.biortech.2009.10.017
20. Das, D.: Microbial Fuel Cell A Bioelectrochemical System that Converts Waste to Watts (Debabrata, D. (ed.)). Springer International, Switzerland (2018)
21. Logan, B.E.: Essential data and techniques for conducting microbial fuel cell and other types of bioelectrochemical system experiments. Chemsuschem **5**(6), 988–994 (2012). https://doi.org/10.1002/cssc.201100604
22. Scott, K., Yu, E.H.: Microbial Electrochemical and Fuel Cells Fundamental and Applications (Scott, K., Yu, E.H. (ed.)). Woodhead Publishing (2016)
23. Behera, B.K., Varma, A.: Microbial resources for sustainable energy. In: Microbial Resources for Sustainable Energy (2016). https://doi.org/10.1007/978-3-319-33778-4
24. Torres, C.I., Kato Marcus, A., Rittmann, B.E.: Kinetics of consumption of fermentation products by anode-respiring bacteria. Appl. Microbiol. Biotechnol. **77**(3), 689–697 (2007). https://doi.org/10.1007/s00253-007-1198-z
25. Logan, B.E., Regan, J.M.: Electricity-producing bacterial communities in microbial fuel cells. Trends Microbiol. **14**(12), 512–518 (2006a). https://doi.org/10.1016/j.tim.2006.10.003
26. Zuo, Y., Xing, D., Regan, J.M., Logan, B.E.: Isolation of the Exoelectrogenic Bacterium Ochrobactrum anthropi YZ-1 by Using a U-Tube Microbial Fuel Cell. Appl. Environ. Microbiol. **74**(10), 3130–3137 (2008). https://doi.org/10.1128/AEM.02732-07
27. Logan, B.E.: Exoelectrogenic bacteria that power microbial fuel cells. Nat. Rev. Microbiol. **7**(5), 375–381 (2009). https://doi.org/10.1038/nrmicro2113
28. Lovley, D.R.: Powering microbes with electricity: direct electron transfer from electrodes to microbes. Environ. Microbiol. Rep. **3**(1), 27–35. https://doi.org/10.1111/j.1758-2229.2010.00211.x
29. Huang, L., Tian, F., Pan, Y., Shan, L., Shi, Y., Logan, B.E.: Mutual benefits of acetate and mixed tungsten and molybdenum for their efficient removal in 40 L microbial electrolysis cells. Water Res. **162**, 358–368 (2019). https://doi.org/10.1016/j.watres.2019.07.003
30. Nancharaiah, Y.V., Mohan, S.V., Lens, P.N.L.: Biological and bioelectrochemical recovery of critical and scarce metals. Trends Biotechnol. **34**(2), 137–155 (2016). https://doi.org/10.1016/j.tibtech.2015.11.003
31. Rinaldi, A., Mecheri, B., Garavaglia, V., Licoccia, S., Di Nardo, P., Traversa, E.: Engineering materials and biology to boost performance of microbial fuel cells: a critical review. Energy Environ. Sci. **1**(4), 417–429 (2008). https://doi.org/10.1039/B806498A
32. Huang, L., Chen, J., Quan, X., Yang, F.: Enhancement of hexavalent chromium reduction and electricity production from a biocathode microbial fuel cell. Bioprocess Biosyst. Eng. **33**(8), 937–945 (2010). https://doi.org/10.1007/s00449-010-0417-7
33. Huang, L., Regan, J.M., Quan, X.: Electron transfer mechanisms, new applications, and performance of biocathode microbial fuel cells. Biores. Technol. **102**, 316–323 (2011). https://doi.org/10.1016/j.biortech.2010.06.096
34. You, S.-J., Ren, N.-Q., Zhao, Q.-L., Wang, J.-Y., Yang, F.-L.: Power generation and electrochemical analysis of biocathode microbial fuel cell using graphite fibre brush as cathode material. Fuel Cells **9**(5), 588–596 (2009). https://doi.org/10.1002/fuce.200900023
35. Cheng, K.Y., Ginige, M.P., Kaksonen, A.H.: Ano-cathodophilic biofilm catalyzes both anodic carbon oxidation and cathodic denitrification. Environ. Sci. Technol. **46**(18), 10372–10378 (2012). https://doi.org/10.1021/es3025066
36. Sun, J., Cao, H., Wang, Z.: Progress in Nitrogen Removal in Bioelectrochemical Systems. Processes **8** (2020)

37. Lovley, D.R.: Erratum: Bug juice: Harvesting electricity with microorganisms. Nat. Rev. Microbiol. **4**(10), 497–508, 797 (2006).https://doi.org/10.1038/nrmicro1442. https://doi.org/10.1038/nrmicro1506
38. Fan, L., Xue, S.: Overview on electricigens for microbial fuel cell. Open Biotechnol. J. **10**, 398–406 (2016). https://doi.org/10.2174/1874070701610010398
39. Scott, K., Rimbu, G.A., Katuri, K.P., Prasad, K.K., Head, I.M.: Application of modified carbon anodes in microbial fuel cells. Process Safety Environ. Protect. **85**(5), 481–488 (2007). https://doi.org/10.1205/psep07018
40. Clauwaert, P., Aelterman, P., Pham, T.H., De Schamphelaire, L., Carballa, M., Rabaey, K., Verstraete, W.: Minimizing losses in bio-electrochemical systems: the road to applications. Appl. Microbiol. Biotechnol. **79**(6), 901–913 (2008). https://doi.org/10.1007/s00253-008-1522-2
41. Abbassi, R., Yadav, A.K., Khan, F., Garaniya, V.: Integrated microbial fuel cells for wastewater treatment. In: Acta Universitatis Agriculturae et Silviculturae Mendelianae Brunensis (Vol. 53, Issue 9). Butterworth Heinemann (2020). http://publications.lib.chalmers.se/records/fulltext/245180/245180.pdf%0Ahttps://hdl.handle.net/20.500.12380/245180%0Ahttp://dx.doi.org/10.1016/j.jsames.2011.03.003%0A10.1016/j.gr.2017.08.001%0Ahttp://dx.doi.org/10.1016/j.precamres.2014.12
42. Shi, L., Squier, T.C., Zachara, J.M., Fredrickson, J.K.: Respiration of metal (hydr)oxides by Shewanella and Geobacter: a key role for multihaem c-type cytochromes. Mol. Microbiol. **65**(1), 12–20 (2007). https://doi.org/10.1111/j.1365-2958.2007.05783.x
43. Carlson, H.K., Iavarone, A.T., Gorur, A., Yeo, B.S., Tran, R., Melnyk, R.A., Mathies, R.A., Auer, M., Coates, J.D.: Surface multiheme c-type cytochromes from Thermincola potens and implications for respiratory metal reduction by Gram-positive bacteria. Proc. Natl. Acad. Sci. **109**(5), 1702 LP–1707 (2012). https://doi.org/10.1073/pnas.1112905109
44. Rabaey, K., Boon, N., Höfte, M., Verstraete, W.: Microbial phenazine production enhances electron transfer in biofuel cells. Environ. Sci. Technol. **39**(9), 3401–3408 (2005). https://doi.org/10.1021/es048563o
45. Liu, J., Qiao, Y., Lu, Z.S., Song, H., Li, C.M.: Enhance electron transfer and performance of microbial fuel cells by perforating the cell membrane. Electrochem. Commun. **15**(1), 50–53 (2012). https://doi.org/10.1016/j.elecom.2011.11.018
46. Parkash, A.: Microbial fuel cells: a source of bioenergy microbial & biochemical technology microbial fuel cells : a source of bioenergy (2016). https://doi.org/10.4172/1948-5948.1000293
47. Du, Z., Li, H., Gu, T.: A state of the art review on microbial fuel cells: A promising technology for wastewater treatment and bioenergy. Biotechnol. Adv. **25**, 464–482 (2007). https://doi.org/10.1016/j.biotechadv.2007.05.004
48. Ye, Y., Ngo, H., Guo, W., Liu, Y., Chang, S., Nguyen, D.D., Ren, J., Liu, Y., Zhang, X.: Feasibility study on a double chamber microbial fuel cell for nutrient recovery from municipal wastewater. Chem. Eng. J. **358**.https://doi.org/10.1016/j.cej.2018.09.215
49. Min, B., Logan, B.E.: Continuous electricity generation from domestic wastewater and organic substrates in a flat plate microbial fuel cell. Environ. Sci. Technol. **38**(21), 5809–5814 (2004). https://doi.org/10.1021/es0491026
50. Fan, Y., Han, S.-K., Liu, H.: Improved performance of CEA microbial fuel cells with increased reactor size. Energy Environ. Sci. **5**(8), 8273–8280 (2012). https://doi.org/10.1039/C2EE21964F
51. Mahdi Mardanpour, M., Nasr Esfahany, M., Behzad, T., Sedaqatvand, R.: Single chamber microbial fuel cell with spiral anode for dairy wastewater treatment. Biosens. Bioelectron. **38**(1), 264–269 (2012). https://doi.org/10.1016/j.bios.2012.05.046
52. He, Z., Minteer, S.D., Angenent, L.T.: Electricity generation from artificial wastewater using an upflow microbial fuel cell. Environ. Sci. Technol. **39**(14), 5262–5267 (2005). https://doi.org/10.1021/es0502876
53. He, Z., Wagner, N., Minteer, S.D., Angenent, L.T.: An upflow microbial fuel cell with an interior cathode: assessment of the internal resistance by impedance spectroscopy. Environ. Sci. Technol. **40**(17), 5212–5217 (2006). https://doi.org/10.1021/es060394f

54. Li, Z., Yao, L., Kong, L., Liu, H.: Electricity generation using a baffled microbial fuel cell convenient for stacking. Biores. Technol. **99**(6), 1650–1655 (2008). https://doi.org/10.1016/j.biortech.2007.04.003
55. Kim, J.R., Premier, G.C., Hawkes, F.R., Dinsdale, R.M., Guwy, A.J.: Development of a tubular microbial fuel cell (MFC) employing a membrane electrode assembly cathode. J. Power Sour. **187**(2), 393–399 (2009). https://doi.org/10.1016/j.jpowsour.2008.11.020
56. Pasupuleti, S.B., Srikanth, S., Venkata Mohan, S., Pant, D.: Continuous mode operation of microbial fuel cell (MFC) stack with dual gas diffusion cathode design for the treatment of dark fermentation effluent. Int. J. Hyd. Energy **40**(36), 12424–12435 (2015). https://doi.org/10.1016/j.ijhydene.2015.07.049
57. Flimban, S.G.A., Ismail, I.M.I., Kim, T., Oh, S.E.: Overview of recent advancements in the microbial fuel cell from fundamentals to applications: Design, major elements, and scalability. Energies **12**(17) (2019). https://doi.org/10.3390/en12173390
58. Virdis, B., Freguia, S., Rozendal, R.A., Rabaey, K., Yuan, Z., Keller, J.: Microbial fuel cells. In: Treatise on Water Science, vol. 4 (2011). https://doi.org/10.1016/B978-0-444-53199-5.00098-1
59. Gezginci, M., Uysal, Y.: The effect of different substrate sources used in microbial fuel cells on microbial community. JSM Environ. Sci. Ecol. **4**(3), 1035 (2016)
60. Jothinathan, D.: Microbial desalination cells: a boon for future generations. In: Microbial Fuel Cell Technology for Bioelectricity (2018).https://doi.org/10.1007/978-3-319-92904-0_12
61. Debabrata, D.: Microbial Fuel Cell A Bioelectrochemical System that Converts Waste to Watts (D. Debabrata (ed.)). Springer International, Switzerland (2018)
62. Gude, V.G.: Wastewater treatment in microbial fuel cells—An overview. J. Clean. Prod. **122**(December 2016), 287–307 (2016). https://doi.org/10.1016/j.jclepro.2016.02.022
63. Rahimnejad, M., Adhami, A.: Microbial fuel cell as new technology for bioelectricity generation: a review. Alex. Eng. J. **54**(3), 745–756 (2015). https://doi.org/10.1016/j.aej.2015.03.031
64. Shanmuganathan, P., Rajasulochana, P., Ramachandra Murthy, A.: Factors affecting the performance of microbial fuel cells. Int. J. Mech. Eng. Technol. **9**(9), 137–148 (2018)
65. Logan, B.E.: Microbial Fuel Cells. Wiley-Interscience (2008)
66. Logan, B., Cheng, S., Watson, V., Estadt, G.: Graphite fiber brush anodes for increased power production in air-cathode microbial fuel cells. Environ. Sci. Technol. **41**(9), 3341–3346 (2007). https://doi.org/10.1021/es062644y
67. Rezaei, F., Richard, T.L., Brennan, R.A., Logan, B.E.: Substrate-enhanced microbial fuel cells for improved remote power generation from sediment-based systems. Environ. Sci. Technol. **41**(11), 4053–4058 (2007). https://doi.org/10.1021/es070426e
68. Kim, J.R., Jung, S.H., Regan, J.M., Logan, B.E.: Electricity generation and microbial community analysis of alcohol powered microbial fuel cells. Biores. Technol. **98**(13), 2568–2577 (2007). https://doi.org/10.1016/j.biortech.2006.09.036
69. Gálvez, A., Greenman, J., Ieropoulos, I.: Landfill leachate treatment with microbial fuel cells; scale-up through plurality. Biores. Technol. **100**(21), 5085–5091 (2009). https://doi.org/10.1016/j.biortech.2009.05.061
70. Niessen, J., Schröder, U., Harnisch, F., Scholz, F.: Gaining electricity from in situ oxidation of hydrogen produced by fermentative cellulose degradation. Lett. Appl. Microbiol. **41**(3), 286–290 (2005). https://doi.org/10.1111/j.1472-765X.2005.01742.x
71. Wang, X., Cheng, S., Feng, Y., Merrill, M.D., Saito, T., Logan, B.E.: Use of carbon mesh anodes and the effect of different pretreatment methods on power production in microbial fuel cells. Environ.Sci. Technol. **43**(17), 6870–6874 (2009). https://doi.org/10.1021/es900997w
72. Pandey, P., Shinde, V.N., Deopurkar, R.L., Kale, S.P., Patil, S.A., Pant, D.: Recent advances in the use of different substrates in microbial fuel cells toward wastewater treatment and simultaneous energy recovery. Appl. Energy **168**, 706–723 (2016). https://doi.org/10.1016/j.apenergy.2016.01.056
73. Kumar, R., Singh, L., Zularisam, A.W., Hai, F.I.: Microbial fuel cell is emerging as a versatile technology : a review on its possible applications , challenges and strategies to improve the performances. June 2017, pp 369–394 (2018). https://doi.org/10.1002/er.3780

74. Tekle, Y., Demeke, A.: Review on microbial fuel cell. Int. J. Eng. Technol. **8**(November), 424–427 (2015)
75. Logan, B.E., Hamelers, B., Rozendal, R., Schröder, U., Keller, J., Freguia, S., Aelterman, P., Verstraete, W., Rabaey, K.: Microbial fuel cells: Methodology and technology. Environ. Sci. Technol. **40**(17), 5181–5192 (2006). https://doi.org/10.1021/es0605016
76. Roy, S., Marzorati, S., Schievano, A., Pant, D.: Author's personal copy, September 2017 (2016). https://doi.org/10.1016/B978-0-12-409548-9.10122-8
77. Moradian, J.M., Fang, Z., Yong, Y.C.: Recent advances on biomass-fueled microbial fuel cell. Bioresour. Bioprocess. **8**(1) (2021). https://doi.org/10.1186/s40643-021-00365-7
78. Logan, B.E., Regan, J.M.: Microbial fuel cells—Challenges and applications. Environ. Sci. Technol. **40**(17), 5172–5180 (2006). https://doi.org/10.1021/es0627592
79. Franks, A.E., Nevin, K.P.: Microbial fuel cells, a current review. Energies **3**(5), 899–919 (2010). https://doi.org/10.3390/en3050899
80. Oh, S.E., Yoon, J.Y., Gurung, A., Kim, D.J.: Evaluation of electricity generation from ultrasonic and heat/alkaline pretreatment of different sludge types using microbial fuel cells. Bioresour. Technol. **165**(C), 21–26 (2014). https://doi.org/10.1016/j.biortech.2014.03.018
81. Fang, W., Zhang, P., Zhang, G., Jin, S., Li, D., Zhang, M., Xu, X.: Effect of alkaline addition on anaerobic sludge digestion with combined pretreatment of alkaline and high pressure homogenization. Biores. Technol. **168**, 167–172 (2014). https://doi.org/10.1016/j.biortech.2014.03.050
82. Yang, G., Zhang, G., Wang, H.: Current state of sludge production, management, treatment and disposal in China. Water Res. **78**, 60–73 (2015). https://doi.org/10.1016/j.watres.2015.04.002
83. Guo, K., Hassett, D.J., Gu, T.: Microbial fuel cells: Electricity generation from organic wastes by microbes. In Microbial Biotechnology: Energy and Environment (2012). https://doi.org/10.1079/9781845939564.0162
84. Koroglu, E.O., Yoruklu, H.C., Demir, A., Ozkaya, B.: Scale-up and commercialization issues of the MFCs: Challenges and implications. In: Biomass, Biofuels, Biochemicals: Microbial Electrochemical Technology: Sustainable Platform for Fuels, Chemicals and Remediation, March 2019, pp. 565–583 (2018). https://doi.org/10.1016/B978-0-444-64052-9.00023-6
85. Gajda, I., Greenman, J., Ieropoulos, I. A.: Recent advancements in real-world Microbial Fuel Cell applications Recent advancements in real-world microbial fuel cell applications. Current Opinion in Electrochemistry, November, 4–10 (2018). https://doi.org/10.1016/j.coelec.2018.09.006

# Waste to Wealth: The Importance of Yeasts in Sustainable Bioethanol Production from Lignocellulosic Biomass

**Akaraphol Watcharawipas, Noreen Suliani Binti Mat Nanyan, and Rika Indri Astuti**

**Abstract** Climate change and global warming are two of the world's most pressing environmental concerns. Bioethanol is considered as an alternative and renewable energy to reduce the usage of fossil fuels due to its environmental-friendly properties. In this chapter, we emphasize on the development of superior yeast strains for the production of bioethanol from lignocellulosic biomass. For instance, the substitution of a threonine at the position of 255 to alanine in ubiquitin ligase Rsp5 could increase the resistance of *Saccharomyces cerevisiae* cells to the fermentation inhibitor acetic acid. Additionally, the development of several *S. cerevisiae* strains that exhibited multiple stress tolerance against high temperature, high ethanol concentration, and a combination of fermentation inhibitors was also discussed. The pentose-utilizing *Pichia* yeast strains have also been constructed using mutagens to improve xylose fermentation and ethanol production. Altogether, these strategies will not only increase our knowledge on yeast stress response and tolerance mechanisms, but they could also be applied in industrial yeast strain engineering to develop more robust yeast strains for higher bioethanol production.

**Keywords** Bioethanol · Biomass · Yeast · *Saccharomyces cerevisiae*

## 1 Introduction

Since the world is currently facing various environmental problems, including climate changes, over-emission of carbon dioxide into the atmosphere and global warming

A. Watcharawipas
National Center for Genetic Engineering and Biotechnology, Pathumthani 12120, Thailand

N. S. B. Mat Nanyan
Bioprocess Technology Division, School of Industrial Technology, Universiti Sains Malaysia, 11900 Penang, Malaysia

R. I. Astuti (✉)
Department of Biology, Faculty of Mathematics and Natural Sciences, IPB University, 16680 Bogor, Indonesia
e-mail: rikaindriastuti@apps.ipb.ac.id



A. Z. Yaser et al. (eds.), *Waste Management, Processing and Valorisation*,
https://doi.org/10.1007/978-981-16-7653-6_14

[1, 2], different approaches have been addressed to manipulate such problems [3–5]. Such an approach is to reduce the use of fossil fuels since they are non-renewable and pollute environments especially when combustion is incomplete. The utilization of bioethanol has gained a lot of interest as alternative energy due to its high-octane number, complete combustion and carbon neutral production process [6–8]. Nowadays, bioethanol is considered as a promising renewable energy that most fits the existing transportation infrastructures. Bioethanol can be produced from both sugar-rich starch and lignocellulosic biomass. First-generation bioethanol production usually relies on sugar-rich feedstocks such as corn, sugarcane and cassava. Second-generation bioethanol production utilizes lignocellulosic biomass as a feedstock. As represented by four different countries including Thailand, China, USA and Brazil, overall bioethanol production and domestic demand have been tentatively increasing over years since 2016 (Fig. 1). This highlights that there is an urgent need for technological development to improve bioethanol production, which in turn ensures its sustainable supply that contributes to human well-being in the bioeconomic society.

## 2 Production of Bioethanol

A variety of biomass has been investigated as resources for bioethanol production. The first generation of bioethanol production employs sugar and starch biomass as fermentation substrate, such as corn, cane and maize. To this date, the majority of ethanol production (84% of the world's ethanol) has been made out of sugar/starch feedstock sources such as corn (USA) and sugarcane (Brazil). However, it has gained serious attention worldwide as those substrates are also food sources which may promote social conflict. Thus, recently, the production of bioethanol has focused on using other potential non-food substrates, such as agricultural, industrial wastes and non-industrial wood, or known as second generation of ethanol production [10]. However, such alternative feedstocks, which are mostly lignocellulosic biomass, require the development of complex biological conversion processes for the ethanol production [11]. The utilization of lignocellulosic wastes promotes zero waste biorefinery technology which brings benefit to the sustainable environment [12, 13].

Production of bioethanol can be conducted via direct fermentation and indirect fermentation with the help of fermentative microbes (Fig. 2). In direct fermentation, biomass sources are pretreated, hydrolysed and then fermented into ethanol [14, 15]. On the other hand, indirect fermentation or also known as syngas fermentation uses gaseous substrate which are obtained from industrial fuel gases of biomass or gasification of biomass, coal and municipal solid waste [16, 17].

Fermentative microbes play pivotal role in the ethanol fermentation. Common industrial fermentative microbes are yeast *Saccharomyces cerevisiae* and bacteria *Zymomonas mobilis* [18, 19]. Further observation and research lead to the application of various other microbes in the future ethanol production including certain species of yeasts such as *Kluyveromyces marxianus* [20], *Candida shehatae*, *Pichia stipitis*

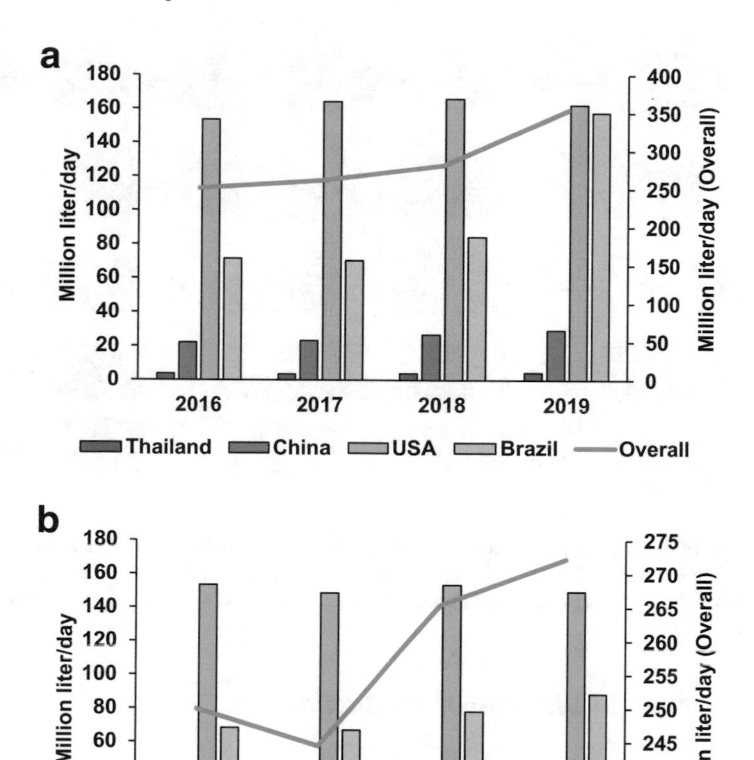


**Fig. 1** Bioethanol production and domestic demand of different countries. **a** Bioethanol production (million litre/day) and **b** domestic demand of ethanol (million litre/day) in Thailand, China, USA and Brazil from 2016 to 2019. The data were originally from: (1) Department of Alternative Energy Development and Efficiency, Thailand; (2) Thai customs; (3) Thai Ethanol Manufacturing Association; (4) Iowa state university and USDA and were previously summarized in annual report by Bank of Thailand [9]

[21] and *Pichia kudriavzevii* [22, 23] and a group of thermophilic bacteria such as *Thermoanaerobacterium saccharolyticum, Thermoanaerobacter ethanolicus* or *Clostridium thermocellum* [24–26]. As in the syngas fermentation, a group of acetogenic bacteria including *Clostridium ljungdahlii and Clostridium autoethanogenum* has been reported to conduct this type of indirect fermentation [27, 28].

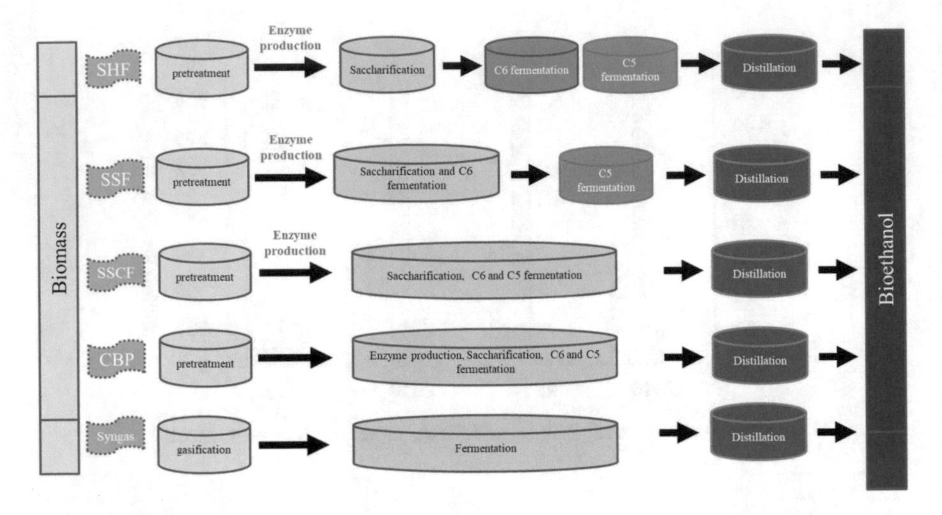


**Fig. 2** Type of biological conversion of substrate biomass into bioethanol including separate hydrolysis and fermentation (SHF), simultaneous saccharification and fermentation (SSF), simultaneous saccharification and co-fermentation (SSCF), consolidated bioprocessing (CBP) and syngas fermentation

## *2.1 Processes in Bioethanol Production*

Generally, bioethanol production involves four main processes consisting of: (1) pretreatment; (2) substrate hydrolysis; (3) fermentation; (4) product separation (Fig. 2) [29, 30]. Pretreatment is performed in order to open several closed structures of the complexed biomass by various methods, e.g. physical treatment by milling and grounding, chemical treatment by acid or alkaline hydrolysis and physicochemical treatment by steam explosion [31]. Substrate hydrolysis is the enzymatic hydrolysis of pretreated biomass to produce sugars, so-called saccharification. Glycoside hydrolase (GH) enzymes are normally used in this process and depend on types of biomass. For examples, amylase and glucoamylase are used for hydrolysis of cassava starch or corn starch to produce glucose [32] while various GH enzyme formulations basically composing of endoglucanase, cellobiohydrolase, beta-glucosidase and xylanase are required for optimal hydrolysis of lignocellulosic biomass such as sugarcane bagasse [33], cassava pulp [34] and rice straw [35]. Then, derived sugars including hexoses, e.g. glucose, fructose and sucrose, and pentoses, e.g. xylose, and arabinose are converted to ethanol by fermentation processes including batch, fed-batch and continuous fermentation, which are mediated by yeasts including *Saccharomyces cerevisiae* for efficient conversion of hexoses to ethanol [36] and *Scheffersomyces stipitis* for effective co-fermentation of hexoses and pentoses [37]. Co-culture of *S. cerevisiae* and *S. stipitis* has been shown to improve ethanol production from rice straw and sugarcane bagasse feedstocks [38, 39]. In addition, overall processes in bioethanol production can be categorized to (1) separate hydrolysis and fermentation (SHF); (2) simultaneous saccharification and fermentation (SSF); (3) simultaneous saccharification and co-fermentation (SSCF) (Fig. 2) [19]. In SHF process, saccharification and fermentation are performed in different bioreactors at different temperatures of about 50 °C and 30 °C, respectively, thereby requiring efficient cooling system with significant cost. In SSF and SSCF processes, both saccharification and fermentation are operated in the same bioreactor at temperature of about 40–50 °C, thereby reducing the cooling cost and number of operational units, preventing bacterial contamination and enhancing enzymatic hydrolysis activity [40]. However, only hexoses are fermented in SSF whereas both hexoses and pentoses are co-fermented in SSCF, thus increasing ethanol productivity [41]. Recently, the more promising technology, consolidated bioprocessing (CBP) (Fig. 2), which is the coupling of saccharification and fermentation into a single step catalysed by a single microorganism or consortium, allowing a reduction of exogenous enzymc cost, has been developed for ethanol production from raw starch [42] and from corn cob-derived hemicellulose [43].

Finally, ethanol produced in the fermentation process needs to be purified from the fermentation broth by conventional distillation processes or by membrane separation such as pervaporation which is an emerging technology allowing purification process to occur at ambient conditions [44]. Considering these processes in bioethanol production, there arc many ways for enhancing bioethanol productivity while mitigating $CO_2$ emission such as process optimization [45], utilization of alternative feedstocks [46] and utilization of appropriate yeasts [47]. Among them, feedstocks and yeasts are not only factors for improving bioethanol productivity, but also major contributors towards sustainable bioethanol production. Thus, the next topics will be particularly focused on the utilization of alternative feedstocks for bioethanol production and the development of yeasts for bioethanol production from lignocellulosic biomass.

### 2.2 *Utilization of Alternative Feedstocks for Bioethanol Production*

In Thailand, first-generation bioethanol production is relied on the utilization of sugar-rich feedstocks including sugarcane molasses [48, 49] and cassava starch [50, 51]. However, this raises food security issues and prompts many researchers to search for alternative feedstocks from non-edible plant parts such as sweet sorghum juice [52], corn stalk juice [53], sugarcane bagasse [23], mixed sugarcane bagasse with dry spent yeast [54] and oil palm empty fruit bunch [55], and from macroalga *Rhizoclonium* [56]. Among different resources for alternative feedstocks, lignocellulosic biomass that includes crop residues and agricultural wastes such as sugarcane bagasse, cassava pulp, rice straw, corn cob and oil palm empty fruit bunch represents the most promising resource for sustainable bioethanol production due to its high abundance and environmental friendliness [57]. Lignocellulosic biomass comprises of approximately 40–50% cellulose, 20–40% hemicellulose, 20–30% lignin and other compositions such as protein and wax [58]. To utilize lignocellulosic biomass as a feedstock, the complexed biomass needs to be pretreated by several methods such as dilute acid hydrolysis and steam explosion and proceeded to enzymatic saccharification to generate hexoses and pentoses such as glucose, fructose, sucrose, xylose and arabinose. Then these sugars can be used as substrates for fermentation to produce ethanol. However, pretreatment of lignocellulosic biomass generates not only sugars but also fermentation inhibitors such as furfural, 5-hydroxymethylfurfural, acetic acid, vanillin and other phenolic compounds that inhibit growth and/or fermentation by yeast cells [59]. In addition to the inhibitors, yeast cells confront various stresses such as high temperature, high osmotic pressure, high ethanol concentration and oxidative stress during fermentation in bioethanol production from lignocellulosic biomass [60, 61]. Therefore, isolation, breeding or genetic engineering to generate yeasts that possess desirable traits to cope with such stresses are the key for sustainable bioethanol production from lignocellulosic biomass.

## 3 Development of Yeasts for Bioethanol Production from Lignocellulosic Biomass

Yeast is the crucial microbial cell factory in fermentation process for converting sugars into bulk and fine chemicals [62]. For bioethanol production, potential yeast strains originally obtained in Thailand are listed (Table 1). These studies highlight that effective ethanol production ranging from 72 to 99% of theoretical yield can be obtained from various types of substrates with batch fermentation. SHF is performed when enzymatic hydrolysis of lignocellulosic biomass is required. Importantly, *S. cerevisiae* TISTR5339 could achieve 90.33% of theoretical yield of ethanol even when complex substrate (sugarcane bagasse mixed with dry spent yeast) is used. Notably, although fermentation temperature is up to 40 °C, *S. cerevisiae* KKU-VN8 could produce ethanol from sweet sorghum juice at 96.32% of theoretical yield. High fermentation temperature is a focus of these studies at least to prevent bacterial contamination. Moreover, *P. kudriavzevii* RZ8-1 exhibited ethanol production from sugarcane bagasse at high yield (77.91% of theoretical yield) and relatively high titer (33.84 g/L), possibly due to its ability to ferment pentose sugars in the hydrolysate. Among several yeast species, *S. cerevisiae* is the well-characterized microbial chassis and has been widely used in fermentation for industrial production of ethanol [63]. In this section, we will discuss how to obtain yeast strains with industrially relevant desirable traits for bioethanol production from lignocellulosic biomass. In general, at least two strategies can be employed: (1) isolation of wild-type yeast strain from natural habitats; (2) breeding or engineering to obtain superior strains.

**Table 1** Potential yeast strains originally obtained in Thailand for bioethanol production

| No. | Strain | Sugar source | Fermentation | Growth temperature (°C) | Ethanol titer (g/L) | Theoretical yield (%) | Reference |
|---|---|---|---|---|---|---|---|
| 1 | *Saccharomyces cerevisiae* UVNR56 | Sugarcane molasses | Batch | 37 | 81.6 | 98.7 | [48] |
| 2 | *S. cerevisiae* KKU-VN8 | Sweet sorghum juice | Batch | 40 | 89.32 | 96.32 | [52] |
| 3 | *S. cerevisiae* TISTR5339 | Sugarcane bagasse-dry spent yeast hydrolysate | Batch, SHF | 30 | 8.98 | 90.33 | [54] |
| 4 | *S. cerevisiae* TISTR5020 | Thailand Hi-brix 53 corn stalk juice | Batch | 36 | 47.39 | 74.85 | [53] |
| | | Macroalga *Rhizoclonium* sp. hydrolysate | Batch, SHF | 30 | 52.34 | 72.3 | [56] |
| 5 | *Pichia kudriavzevii* DMKU 3-ET15 | Cassava starch hydrolysate | Batch, SHF | 40 | 78.6 | 85.5 | [50] |
| 6 | *P. kudriavzevii* RZ8-1 | Sugarcane bagasse hydrolysate | Batch, SHF | 40 | 33.84 | 77.91 | [23] |
| 7 | *Kluyveromyces marxianus* TISTR5116 | Oil palm empty fruit bunch hydrolysate | Batch, SHF | 37.5 | 28.1 | 76.28 | [55] |

### 3.1 Acetate-Tolerant Yeasts

Among several fermentation inhibitors, of which the formation can be reduced by the pretreatment process optimization, acetic acid which is generated from acetylated hemicellulose cannot be subsidized and is always present in lignocellulosic hydrolysate over 10 g/L [64]. Therefore, we first focused on acetic acid as a fermentation inhibitor and the stress responses that it induces. At pH < 4.75 (pKa of acetic acid), acetic acid mainly exists in a protonated form that can enter the cells by passive diffusion across the plasma membrane and facilitated diffusion mediated by the aquaglyceroporin channel Fps1p [65]. In the cytosol with near neutral pH, acetic acid dissociates to acetate and proton. The dissociated protons cause intracellular acidification that negatively affects cellular metabolisms. At high concentration, acetic acid triggers generation of reactive oxygen species and mitochondrial dysfunction, leading to program cell death [66]. The dissociated acetate anions also disturb intracellular turgor pressure [67]. To cope with the acetic acid stress, yeast cells mainly relied on: (1) transcriptional changes governed by the weak acid stress responsive transcription factor Haa1p [68]; (2) extrusion of excess protons out of the cytosol by activity of the plasma membrane P2-type $H^+$-ATPase Pma1p and sequestering of excess protons in the vacuole by activity of the vacuolar membrane $H^+$-ATPase (V-ATPase) [69, 70]; (3) export of excess acetate anions through putative transporters such as the drug:$H^+$ antiporters Tpo2p and Tpo3p [71]; and (4) block of acetic acid entry by the downregulation of the aquaglyceroporin channel Fps1p mediated by the mitogen-activated protein kinase (MAPK) Hog1 in a ubiquitin dependent manner [65]. However, in neutralized lignocellulosic hydrolysate (pH > 4.75), acetic acid mostly exists in the form of acetate anions conjugated with sodium or potassium cations, depending on alkaline base used for neutralization. Recent studies show that sodium acetate is more toxic to yeast cells than potassium acetate [72] and sodium acetate also induces more intracellular sodium accumulation than sodium chloride at equal molar concentration [73], suggesting that sodium acetate is a composite stress between sodium stress and acetic acid stress [74]. Taken together, both acetic acid and sodium acetate inhibit growth and fermentation, thereby meliorating bioethanol production from lignocellulosic hydrolysate.

To improve bioethanol productivity, yeast cells that exhibit tolerance to acetic acid or sodium acetate have been developed. Overexpression of the weak acid stress-responsive transcription factor Haa1p in the industrial *S. cerevisiae* Ethanol Red conferred acetic acid tolerance and improved ethanol production under acetic acid containing conditions [75]. Substitution of the Thr255Ala mutant allele of the E3 ubiquitin ligase *RSP5* in the genome caused sodium acetate tolerance to Ethanol Red and enhanced initial fermentation rate under the sodium acetate containing condition [73] (Fig. 3). In addition, [64] recently reported that the acetate-tolerant xylose-utilizing *S. cerevisiae* XUSAE57 obtained from adaptive evolution was able to efficiently convert glucose and xylose in sugarcane bagasse hydrolysate containing 2.5 g/L acetic acid and 0.8 g/L phenolic compounds to ethanol at the yield of 0.49 g

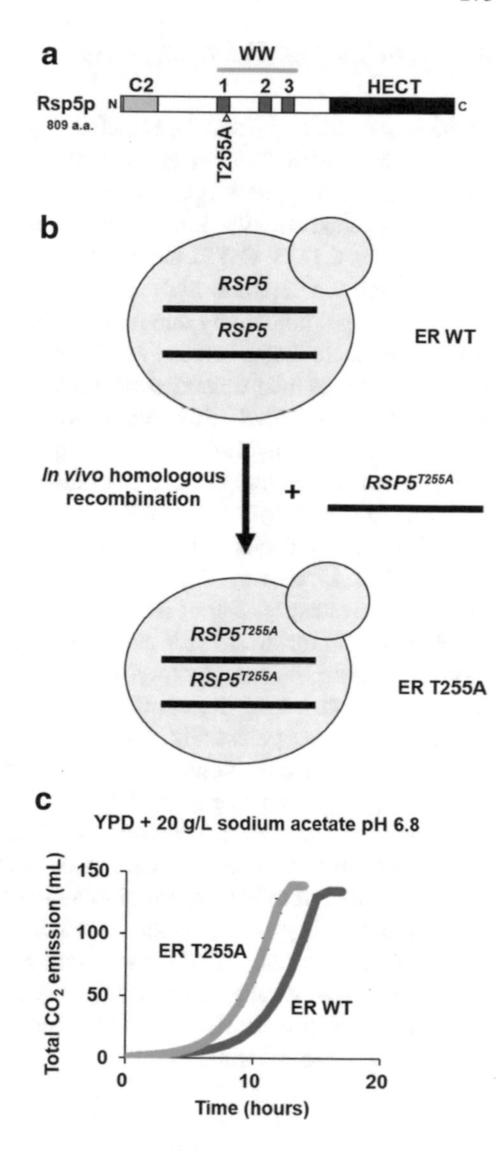


**Fig. 3** Construction of the sodium acetate-tolerant *S. cerevisiae* ER T255A and fermentation. **a** The Thr255Ala or T255A mutation is located in the substrate recognition domain (WW1 domain) of Rsp5p. **b** *S. cerevisiae* ER T255A was constructed by replacement of the wild-type *RSP5* with the mutant $RSP5^{T255A}$ allele via in vivo homologous recombination. **c** *S. cerevisiae* ER T255A showed higher initial fermentation rate in the sodium acetate containing condition. The information was partially obtained from the previous study [73]

ethanol/ g sugars, the highest yield reported to date. Altogether, these studies underline the importance of genetic engineering approach for developing acetic acid or sodium acetate-tolerant industrial yeast strains to improve bioethanol production from lignocellulosic biomass.

### 3.2 Multistress Tolerant Yeasts

In the fermentation process for bioethanol production from lignocellulosic biomass, yeast cells face multiple stresses at the same time instead of individual stress. This indicates that development of yeast strains capable of tolerating multiple stresses is indispensable. Recent study revealed that multistress-tolerant *S. cerevisiae* strains C3253, C3751 and C4377, isolated from Thai fruits, displayed tolerance to high temperature, high ethanol concentration, high osmotic pressure and oxidative stress, potentially due to continuous expression of the heat shock protein *HSP* genes including *SSA4* and *HSP82* under long-term exposure to such stresses, increased intracellular trehalose accumulation, improved cell wall remodelling and redox balancing [76]. Their previous work also interestingly demonstrated that by using spore to cell hybridization technique, multistress-tolerant *S. cerevisiae* TJ14 that possessed high temperature tolerance (up to 41 °C), high ethanol productivity and acid tolerance (pH 3.5) could be developed, and this strain exhibited ethanol production of 46.0 g/L with 90% theoretical yield at 41 °C, pH 3.5 from 10% glucose [77]. Moreover, [78] recently showed that the mutant *S. cerevisiae* Afb.01 possessing overexpression of the thioredoxin *TRX1* gene exhibited tolerance to the combined inhibitor stress (2 g/L furfural and 10 g/L acetic acid) and produced ethanol from sugarcane bagasse hydrolysate at the titer of 10.5 g/L with 58.7% theoretical yield, higher than the control strain. Based on proteomic and metabolomic analysis, this was potentially due to: (1) upregulation of the redox homeostasis system including the thioredoxin Trx1p and the 6-phosphogluconate dehydrogenase Gnd1p; (2) enhancement of the pentose phosphate pathway, glycolysis and TCA cycle; (3) increased accumulation of the stress response proteins including the heat shock proteins Hsp26p, Hsp82p, Hsp104p and Ssa4p, and the stress protectants consisting of trehalose, 4-aminobutyrate (GABA), putrescine and various fatty acids, and possibly; (4) increased response to DNA damage via the upregulation of the putative DNA replicative stress response protein Fmp16p. Taken together, these studies highlight various possible targets for future strain engineering to generate a robust multistress-tolerant yeast strain useful for sustainable bioethanol production from lignocellulosic biomass.

### 3.3 Ethanologenic Microbes and Their Strain Development

Industrial ethanologenic microbes either *S. cerevisiae* or *Z. mobilis* cannot ferment pentose sugars to ethanol, they only utilize hexose sugars instead [19]. Thus, lignocellulosic ethanol production will not be efficient, as biomass is needed in large amount while the excess pentose sugars are not utilized but wasted instead. Strain development has been employed to optimize the ethanol production from lignocellulosic substrate from those industrial yeasts via genetic and physiology engineering approaches [79–81].

Certain approaches have been conducted to optimize lignocellulosic ethanol production by using co-culture of pentose-utilizing bacteria and yeast [82, 83]. In this regard, bacteria employ the isomerase pathway to direct xylose to their central metabolism [84], whereas yeast uses the reductase and dehydrogenase pathways to convert xylose to xylulose via the intermediate xylitol [85]. Xylulose then enters pentose phosphate pathway that supplies certain intermediates in glycolysis pathway. Thus, as a result, pyruvic acid from glycolysis is used for ethanol fermentation via intermediate, acetaldehyde [85].

*S. cerevisiae* has also been engineered to be able to use pentose sugars, cellobiose and xylose. Numbers of pentose-utilization genes from other microbes were introduced to *S. cerevisiae* cells [86]. The cellodextrin assimilation pathway consists of a cellodextrin transporter (*cdt-1*) and an intracellular β-glucosidase (*ghl-1*) from the filamentous fungus *N. crassa.* The modified xylose metabolic pathway utilizes xylose reductase isoenzymes (wild-type XR and a mutant XRR276H), xylitol dehydrogenase (*XYL2*) and xylulokinase (*XKS1*) from the xylose-fermenting yeast *P. stipitis*. This particular engineered *S. cerevisiae* was capable to simultaneously utilize cellobiose and xylose, as neither carbon source represses consumption of the other, including glucose.

## *3.4* Pichia *as Emerging Ethanologenic Yeast*

Yeast genera of *Pichia* has been reported to produce ethanol from various lignocellulosic biomass (Table 2). The ability of *Pichia* is to utilize pentose sugar, suggesting its potential application in the second generation of ethanol production [22, 87, 88].

*P. kudriavzevii* has been reported as thermo-tolerant yeast which therefore is suitable for high-temperature ethanol production [92, 93]. Yeast *P. kudriavzevii* is capable to grow at 41 °C, while other strains are capable to grow at 43 °C [94]. Indeed, certain strains of *P. kudriavzevii* have been reported to produce ethanol such as high temperature including *P. kudriavzevii* MF-121 [94], *P. kudriavzevii* DMKU 3-ET15 [50] and *P. kudriavzevii* RZ8-1 [23].

*P. kudriavezii* has also been reported as pentose-utilizing yeast. In addition, *P. kudriavzevii* was found to be tolerant against various fermentation-related stresses

**Table 2** Yeast *Pichia* spp. strains which were used for bioethanol production by using different substrates

| No. | Strain | Substrate | Ethanol yield (g/L) | References |
|---|---|---|---|---|
| 1 | *Pichia stipitis* CBS 6054 | Corn stover | 15.0 | [89] |
| 2 | *Pichia stipitis* | Rice straw | 21.0 | [21] |
| 3 | *Pichia stipitis* | Wheat straw | 12.2 | [90] |
| 4 | *Pichia kudriavzevii* | Orange peel | 54.0 | [91] |
| 5 | *Pichia kudriavzevii* 1P4 | Glucose | 32.05 | [88] |

including high glucose, ethanol, pH, lignocellulose inhibitors such as acetic acid, formic acid and furfural [95–98]. Such characters are important for efficient conversion of lignocellulosic biomass leading to the high ethanol productivity [99].

Strain development of *Pichia* spp. has also been conducted in order to increase ethanol production. Mutant *P. stipitis* NRRL Y-7124 was constructed using UV-modification thus capable in using xylose/glucose anaerobically with higher ethanol production than industrial yeast *S. cerevisiae* [100]. UV-irradiation has also been used to develop mutant of *Pichia stipitis* with increased ethanol tolerance and production using xylose as substrate [101] which is capable to ferment wheat straw hemicellulose with enhanced ethanol yield [102] and improves xylose fermentation and ethanol yield [103]. Mutagen such as ethyl methanesulfonate (EMS) has also been used to construct mutant *P. kudriavzevii* with increased intracellular proline which is associated with high temperature and ethanol stress tolerance phenotype [22]. Further techniques in strain development such as protoplast fusion, breeding, genome editing and shuffling, directed evolution can be applied to construct ethanologenic *Pichia* isolates that are feasible for industrial applications.

## 4 Conclusion

Although our existing technology for sustainable bioethanol production from lignocellulosic hydrolysate has not yet achieved economic feasibility in terms of process hurdles and production cost, newer approaches are emerging and could be a game-changer. For instance, CBP is one of the newly promising technologies to solve such constraints regarding number of operational units, cooling cost, bacterial contamination and exogenous enzymes required for substrate hydrolysis, thus being useful for production of not only bioethanol but also a variety of biochemicals demanding in the bioeconomic era. When combined with the genetic modifications of bioethanol yeast producers, these technologies are expected to further boost and improve the production of bioethanol from lignocellulosic biomass.

## References

1. Wei, Y.M., Han, R., Wang, C., Yu, B., Liang, Q.M., Yuan, X.C., Chang, J., Zhao, Q., Liao, H., Tang, B., Yan, J., Cheng, L., Yang, Z.: Self-preservation strategy for approaching global warming targets in the post-Paris Agreement era. Nat. Commun. **11**, 1–13 (2020). https://doi.org/10.1038/s41467-020-15453-z
2. Hari, V., Rakovec, O., Markonis, Y., Hanel, M., Kumar, R.: Increased future occurrences of the exceptional 2018–2019 Central European drought under global warming. Sci. Rep. **10**, 1–10 (2020). https://doi.org/10.1038/s41598-020-68872-9
3. Kunanuntakij, K., Varabuntoonvit, V., Vorayos, N., Panjapornpon, C., Mungcharoen, T.: Thailand Green GDP assessment based on environmentally extended input-output model. J. Clean. Prod. **167**, 970–977 (2017). https://doi.org/10.1016/j.jclepro.2017.02.106

4. Hosseini-Motlagh, S.M., Johari, M., Zirakpourdehkordi, R.: Grain production management to reduce global warming potential under financial constraints and time value of money using evolutionary game theory. Int. J. Prod. Res. **59**, 1–22 (2020). https://doi.org/10.1080/00207543.2020.1773562
5. Wang, Q., Wang, S.: Preventing carbon emission retaliatory rebound post-COVID-19 requires expanding free trade and improving energy efficiency. Sci. Total Environ. **746**, 141158 (2020). https://doi.org/10.1016/j.scitotenv.2020.141158
6. Kaenchan, P., Puttanapong, N., Bowonthumrongchai, T., Limskul, K., Gheewala, S.H.: Macroeconomic modeling for assessing sustainability of bioethanol production in Thailand. Energy Policy **127**, 361–373 (2019). https://doi.org/10.1016/j.enpol.2018.12.026
7. Haputta, P., Puttanapong, N., Silalertruksa, T., Bangviwat, A., Prapaspongsa, T., Gheewala, S.H.: Sustainability analysis of bioethanol promotion in Thailand using a cost-benefit approach. J. Clean. Prod. **251**, 119756 (2020). https://doi.org/10.1016/j.jclepro.2019.119756
8. Kuittinen, S., Hietaharju, J., Kupiainen, L., Hassan, M.K., Yang, M., Kaipiainen, E., Villa, A., Kangas, J., Keinänen, M., Vepsäläinen, J., Pappinen, A.: Bioethanol production from short rotation *S. schwerinii* E. Wolf is carbon neutral with utilization of waste-based organic fertilizer and process carbon dioxide capture. J. Clean. Prod. **293** (2021). https://doi.org/10.1016/j.jclepro.2021.126088
9. Preecha, S., Nuanyai, P.: Report on situation of ethanol price in Thailand and foreign countries (3rd quarter) (2019)
10. Zabed, H., Sahu, J.N., Suely, A., Boyce, A.N., Faruq, G.: Bioethanol production from renewable sources: current perspectives and technological progress. Renew. Sustain. Energy Rev. **71**, 475–501 (2017). https://doi.org/10.1016/j.rser.2016.12.076
11. Chiaramonti, D., Prussi, M., Ferrero, S., Oriani, L., Ottonello, P., Torre, P., Cherchi, F.: Review of pretreatment processes for lignocellulosic ethanol production, and development of an innovative method. Biomass Bioenerg. **46**, 25–35 (2012). https://doi.org/10.1016/j.biombioe.2012.04.020
12. Branco, R.H.R., Serafim, L.S., Xavier, A.M.R.B.: Second generation bioethanol production: on the use of pulp and paper industry wastes as feedstock. Fermentation **5**, 1–30 (2019). https://doi.org/10.3390/fermentation5010004
13. Miliotti, E., Dell'Orco, S., Lotti, G., Rizzo, A.M., Rosi, L., Chiaramonti, D.: Lignocellulosic ethanol biorefinery: Valorization of lignin-rich stream through hydrothermal liquefaction. Energies **12** (2019). https://doi.org/10.3390/en12040723
14. Huang, J., Chen, D., Wei, Y., Wang, Q., Li, Z., Chen, Y., Huang, R.: Direct ethanol production from lignocellulosic sugars and sugarcane bagasse by a recombinant *Trichoderma reesei* strain HJ48. Sci. World J. **2014** (2014). https://doi.org/10.1155/2014/798683
15. Tanimura, A., Kikukawa, M., Yamaguchi, S., Kishino, S., Ogawa, J., Shima, J.: Direct ethanol production from starch using a natural isolate, *Scheffersomyces shehatae*: toward consolidated bioprocessing. Sci. Rep. **5** (2015). https://doi.org/10.1038/srep09593
16. Devarapalli, M., Atiyeh, H.K.: A review of conversion processes for bioethanol production with a focus on syngas fermentation. Biofuel Res. J. **2**, 268–280 (2015). https://doi.org/10.18331/BRJ2015.2.3.5
17. Ciliberti, C., Biundo, A., Albergo, R., Agrimi, G., Braccio, G., de Bari, I., Pisano, I.: Syngas derived from lignocellulosic biomass gasification as an alternative resource for innovative bioprocesses. Processes. **8**, 1–38 (2020). https://doi.org/10.3390/pr8121567
18. Li, Y., Zhai, R., Jiang, X., Chen, X., Yuan, X., Liu, Z., Jin, M.: Boosting ethanol productivity of *Zymomonas mobilis* 8b in enzymatic hydrolysate of dilute acid and ammonia pretreated corn stover through medium optimization, high cell density fermentation and cell recycling. Front. Microbiol. **10**, 1–10 (2019). https://doi.org/10.3389/fmicb.2019.02316
19. Mohd Azhar, S.H., Abdulla, R., Jambo, S.A., Marbawi, H., Gansau, J.A., Mohd Faik, A.A., Rodrigues, K.F.: Yeasts in sustainable bioethanol production: a review. Biochem. Biophys. Rep. **10**, 52–61 (2017). https://doi.org/10.1016/j.bbrep.2017.03.003
20. de Sousa, C.C., Gonçalves, G.T.I., Falleiros, L.N.S.S.: Ethanol production using agroindustrial residues as fermentation substrates by *Kluyveromyces marxianus*. Ind. Biotechnol. **14**, 308–314 (2018). https://doi.org/10.1089/ind.2018.0023

21. Li, Y., Park, J. Yil, Shiroma, R., Tokuyasu, K.: Bioethanol production from rice straw by a sequential use of *Saccharomyces cerevisiae* and *Pichia stipitis* with heat inactivation of *Saccharomyces cerevisiae* cells prior to xylose fermentation. J. Biosci. Bioeng. **111**, 682–686 (2011). https://doi.org/10.1016/j.jbiosc.2011.01.018
22. Astuti, R.I., Alifianti, S., Maisyitoh, R.N., Mubarik, N.R., Meryandini, A.: Ethanol production by novel proline accumulating *Pichia kudriavzevii* mutants strains tolerant to high temperature and ethanol stresses. Online J. Biol. Sci. **18**, 349–357 (2018). https://doi.org/10.3844/ojbsci.2018.349.357
23. Chamnipa, N., Thanonkeo, S., Klanrit, P., Thanonkeo, P.: The potential of the newly isolated thermotolerant yeast *Pichia kudriavzevii* RZ8-1 for high-temperature ethanol production. Brazilian J. Microbiol. **49**, 378–391 (2018). https://doi.org/10.1016/j.bjm.2017.09.002
24. Chang, T., Yao, S.: Thermophilic, lignocellulolytic bacteria for ethanol production: current state and perspectives. Appl. Microbiol. Biotechnol. **92**, 13–27 (2011). https://doi.org/10.1007/s00253-011-3456-3
25. Scully, S.M., Orlygsson, J.: Recent advances in second generation ethanol production by thermophilic bacteria. Energies **8**, 1–30 (2015). https://doi.org/10.3390/en8010001
26. Taylor, M.P., Eley, K.L., Martin, S., Tuffin, M.I., Burton, S.G., Cowan, D.A.: Thermophilic ethanologenesis: future prospects for second-generation bioethanol production. Trends Biotechnol. **27**, 398–405 (2009). https://doi.org/10.1016/j.tibtech.2009.03.006
27. Bengelsdorf, F.R., Straub, M., Dürre, P.: Bacterial synthesis gas (syngas) fermentation. Environ. Technol. (United Kingdom) **34**, 1639–1651 (2013). https://doi.org/10.1080/09593330.2013.827747
28. Phillips, J.R., Huhnke, R.L., Atiyeh, H.K.: Syngas fermentation: A microbial conversion process of gaseous substrates to various products. Fermentation. **3** (2017). https://doi.org/10.3390/fermentation3020028
29. Dias, M.O.S., Ensinas, A. V., Nebra, S.A., Maciel Filho, R., Rossell, C.E.V., Maciel, M.R.W.: Production of bioethanol and other bio-based materials from sugarcane bagasse: Integration to conventional bioethanol production process. Chem. Eng. Res. Des. **87**, 1206–1216 (2009). https://doi.org/10.1016/j.cherd.2009.06.020
30. Dey, P., Pal, P., Kevin, J.D., Das, D.B.: Lignocellulosic bioethanol production: prospects of emerging membrane technologies to improve the process—a critical review. Rev. Chem. Eng. **36**, 333–367 (2020). https://doi.org/10.1515/revce-2018-0014
31. Rocha, G.J.M., Martín, C., da Silva, V.F.N., Gómez, E.O., Gonçalves, A.R.: Mass balance of pilot-scale pretreatment of sugarcane bagasse by steam explosion followed by alkaline delignification. Bioresour. Technol. **111**, 447–452 (2012). https://doi.org/10.1016/j.biortech.2012.02.005
32. Pradyawong, S., Juneja, A., Bilal Sadiq, M., Noomhorm, A., Singh, V.: Comparison of cassava starch with corn as a feedstock for bioethanol production. Energies **11**, 1–11 (2018). https://doi.org/10.3390/en11123476
33. De Guilherme, A.A., Dantas, P.V.F., Soares, J.C.J., Dos Santos, E.S., Fernandes, F.A.N., De Macedo, G.R.: Pretreatments and enzymatic hydrolysis of sugarcane bagasse aiming at the enhancement of the yield of glucose and xylose. Brazilian J. Chem. Eng. **34**, 937–947 (2017). https://doi.org/10.1590/0104-6632.20170344s20160225
34. Bunterngsook, B., Laothanachareon, T., Natrchalayuth, S., Lertphanich, S., Fujii, T., Inoue, H., Youngthong, C., Chantasingh, D., Eurwilaichitr, L., Champreda, V.: Optimization of a minimal synergistic enzyme system for hydrolysis of raw cassava pulp. RSC Adv. **7**, 48444–48453 (2017). https://doi.org/10.1039/c7ra08472b
35. Suwannarangsee, S., Bunterngsook, B., Arnthong, J., Paemanee, A., Thamchaipenet, A., Eurwilaichitr, L., Laosiripojana, N., Champreda, V.: Optimisation of synergistic biomass-degrading enzyme systems for efficient rice straw hydrolysis using an experimental mixture design. Bioresour. Technol. **119**, 252–261 (2012). https://doi.org/10.1016/j.biortech.2012.05.098
36. Gao, Y., Xu, J., Yuan, Z., Jiang, J., Zhang, Z., Li, C.: Ethanol production from sugarcane bagasse by fed-batch simultaneous saccharification and fermentation at high solids loading. Energy Sci. Eng. **6**, 810–818 (2018). https://doi.org/10.1002/ese3.257

37. Santos, S.C., de Sousa, A.S., Dionísio, S.R., Tramontina, R., Ruller, R., Squina, F.M., Vaz Rossell, C.E., da Costa, A.C., Ienczak, J.L.: Bioethanol production by recycled *Scheffersomyces stipitis* in sequential batch fermentations with high cell density using xylose and glucose mixture. Bioresour. Technol. **219**, 319–329 (2016). https://doi.org/10.1016/j.biortech.2016.07.102
38. Unrean, P., Khajeeram, S.: Model-based optimization of *Scheffersomyces stipitis* and *Saccharomyces cerevisiae* co-culture for efficient lignocellulosic ethanol production. Bioresour. Bioprocess. **2**, 1–11 (2015). https://doi.org/10.1186/s40643-015-0069-1
39. Khajeeram, S., Puseenam, A., Roongsawang, N., Unrean, P.: Optimal design of cost-effective simultaneous saccharification and co-fermentation through integrated process optimization. Bioenergy Res. **10**, 891–902 (2017). https://doi.org/10.1007/s12155-017-9851-6
40. Olofsson, K., Bertilsson, M., Lidén, G.: A short review on SSF—an interesting process option for ethanol production from lignocellulosic feedstocks. Biotechnol. Biofuels. **1**, 1–14 (2008). https://doi.org/10.1186/1754-6834-1-7
41. Koppram, R., Nielsen, F., Albers, E., Lambert, A., Wännström, S., Welin, L., Zacchi, G., Olsson, L.: Simultaneous saccharification and co-fermentation for bioethanol production using corncobs at lab PDU and demo scales. Biotechnol. Biofuels **6**, 2–11 (2013). https://doi.org/10.1186/1754-6834-6-2
42. Cripwell, R.A., Favaro, L., Viljoen-Bloom, M., van Zyl, W.H.: Consolidated bioprocessing of raw starch to ethanol by *Saccharomyces cerevisiae*: achievements and challenges. Biotechnol. Adv. **42**, 107579 (2020). https://doi.org/10.1016/j.biotechadv.2020.107579
43. Cunha, J.T., Romaní, A., Inokuma, K., Johansson, B., Hasunuma, T., Kondo, A., Domingues, L.: Consolidated bioprocessing of corn cob-derived hemicellulose: engineered industrial *Saccharomyces cerevisiae* as efficient whole cell biocatalysts. Biotechnol. Biofuels **13**, 1–15 (2020). https://doi.org/10.1186/s13068-020-01780-2
44. Peng, P., Lan, Y., Liang, L., Jia, K.: Membranes for bioethanol production by pervaporation. Biotechnol. Biofuels **14**, 1–33 (2021). https://doi.org/10.1186/s13068-020-01857-y
45. Dias, M.O.S., Junqueira, T.L., Rossell, C.E. V., MacIel Filho, R., Bonomi, A.: Evaluation of process configurations for second generation integrated with first generation bioethanol production from sugarcane. Fuel Process. Technol. **109**, 84–89 (2013). https://doi.org/10.1016/j.fuproc.2012.09.041
46. García, A., González Alriols, M., Labidi, J.: Evaluation of different lignocellulosic raw materials as potential alternative feedstocks in biorefinery processes. Ind. Crops Prod. **53**, 102–110 (2014). https://doi.org/10.1016/j.indcrop.2013.12.019
47. Kasavi, C., Finore, I., Lama, L., Nicolaus, B., Oliver, S.G., Toksoy Oner, E., Kirdar, B.: Evaluation of industrial *Saccharomyces cerevisiae* strains for ethanol production from biomass. Biomass Bioenerg. **45**, 230–238 (2012). https://doi.org/10.1016/j.biombioe.2012.06.013
48. Thammasittirong, S.N.R., Thirasaktana, T., Thammasittirong, A., Srisodsuk, M.: Improvement of ethanol production by ethanoltolerant *Saccharomyces cerevisiae* UVNR56. Springerplus **2**, 1–5 (2013). https://doi.org/10.1186/2193-1801-2-583
49. Silalertruksa, T., Gheewala, S.H., Pongpat, P.: Sustainability assessment of sugarcane biorefinery and molasses ethanol production in Thailand using eco-efficiency indicator. Appl. Energy. **160**, 603–609 (2015). https://doi.org/10.1016/j.apenergy.2015.08.087
50. Yuangsaard, N., Yongmanitchai, W., Yamada, M., Limtong, S.: Selection and characterization of a newly isolated thermotolerant *Pichia kudriavzevii* strain for ethanol production at high temperature from cassava starch hydrolysate. Antonie van Leeuwenhoek, Int. J. Gen. Mol. Microbiol. **103**, 577–588 (2013). https://doi.org/10.1007/s10482-012-9842-8
51. Jakrawatana, N., Pingmuangleka, P., Gheewala, S.H.: Material flow management and cleaner production of cassava processing for future food, feed and fuel in Thailand. J. Clean. Prod. **134**, 633–641 (2016). https://doi.org/10.1016/j.jclepro.2015.06.139
52. Techaparin, A., Thanonkeo, P., Klanrit, P.: High-temperature ethanol production using thermotolerant yeast newly isolated from Greater Mekong Subregion. Brazilian J. Microbiol. **48**, 461–475 (2017). https://doi.org/10.1016/j.bjm.2017.01.006

53. Bautista, K., Unpaprom, Y., Ramaraj, R.: Bioethanol production from corn stalk juice using *Saccharomyces cerevisiae* TISTR 5020. Energy Sources, Part A Recover. Util. Environ. Eff. **41**, 1615–1621 (2019). https://doi.org/10.1080/15567036.2018.1549136
54. Khonngam, T., Salakkam, A.: Bioconversion of sugarcane bagasse and dry spent yeast to ethanol through a sequential process consisting of solid-state fermentation, hydrolysis, and submerged fermentation. Biochem. Eng. J. **150**, 107284 (2019). https://doi.org/10.1016/j.bej.2019.107284
55. Sukhang, S., Choojit, S., Reungpeerakul, T., Sangwichien, C.: Bioethanol production from oil palm empty fruit bunch with SSF and SHF processes using *Kluyveromyces marxianus* yeast. Cellulose **27**, 301–314 (2020). https://doi.org/10.1007/s10570-019-02778-2
56. Khammee, P., Ramaraj, R., Whangchai, N., Bhuyar, P., Unpaprom, Y.: The immobilization of yeast for fermentation of macroalgae *Rhizoclonium* sp. for efficient conversion into bioethanol. Biomass Convers. Biorefinery. **11**, 827–835 (2021). https://doi.org/10.1007/s13399-020-00786-y
57. Limayem, A., Ricke, S.C.: Lignocellulosic biomass for bioethanol production: current perspectives, potential issues and future prospects. Prog. Energy Combust. Sci. **38**, 449–467 (2012). https://doi.org/10.1016/j.pecs.2012.03.002
58. Kumar, R., Strezov, V., Weldekidan, H., He, J., Singh, S., Kan, T., Dastjerdi, B.: Lignocellulose biomass pyrolysis for bio-oil production: A review of biomass pre-treatment methods for production of drop-in fuels. Renew. Sustain. Energy Rev. **123** (2020). https://doi.org/10.1016/j.rser.2020.109763
59. Kumar, V., Yadav, S.K., Kumar, J., Ahluwalia, V.: A critical review on current strategies and trends employed for removal of inhibitors and toxic materials generated during biomass pretreatment. Bioresour. Technol. **299**, 122633 (2020). https://doi.org/10.1016/j.biortech.2019.122633
60. Auesukaree, C.: Molecular mechanisms of the yeast adaptive response and tolerance to stresses encountered during ethanol fermentation. J. Biosci. Bioeng. **124**, 133–142 (2017). https://doi.org/10.1016/j.jbiosc.2017.03.009
61. Takagi, H.: Molecular mechanisms and highly functional development for stress tolerance of the yeast *Saccharomyces cerevisiae*. Biosci. Biotechnol. Biochem. **85**, 1017–1037 (2021). https://doi.org/10.1093/bbb/zbab022
62. Kavšček, M., Stražar, M., Curk, T., Natter, K., Petrovič, U.: Yeast as a cell factory: current state and perspectives. Microb. Cell Fact. **14**, 1–10 (2015). https://doi.org/10.1186/s12934-015-0281-x
63. Li, H., Wu, M., Xu, L., Hou, J., Guo, T., Bao, X., Shen, Y.: Evaluation of industrial *Saccharomyces cerevisiae* strains as the chassis cell for second-generation bioethanol production. Microb. Biotechnol. **8**, 266–274 (2015). https://doi.org/10.1111/1751-7915.12245
64. Ko, J.K., Enkh-Amgalan, T., Gong, G., Um, Y., Lee, S.M.: Improved bioconversion of lignocellulosic biomass by *Saccharomyces cerevisiae* engineered for tolerance to acetic acid. GCB Bioenergy. **12**, 90–100 (2020). https://doi.org/10.1111/gcbb.12656
65. Mollapour, M., Piper, P.W.: Hog1 mitogen-activated protein kinase phosphorylation targets the yeast Fps1 Aquaglyceroporin for endocytosis, thereby rendering cells resistant to acetic acid. Mol. Cell. Biol. **27**, 6446–6456 (2007). https://doi.org/10.1128/mcb.02205-06
66. Giannattasio, S., Guaragnella, N., Ždralević, M., Marra, E.: Molecular mechanisms of *Saccharomyces cerevisiae* stress adaptation and programmed cell death in response to acetic acid. Front. Microbiol. **4**, 1–7 (2013). https://doi.org/10.3389/fmicb.2013.00033
67. Mollapour, M., Shepherd, A., Piper, P.W.: Presence of the Fps1p aquaglyceroporin channel is essential for Hog1p activation, but suppresses Slt2(Mpk1)p activation, with acetic acid stress of yeast. Microbiology **155**, 3304–3311 (2009). https://doi.org/10.1099/mic.0.030502-0
68. Swinnen, S., Henriques, S.F., Shrestha, R., Ho, P.W., Sá-Correia, I., Nevoigt, E.: Improvement of yeast tolerance to acetic acid through Haa1 transcription factor engineering: towards the underlying mechanisms. Microb. Cell Fact. **16**, 1–15 (2017). https://doi.org/10.1186/s12934-016-0621-5

69. Ullah, A., Orij, R., Brul, S., Smits, G.J.: Quantitative analysis of the modes of growth inhibition by weak organic acids in *Saccharomyces cerevisiae*. Appl. Environ. Microbiol. **78**, 8377–8387 (2012). https://doi.org/10.1128/AEM.02126-12
70. Smardon, A.M., Kane, P.M.: Loss of vacuolar $H^+$-ATPase activity in organelles signals ubiquitination and endocytosis of the yeast plasma membrane proton pump Pma1p. J. Biol. Chem. **289**, 32316–32326 (2014). https://doi.org/10.1074/jbc.M114.574442
71. Mira, N.P., Palma, M., Guerreiro, J.F., Sá-Correia, I.: Genome-wide identification of *Saccharomyces cerevisiae* genes required for tolerance to acetic acid. Microb. Cell Fact. **9**, 1–13 (2010). https://doi.org/10.1186/1475-2859-9-79
72. Peña, P.V., Glasker, S., Srienc, F.: Genome-wide overexpression screen for sodium acetate resistance in *Saccharomyces cerevisiae*. J. Biotechnol. **164**, 26–33 (2013). https://doi.org/10.1016/j.jbiotec.2012.12.005
73. Watcharawipas, A., Watanabe, D., Takagi, H.: Enhanced sodium acetate tolerance in *Saccharomyces cerevisiae* by the Thr255Ala mutation of the ubiquitin ligase Rsp5. FEMS Yeast Res. **17**, 1–10 (2017). https://doi.org/10.1093/femsyr/fox083
74. Watcharawipas, A., Watanabe, D., Takagi, H.: Sodium acetate responses in *Saccharomyces cerevisiae* and the ubiquitin ligase Rsp5. Front. Microbiol. **9**, 1–9 (2018). https://doi.org/10.3389/fmicb.2018.02495
75. Inaba, T., Watanabe, D., Yoshiyama, Y., Tanaka, K., Ogawa, J., Takagi, H., Shimoi, H., Shima, J.: An organic acid-tolerant HAA1-overexpression mutant of an industrial bioethanol strain of *Saccharomyces cerevisiae* and its application to the production of bioethanol from sugarcane molasses. AMB Express **3**, 1–7 (2013). https://doi.org/10.1186/2191-0855-3-74
76. Kitichantaropas, Y., Boonchird, C., Sugiyama, M., Kaneko, Y., Harashima, S., Auesukaree, C.: Cellular mechanisms contributing to multiple stress tolerance in *Saccharomyces cerevisiae* strains with potential use in high-temperature ethanol fermentation. AMB Express. **6** (2016). https://doi.org/10.1186/s13568-016-0285-x
77. Benjaphokee, S., Hasegawa, D., Yokota, D., Asvarak, T., Auesukaree, C., Sugiyama, M., Kaneko, Y., Boonchird, C., Harashima, S.: Highly efficient bioethanol production by a *Saccharomyces cerevisiae* strain with multiple stress tolerance to high temperature, acid and ethanol. N. Biotechnol. **29**, 379–386 (2012). https://doi.org/10.1016/j.nbt.2011.07.002
78. Unrean, P., Gätgens, J., Klein, B., Noack, S., Champreda, V.: Elucidating cellular mechanisms of *Saccharomyces cerevisiae* tolerant to combined lignocellulosic-derived inhibitors using high-throughput phenotyping and multiomics analyses. FEMS Yeast Res. **18**, 1–10 (2018). https://doi.org/10.1093/femsyr/foy106
79. Fernandes, S., Murray, P.: Metabolic engineering for improved microbial pentose fermentation. Bioeng. Bugs. **1** (2010). https://doi.org/10.4161/bbug.1.6.12724
80. Hahn-Hägerdal, B., Karhumaa, K., Fonseca, C., Spencer-Martins, I., Gorwa-Grauslund, M.F.: Towards industrial pentose-fermenting yeast strains. Appl. Microbiol. Biotechnol. **74**, 937–953 (2007). https://doi.org/10.1007/s00253-006-0827-2
81. Young, E., Lee, S.M., Alper, H.: Optimizing pentose utilization in yeast: The need for novel tools and approaches. Biotechnol. Biofuels **3**, 1–12 (2010). https://doi.org/10.1186/1754-6834-3-24
82. Ire, F.S., Ezebuiro, V., Ogugbue, C.J.: Production of bioethanol by bacterial co-culture from agro-waste-impacted soil through simultaneous saccharification and co-fermentation of steam-exploded bagasse. Bioresour. Bioprocess. **3** (2016). https://doi.org/10.1186/s40643-016-0104-x
83. Nehme, N., Mathieu, F., Taillandier, P.: Impact of the co-culture of *Saccharomyces cerevisiae-Oenococcus oeni* on malolactic fermentation and partial characterization of a yeast-derived inhibitory peptidic fraction. Food Microbiol. **27**, 150–157 (2010). https://doi.org/10.1016/j.fm.2009.09.008
84. Zhao, Z., Xian, M., Liu, M., Zhao, G.: Biochemical routes for uptake and conversion of xylose by microorganisms. Biotechnol. Biofuels **13**, 1–12 (2020). https://doi.org/10.1186/s13068-020-1662-x

85. Cunha, J.T., Soares, P.O., Romaní, A., Thevelein, J.M., Domingues, L.: Xylose fermentation efficiency of industrial *Saccharomyces cerevisiae* yeast with separate or combined xylose reductase/xylitol dehydrogenase and xylose isomerase pathways. Biotechnol. Biofuels **12**, 1–14 (2019). https://doi.org/10.1186/s13068-019-1360-8
86. Ha, S.J., Galazka, J.M., Kim, S.R., Choi, J.H., Yang, X., Seo, J.H., Glass, N.L., Cate, J.H.D., Jin, Y.S.: Engineered *Saccharomyces cerevisiae* capable of simultaneous cellobiose and xylose fermentation. Proc. Natl. Acad. Sci. U. S. A. **108**, 504–509 (2011). https://doi.org/10.1073/pnas.1010456108
87. Martha, M.I., Astuti, R.I., Wahyuni, W.T.: Enhanced ethanol production by high temperature-tolerance mutant. Microbiol. Indones. **14**, 66–72 (2020). https://doi.org/10.5454/mi.14.2.x
88. Ulya, D., Indri Astuti, R., Meryandini, A.: The ethanol production activity of indigenous thermotolerant yeast *Pichia kudriavzevii* 1P4. Microbiol. Indones. **14**, 1 (2021). https://doi.org/10.5454/mi.14.4.1
89. Agbogbo, F.K., Wenger, K.S.: Production of ethanol from corn stover hemicellulose hydrolyzate using *Pichia stipitis*. J. Ind. Microbiol. Biotechnol. **34**, 723–727 (2007). https://doi.org/10.1007/s10295-007-0247-z
90. Bellido, C., Bolado, S., Coca, M., Lucas, S., González-Benito, G., García-Cubero, M.T.: Effect of inhibitors formed during wheat straw pretreatment on ethanol fermentation by *Pichia stipitis*. Bioresour. Technol. **102**, 10868–10874 (2011). https://doi.org/10.1016/j.biortech.2011.08.128
91. Koutinas, M., Patsalou, M., Stavrinou, S., Vyrides, I.: High temperature alcoholic fermentation of orange peel by the newly isolated thermotolerant *Pichia kudriavzevii* KVMP10. Lett. Appl. Microbiol. **62**, 75–83 (2016). https://doi.org/10.1111/lam.12514
92. Kitagawa, T., Tokuhiro, K., Sugiyama, H., Kohda, K., Isono, N., Hisamatsu, M., Takahashi, H., Imaeda, T.: Construction of a β-glucosidase expression system using the multistress-tolerant yeast *Issatchenkia orientalis*. Appl. Microbiol. Biotechnol. **87**, 1841–1853 (2010). https://doi.org/10.1007/s00253-010-2629-9
93. Kwon, Y.J., Ma, A.Z., Li, Q., Wang, F., Zhuang, G.Q., Liu, C.Z.: Effect of lignocellulosic inhibitory compounds on growth and ethanol fermentation of newly-isolated thermotolerant *Issatchenkia orientalis*. Bioresour. Technol. **102**, 8099–8104 (2011). https://doi.org/10.1016/j.biortech.2011.06.035
94. Isono, N., Hayakawa, H., Usami, A., Mishima, T., Hisamatsu, M.: A comparative study of ethanol production by *Issatchenkia orientalis* strains under stress conditions. J. Biosci. Bioeng. **113**, 76–78 (2012). https://doi.org/10.1016/j.jbiosc.2011.09.004
95. Astuti, R.I., Nurhayati, N., Ukit, Alifiyanti, S., Sunarti, T.C., Meryandini, A.: Exogenous L-proline increases stress tolerance of yeast *Pichia kudriavzevii* against inhibitors in lignocellulose hydrolysates and enhances its ethanol production. IOP Conf. Ser. Earth Environ. Sci. **197** (2018). https://doi.org/10.1088/1755-1315/197/1/012052
96. Ndubuisi, I.A., Qin, Q., Liao, G., Wang, B., Moneke, A.N., Ogbonna, J.C., Jin, C., Fang, W.: Effects of various inhibitory substances and immobilization on ethanol production efficiency of a thermotolerant *Pichia kudriavzevii*. Biotechnol. Biofuels. **13**, 1–12 (2020). https://doi.org/10.1186/s13068-020-01729-5
97. Nweze, J.E., Ndubuisi, I., Murata, Y., Omae, H., Ogbonna, J.C.: Isolation and evaluation of xylose-fermenting thermotolerant yeasts for bioethanol production. Biofuels **12**, 1–10 (2019). https://doi.org/10.1080/17597269.2018.1564480
98. Phong, H.X., Klanrit, P., Dung, N.T.P., Yamada, M., Thanonkeo, P.: Isolation and characterization of thermotolerant yeasts for the production of second-generation bioethanol. Ann. Microbiol. **69**, 765–776 (2019). https://doi.org/10.1007/s13213-019-01468-5
99. Costa, D.A., De Souza, C.J.A., Costa, P.S., Rodrigues, M.Q.R.B., Dos Santos, A.F., Lopes, M.R., Genier, H.L.A., Silveira, W.B., Fietto, L.G.: Physiological characterization of thermotolerant yeast for cellulosic ethanol production. Appl. Microbiol. Biotechnol. **98**, 3829–3840 (2014). https://doi.org/10.1007/s00253-014-5580-3
100. Hughes, S.R., Gibbons, W.R., Bang, S.S., Pinkelman, R., BischoV, K.M., Slininger, P.J., Qureshi, N., Kurtzman, C.P., Liu, S., Saha, B.C., Jackson, J.S., Cotta, M.A., Rich, J.O.,

Javers, J.E.: Random UV-C mutagenesis of *Scheffersomyces* (formerly *Pichia*) *stipitis* NRRL Y-7124 to improve anaerobic growth on lignocellulosic sugars. J. Ind. Microbiol. Biotechnol. **39**, 163–173 (2012). https://doi.org/10.1007/s10295-011-1012-x

101. Watanabe, T., Watanabe, I., Yamamoto, M., Ando, A., Nakamura, T.: A UV-induced mutant of *Pichia stipitis* with increased ethanol production from xylose and selection of a spontaneous mutant with increased ethanol tolerance. Bioresour. Technol. **102**, 1844–1848 (2011). https://doi.org/10.1016/j.biortech.2010.09.087
102. Koti, S., Govumoni, S.P., Gentela, J., Venkateswar Rao, L.: Enhanced bioethanol production from wheat straw hemicellulose by mutant strains of pentose fermenting organisms *Pichia stipitis* and *Candida shehatae*. Springerplus **5**, 1–9 (2016). https://doi.org/10.1186/s40064-016-3222-1
103. Geiger, M., Gibbons, J., West, T., Hughes, S.R., Gibbons, W.: Evaluation of UV-C mutagenized *Scheffersomyces stipitis* strains for ethanol production. J. Lab. Autom. **17**, 417–424 (2012). https://doi.org/10.1177/2211068212452873

# Achieving Sustainable Solid Waste Management in Sub-Saharan Africa: The Option of Valorisation and Circular Economy Model

**Folarin Olawale Saburi**

**Abstract** Poor solid waste management practices have remained a major form problem in Sub-Saharan Africa. Improper waste disposal such as open dumping is a deliberate, albeit unconscious squandering of valuable feedstock for industrial production. Most wastes are reusable or recyclable, serving as a source of livelihood for the vulnerable and as a source of cheaper raw materials for industries. Although there are several waste management options, the application of life-cycle assessment to waste management practices favours a circular economy and waste valorisation. The developed economies have long embraced circular economy as a sustainable waste management option. There is a need for the developing nations in Sub-Saharan Africa to also move quickly in this direction. The contribution of this waste management approach to the efforts of low-income countries in achieving some of their climate change action obligations, and more importantly, most of their United Nations Sustainable Development Goals (SDGs) encapsulates the remarkable benefits of circular economy. Although there are different subcategories of solid wastes, the application of circular economy to vehicle tyres and electronic wastes has been considered in this section due to the huge challenges they pose to waste management in developing countries and the invaluable materials that can be recovered from them as feedstocks.

**Keywords** Solid waste · Life-cycle assessment · Circular economy · Waste valorisation · Sub-Saharan Africa

## 1 Introduction

Every year, the world generates over 2 billion tonnes of solid waste, and about 33 per cent of this waste is managed in an unsustainable manner [1]. Solid wastes include wastes generated from homes, offices and a wide range of industrial and commercial activities. These may include discarded plastic containers and bags, food wastes,

Folarin O. S. (✉)
Centre for Environmental Research Education Awareness and Strategies (CEREAS), 15B, Ajayi Osungbekun Street, Agodi GRA, Ibadan, Oyo State, Nigeria
e-mail: wale@cereas.org

 
A. Z. Yaser et al. (eds.), *Waste Management, Processing and Valorisation*,
https://doi.org/10.1007/978-981-16-7653-6_15

electronic wastes, agricultural wastes, construction wastes, sludge from industrial effluent treatment plants, discarded industrial packaging materials, regulated medical wastes, expired pharmaceuticals, waste tyres and a host of others. Some solid wastes are in semi-solid or liquid form, and some of them have gaseous materials, and proper management of these discarded materials is an important element of public and environmental health [2].

It has been estimated that the quantity of solid waste will continue to rise as the world population increases with increase in the quantity of goods produced and waste generated. Specifically, by the year 2050, the global waste volume generated per annum would have risen to 3.4 billion tonnes, mostly coming from East Asia and Pacific, South Asia and Sub-Saharan Africa [3] (Fig. 1).

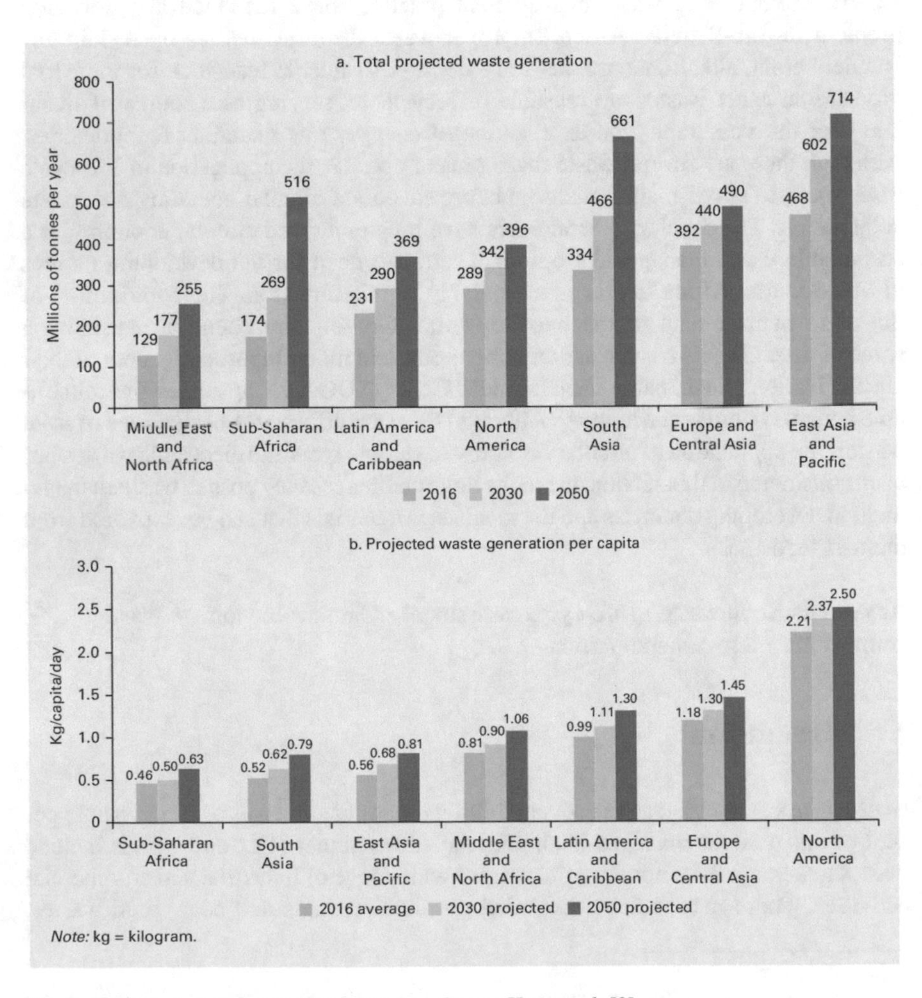


**Fig. 1** Projected waste generation by region. *Source* Kaza et al. [3]

Unlike in developed countries where the challenges of waste management are complicated by lack of robust taxing and loan and debt service regimes to support its management infrastructure, solid waste management is more challenging in developing countries. This is due to the preference for cheaper disposal alternatives such as open dumps and landfills [4]. As a result, developing countries face impending public health crises associated with poor solid waste management practices. The perennial challenges of solid waste disposal and associated problems in Sub-Saharan Africa require urgent and appropriate actions. One of the most appropriate strategies is the adoption of circular economy model in solid waste management. This guarantees attainment of some sustainable development goals of the United Nations while effectively addressing the problem of solid wastes disposal.

## 2 Waste Management and Sustainable Development Goals

The Sustainable Development Goals (SDGs), also known as the Global Goals, adopted by the United Nations in 2015, are a universal call to action aimed at ending poverty, protect the planet and ensure that by 2030 all people enjoy peace and prosperity. A closer look at the 17 SDGs provides insight into the vital roles of proper solid waste management towards the attainment of some of the goals (Table 1). For example, improper solid waste disposal is a risk factor for vector-borne diseases and sustainable solid waste management is of utmost importance to their prevention [5]. In addition, the common practice of burning wastes increases public health risk due to the release of toxic fumes and gases from the materials. In recent times, the coronavirus disease outbreak has drastically increased the volume of solid waste generation [6], and improper disposal of discarded personal protective equipment such as non-surgical face masks, face shields, gloves and gowns may increase the risk of COVID-19 spread [7, 8]. Therefore, it is practically impossible to achieve some of the SDGs without sustainable solid waste management.

## 3 Solid Waste Management in Developing Countries

Urbanisation in developing countries is becoming rapid with disproportionately high population in large cities [9]; and this rural-to-urban shift in population has been associated with rising volume of waste in countries with poorly developed industrial bases and developing economies [10]. Developing economies consist of seven North African countries, seven Central African countries, thirteen nations in East Africa, eleven Southern Africa countries, all the fifteen members of the Economic Community of West African States. In East Asia, twenty-three countries are classified as developing economies, nine in South Asia, 14 in Western Asia, 7 in Latin America and the Caribbean, 10 each in Central and South America [11]. In these

**Table 1** Relationship between solid waste management and sustainable development goals

| SN | Sustainable development goal | Waste management and specific target |
|---|---|---|
| 1 | SDG3 | SDG3 is to substantially reduce the number of deaths and illnesses from hazardous chemicals and air, water and soil pollution and contamination. In addition, the goal is to minimise deaths resulting from non-communicable diseases such as malaria. Poor solid waste management usually results in pollution of water and soil while also providing breeding site for vectors of non-communicable diseases |
| 2 | SDG6 | To ensure availability of clean water and sanitation for all, as specified under SDG6, sound solid waste management programme is essential |
| 1 | SDG 11 | The specific target of SDG11 is for everyone to have healthy, safe communities and sustainable cities. This is not achievable without sustainable solid waste management |
| | SDG12 | SDG12 is aimed at achieving sustainable management and efficient use of natural resources; environmentally sound management of chemicals and all wastes throughout their life cycle; and substantial reduction in waste generation through prevention, reduction, recycling and reuse |
| 2 | SDG13 | The Climate Action (SDG 13) is geared towards reducing the emission of greenhouse gases like methane and carbon dioxide ($CO_2$), which are released from, among other sources, open dumping and or burning of refuse |
| 3 | SDG14 | Waste plastics, often improperly disposed in developing countries, have been found in abundance in the marine environment. This is a threat to the achievement of SDG14, which is aimed at protecting the marine ecological system |
| 4 | SDG15 | SDG 15 is a drive towards minimising pollution in the terrestrial ecosystem for a healthier environment |

low-income countries across different continents, improper waste disposal is an age-long problem. This may further complicate environmental degradation and public health crisis such as the spread of diseases [12].

Open dumping is an age-long practice in developing countries across the world [13]. In Sub-Sahara Africa where most countries are in the developing economies category, the practice is widespread, accounting for about 75% of the disposal method for the total volume of wastes generated [14]. The prevalent unsustainable solid waste management practices may be attributed to the common features associated with developing countries such as poor governance and lack of capability to finance critical infrastructure without foreign aids. Apparently, poor governance reflects in weak environmental regulations and enforcement, which by extension has been the primary contributor to poor waste management culture among the populace. Lack of finance for the development of standard waste management infrastructure has been a factor contributing to problems associated with the disposal of products at their end of life. As a result, the common waste management practices are open dumping of waste that could have served as a source of raw materials for new products.

Some solid wastes such as discarded plastic containers of pesticides are hazardous in nature. They are generated from diverse sources and discarded in waste bins or at dumpsites. Their chemical composition necessitated special handling procedures and disposal methods to safeguard the environment and human health. The lack of infrastructure to manage such wastes in developing countries makes it a huge challenge and threat to public health. Where such infrastructures are available, the identified existing or potential impacts of deploying facilities for waste management are of concern. This applies in general to the valorisation and circular economy of solid wastes in developing countries. As a result, the focus is on life-cycle assessment approach to solid waste management and the circular economy of hazardous solid wastes.

## 4 Life-Cycle Assessment Approach to Solid Waste Management

Life-cycle assessment (LCA) is a quantitative approach to evaluate the potential adverse environmental effects of the product(s) across raw materials extraction and processing, products manufacturing, distribution and utilisation, thereby identifying specific points of significant impacts. The methodologies and tools of LCA are being used more in recent times for solid waste management as they evaluate and compare the environmental aspects and impacts of different waste management systems and practices. LCA is applied in the evaluation of waste management practices [15]. This approach to solid waste management is an important decision-making tool for public policymakers, especially waste management authorities for more informed, thoughtful decisions [16].

The application of LCA in waste management entails the cradle-to-grave environmental analysis of discarded waste materials, and the product is produced from waste materials. At the stage of materials recovery from waste materials, wastes are usually generated as part of the remote, indirect adverse environmental consequences of new products derived from recycled wastes. The new products, at their end of life, may be managed further in many ways through reuse, recycling, energy recovery or disposal, and each with varying environmental impacts [17]. Therefore, LCA can serve as an effective and useful tool in evaluating the environmental impacts or performance of a proposed or existing solid waste management programme in developing countries.

### 4.1 Social Impacts of Life-Cycle Assessment

Social life-cycle assessment (S-LCA) is a procedure for product information acquisition, analysis and communication with respect to its existing and possible socioeconomic impacts throughout the product's life cycle [18]. This technique focuses on the current and potential social effects of processes, products and services from cradle to grave [19]. In general, the outcome of LCA suggests that solving one problem almost always leads to other problems [20].

### 4.2 Material Recovery and Social Risks

The importance of raw materials has been underscored as indispensable to the attainment of some United Nations SDGs [21]. They play crucial roles in production processes, contribute to the drive of any society or nation towards industrialisation and assist in the provision of employment opportunities, improved gross domestic products, among development goals. However, the extraction or production of these crucial components of manufacturing inventory has potential and significant socioeconomic impacts, generally in relation to poor governance and weak institutional and legal frameworks [22].

Production of raw materials from wastes starts with the acquisition of site where adequate land is available for such recycling facility to be sited. The most critical potential socio-economic impact associated with land acquisition is physical or economic displacement. Some World Bank Environmental and Social Standards as well as the International Finance Corporation Performance Standards are usually triggered by land acquisition; and the social impacts associated with it require appropriate mitigation measures that are in line with international best practice.

In addition, the life-cycle assessment of the raw materials extraction process comes with a generation of wastes at different phases, from the preconstruction stage, through construction or installation of extraction facilities through the production phase and facility decommissioning phase. Wastes generated at recycling site become a social and environmental burden if no proper waste management plan is in place. Emissions, which may result from waste recycling processes, may also be a social nuisance to the host communities of the recycling facility.

## 5 Life-Cycle Assessment and Circular Economy of Solid Wastes

The life-cycle assessment of solid waste management options favours material recovery, which led to the widespread acceptance of circular economy. Circular economy (CE) is an archetypical shift away from the prevailing unsustainable linear economic model towards environmental sustainability and viable waste management practices. The economic model was conceptualised by Pearce and Turner as an economy based on the conversion of wastes into material resources through a natural ecosystem feedback mechanisms or technological feedback mechanism. It is an important model that is key to achieving effective handling of solid waste crisis prevalent in developing countries while also creating job opportunities and cheaper raw materials for industries particularly in emerging economies. Moreover, the waste management hierarchy underscores the importance of recycling as an effective waste management strategy, and circular enterprises play an important role in creating higher value chains and environmental preservation [23]. This is the only right direction to go for waste management authorities in developing countries, and starting with investment in national public awareness and provision of technical skills would play a key role in achieving successful implementation of circular enterprise initiatives.

Solid wastes generated at every stage of a product life cycle are not fully utilised as raw materials for the production of another product, especially in developing countries. On the other hand, advanced economies have embraced the CE approach as one of the best waste management options. Member states in the European Union have concentrated their interest in the implementation of CE, and especially small- and medium-sized enterprises [24]. Full integration and implementation of this circular business model into the economic plan of developing countries is capable of ameliorating many social and economic crisis such as poor environmental health, pollution problems, health crisis and a high rate of unemployment, which are common in less-developed countries.

## *5.1 Circular Economy in Sub-Saharan Africa*

The recycling sector of solid waste management in Sub-Saharan Africa is dominated by informal recyclers. As a result, accurate statistics on the volume of waste recovered, recycled or valorised are not readily available and difficult to project [25]. In Sub-Saharan Africa using Nigeria as a case study, public awareness about circular economy is still poor, and the perception that material recovery from waste is for the underprivileged and scavengers is prevalent. The solid waste management system in most part of the country is still elementary, unsustainable, largely inefficient and disorganised. This is due to lack of adequate waste management facilities and deprived access to solid waste collection and management services [26]. It is noteworthy, that concerted efforts are being made by government parastatals and agencies to catch up with the exacting realities and proffer solution to the solid waste crisis in Nigeria—Africa's most populous country.

Recently, the federal government of Nigeria developed the National Policy on Solid Waste Management (2020) with a supporting implementation strategy, aimed at sustainably managing the severe solid waste problem as it touches all categories of solid wastes generated in the country. Prior to development of the policy, the National

Environmental Standards and Regulations Enforcement Agency (NESREA) initiated Extended Producers Responsibility (EPR) in line with international guidance to minimise solid wastes in dumpsites and most of the stakeholders in the manufacturing sectors of the Nigerian economy have subscribed to the initiative, thereby enhancing the supply chain for some of their raw materials. The key players in the implementation of EPR in Nigeria are members of the Food and Beverages Recycling Alliance (FBRA).

Adopting the circular economic models can help countries in Sub-Saharan Africa to attain some of their climate action obligations and sustainable development goals. For instance, recovery of waste polyethylene terephthalate (PET) and plastic recycling has created thousands of job opportunities for the poor and unemployed groups, including youths and women groups in many African countries. In Ghana, for example, a newly established waste and plastics recycling plant has the potential to generate jobs for 2300 people [27].

Waste management authorities in Sub-Saharan African countries can take a cue from Latin America and the Caribbean (LAC) countries. They have fast-growing cities and rising volume of their solid wastes, which are mostly organic in nature [28, 29]; in spite of poor evolution of solid waste management practices among member countries, circular economy has received significant attention in LAC, most of which are equally in the category of developing countries [30].

Although there are different subcategories of solid wastes, the application of circular economy to vehicle tyres and electronic wastes has been considered in this section due to the massive challenges they pose to waste management in less advanced countries. In addition, they are solid waste materials with high investment prospects in terms of material recovery.

At the products' end of life, they are being converted to energy source, in form of electricity or as fuels. This is an emerging practice in developing countries while the developed countries are at the advanced stage, which has helped them in achieving cleaner cities and sustainable solid waste management. There are reliable and cost-effective technologies required to explore renewable energy resources in Africa through the use of organic solid wastes. One of such affordable technology is the use of organic waste to generate biogas for power production either for off-grid communities or for integration into the national grid. In addition, agro-allied industries in Africa make use of their waste materials to generate process heat and power, which is used locally [31].

## 5.2 Solid Waste Valorisation

Waste valorisation is an integral component of circular business model. It involves reuse, recycling or conversion of waste materials into more useful products or fuels. Apart from organic waste, which is non-hazardous and biodegradable; waste vehicle tyres and electronic wastes are two major wastes materials with grave environmental consequences that should be considered in the context of circular economy.

### 5.2.1 Sustainable Waste Tyres Management

Waste tyres are one of the most challenging solid wastes crises with an estimated 1.5 billion waste tyres generated annually [32]. Their resistance to biodegradation and release of toxic fumes if burnt are factors that make the waste tyre a threat to the environment and human health. The difficulties associated with their disposal and the need for sustainable waste management made waste tyres valorisation an attractive option [33]. In the last 10 years, some studies in developed and developing countries have employed LCA to assess the environmental impact of waste tyres and disposal options. They are been reused for many industrial and commercial purposes, in civil engineering, and as fuel in cement production plants (Fig. 2).

Waste tyre recycling is gaining significant attention and acceptance due to the increase in raw material prices and growing environmental awareness among governments, manufacturers and consumers [35]. Pyrolysis and hydrothermal liquefaction are two major environment-friendly thermochemical conversion technologies for the effective conversion of waste tyre into energy. The oil from the process is a viable raw material for the production of hydrocarbon compounds such as benzene and xylene [32], which are further used as raw materials in petrochemical industries.

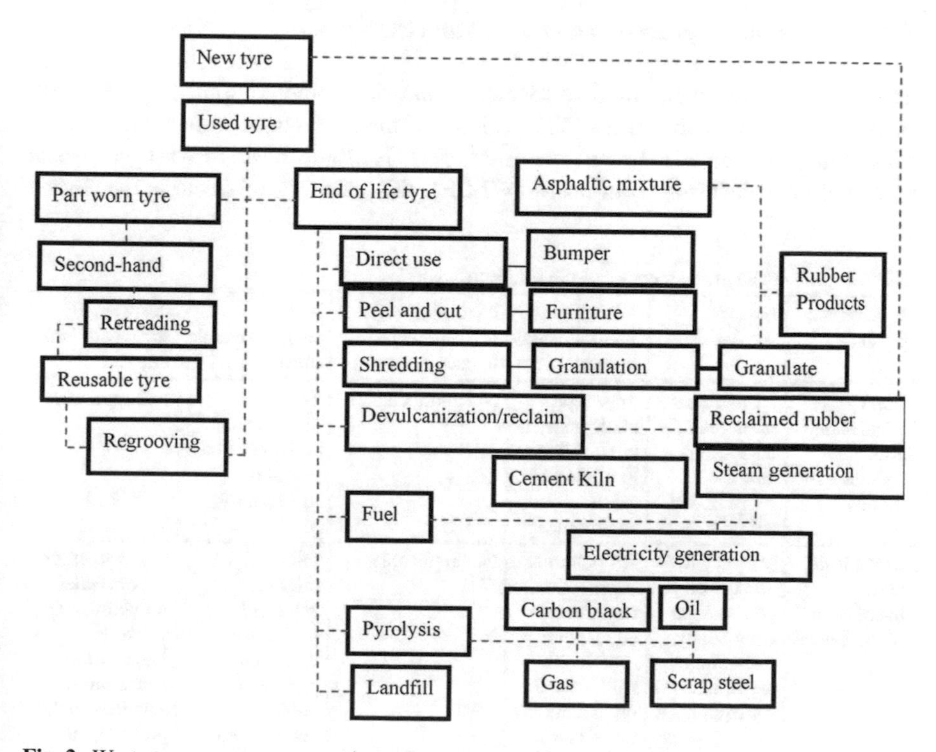


**Fig. 2** Waste tyres management options. *Source* Mmereki et al. [34]

There is also a rising demand for pyrolysis oil particularly for use in diesel engines. In addition, the growing rate of infrastructure development projects across the world and the drive towards industrialisation in developing countries would play a major role in the demand for pyrolysis oil [36].

### 5.2.2 Environmental and Social Impacts of Waste Tyre Valorisation

There are positive and adverse potential impacts of tyre valorisation on the biophysical and human environment. The use of scrap tyres for fuel in cement plants provides more reductions in most environmental impact categories compared to other scrap tyre applications, excluding applications in artificial turf. For example, every metric ton of tyre-derived fuel substituted for coal in cement kilns eliminates an estimated 0.5 tonnes ($CO_2$ equivalent) of direct and indirect greenhouse gas emissions [37]. The other benefits of waste tyre valorisation are more than a reduction in greenhouse gases and global warming. Other benefits include resource conservation and reduction in the use of fossil fuels (Table 2).

### 5.2.3 Circular Approach to E-Waste Valorisation

The surging demand for modern electrical and electronic equipment (EEE) with increasingly shorter life cycle [38], which has made electronic waste an issue of critical global interest. Generally referred to as electronic waste (e-waste) or waste electrical and electronic equipment (WEEE), this category of waste is the fastest

**Table 2** Life-cycle assessment of waste tyre pyrolysis

| Process and products | Impact sources | Adverse environmental impact | Positive environmental impact | Adverse socio-economic impact | Positive socio-economic impact |
|---|---|---|---|---|---|
| Land acquisition and clearance for facility | Displacement of original landowners and restriction of access to land | Impact on biodiversity if the facility is located in a green environment | Not applicable | Economic displacement and impact on means of livelihood | Not applicable |
| Construction and installation of equipment | Movement and use of construction equipment; excavation; use of lube oil for equipment | Air pollution from construction equipment, Accidental spill of lube oil and impact on soil and water resources | Not applicable | Influx of migrant workers to project site; increased pressure on public facilities, etc. | Creation of job opportunities for construction workers; improved economic activities in the project area |

(continued)

**Table 2** (continued)

| Process and products | Impact sources | Adverse environmental impact | Positive environmental impact | Adverse socio-economic impact | Positive socio-economic impact |
|---|---|---|---|---|---|
| Raw materials acquisition | Material sourcing and transportation | Emission from haulage vehicles<br>Impact on road traffic and accident | Reduction in volume of waste going to landfill and its associated benefits | Visual impact of waste materials at storage section | Source of employment and income for the vulnerable groups |
| Pyrolysis | Material recovery, heat application and water for cooling | Emission of gases, Residual wastes; high energy consumption, Water consumption for cooling | Recovery of non-biodegradable waste materials and environmental conservation; Sustainable production | Not applicable | Job creation and skills transfer |
| Pyrolytic oil and carbon black | Spills due to mishandling or leaks<br>Improper storage or movement | Soil contamination with hazardous hydrocarbon substance; Pollution of water resources; Inhalation of fugitive carbon dust | Not applicable | Not applicable | Oil and carbon black are cheaper raw materials for industries and road construction<br>Source of income for investors and revenues for government |

growing waste stream globally due to the growing human population and rapid technological advancement according to the United Nations. Annually, the volume of WEE generated worldwide is estimated at 50 million, and this is likely to increase to 120 million tonnes per annum by the year 2050 [39]. This is due to the volume of WEEE generated, the diversity and toxicity of some of its components, their potential environmental and health hazards and the high added value in their components, it is one of the most critical waste categories with respect to public policy and decision-making for waste management. [40].

The problem of WEEE is more complicated generally in developing countries due to the inability to afford new EEE by a significant population, which makes them the target destination for shipment of near end-of-life appliances from advanced countries. People with lower income and poor purchasing power can only afford to buy used EEE [41]. Moreover, the transboundary movement of used electronic appliances alongside tonnes of end-of-life EEE to less developed countries has been attributed to stringent environmental regulations, rising cost of manpower and poor returns on investment in recycling them in the advanced countries [42]. Consequently, more electrical and electronic wastes are generated rapidly in developing countries, which are being improperly handled through rudimentary methods without appropriate tools and poor facilities [43]. Nonetheless, the significant roles of informal handlers of e-waste in the overall WEEE management among third world countries with poor or no formal waste management structures have been well-acknowledged [44]. Considering the prevalent problems of this type of waste and the associated impacts of its improper disposal, “it is of crucial importance to develop circular economy solutions that prioritise reuse and recycling, as well as reducing the amount of waste that is disposed of at landfill” [45]. The reuse option is more attractive under the circular economy concept for electronic waste than the recycling option.

Wastes categorised as WEE are diverse in nature and components. As a result, there is no specific valorisation process that fits all. However, the general material recovery and recycling process includes e-waste collection, categorisation and sorting, module separation, disassembly and segregation. Further material recovery procedures are based on the outcomes of material segregation (Fig. 3).

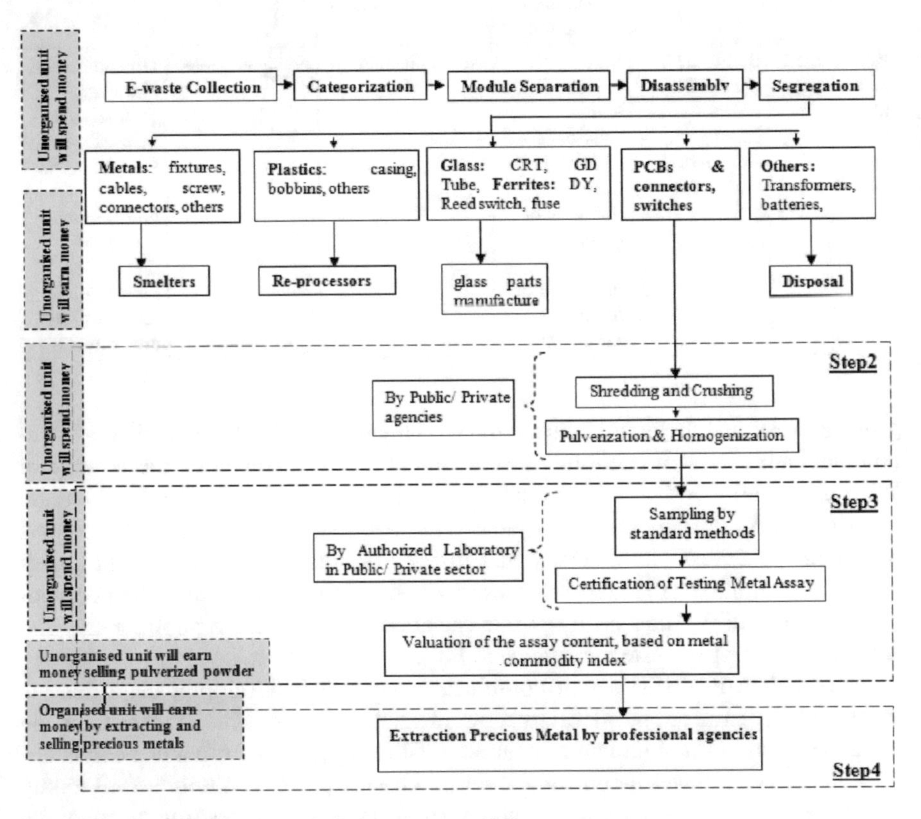


**Fig. 3** Process flowchart for e-waste management [46]

## 6 Conclusion

Increase in generation of solid waste remains inevitable as population rises leading to increase in purchase of products, use and disposal. In addition, the environmental, health and social problems associated with solid waste disposal in developing countries will continue to be an economic burden without a sustainable waste management programme. Clearly, countries in Sub-Saharan Africa will continue to struggle with poor economic growth until significant steps are taken to drive circular enterprise. The integration of circular business model into the economic programme of developing countries and its effective implementation will significantly reduce the burden of solid waste. Also, employment opportunities and means of livelihood are created through this waste management approach while sustainable development goals and targets are achieved. Undoubtedly, the challenges of rising volume of waste due to the rising population, shorter lifespan of products, poor disposal attitude of most people in developing countries can be daunting. Nonetheless, these challenges confronting developing nations come with immense economic potentials with significant positive social impacts. This is underscored by the highlighted benefits of circular economy which encourages material recovery and resource efficiency, cheaper feedstock for the manufacturing sector, lesser volume of waste at landfills, a cleaner city and its associated public health benefits and job opportunities for the vulnerable members of the society. Although general life-cycle assessment favours circular economy, specific LCA of each category of solid wastes remains an indispensable, important decision-making tool to ensure that public policies on solid waste management are formulated based on adequate knowledge of their adverse and or positive environmental and socio-economic impacts.

## References


1. World Bank: Solid Waste Management. Urban Development Brief (2019). https://www.worldbank.org/en/topic/urbandevelopment/brief/solid-waste-management; Retrieved April 16 Apr 2021
2. United State Environmental Protection Agency: Criteria for the Definition of Solid Waste and Solid and Hazardous Waste Exclusions (2020). https://www.epa.gov/hw/criteria-definition-solid-waste-and-solid-and-hazardous-waste-exclusions. Retrieved 12 May 2021
3. Kaza, S., Yao, L., Bhada-Tata, P., Woerden, F.: What a Waste 2.0: A Global Snapshot of Solid Waste Management to 2050 (2018)
4. Ali, S.M., Pervaiz, A., Afzal, B., Hamid, N., Yasmin, A.: Open dumping of municipal solid waste and its hazardous impacts on soil and vegetation diversity at waste dumping sites of Islamabad city. J. King Saud Univ. Sci. **26**(1), 59–65 (2014). https://doi.org/10.1016/j.jksus.2013.08.003
5. Krystosik, A., Njoroge, G., Odhiambo, L., Forsyth, J., Mutuku, F., LaBeaud, A.D.: Solid wastes provide breeding sites, burrows, and food for biological disease vectors, and urban zoonotic reservoirs: a call to action for solutions-based research. Front. Public Health (2020). https://doi.org/10.3389/fpubh.2019.00405

6. Torres, F.G., De-la-Torre, G.E.: Face mask waste generation and management during the COVID-19 pandemic: an overview and the Peruvian case. Sci. Total Environ. **786**, 147628 (2021). https://doi.org/10.1016/j.scitotenv.2021.147628
7. Nzediegwu, C., Chang, S.X.: Improper solid waste management increases potential for COVID-19 spread in deve. Resour. Conserv. Recycl. **161**, 104947 (2020). https://doi.org/10.1016/j.resconrec.2020.104947. https://www.afdb.org/en/news-and-events/press-releases/new-report-african-circular-economy-alliance-shines-light-five-big-bets-circular-economy
8. Yousefi, M., Oskoei, V., Jafari, A.J., Farzadkia, M., Firooz, M.H., Abdollahinejad, B., Torkashvand, J.: Municipal solid waste management during COVID-19 pandemic: effects and repercussions. Environ Sci Pollut Res Int **5**(3), 1–10 (2021). https://doi.org/10.1007/s11356-021-14214-9
9. Henderson, V.: Urbanization in developing countries. World Bank. © World Bank (2002). https://openknowledge.worldbank.org/handle/10986/16420 License: CC BY-NC-ND 3.0 IGO
10. Lagerkvist, A., Dahlén, L.: Solid waste generation and characterization. In: Meyers, R.A. (ed.) Encyclopedia of Sustainability Science and Technology. Springer, New York, NY (2012). https://doi.org/10.1007/978-1-4419-0851-3_110
11. World Economic Situation and Prospects Report: Country classification, p. 166. United Nations, New York (2020)
12. Ferronato, N., Torretta, V.: Waste mismanagement in developing countries: a review of global issues. Int. J. Environ. Res. Public Health **16**(6), 1060 (2019). https://doi.org/10.3390/ijerph16061060
13. Yadav H., Kumar P., Singh V.P.: Hazards from the municipal solid waste dumpsites: a review. In: Singh, H., Garg, P., Kaur, I. (eds) Proceedings of the 1st International Conference on Sustainable Waste Management through Design. ICSWMD 2018. Lecture Notes in Civil Engineering, vol 21. Springer, Cham (2019). https://doi.org/10.1007/978-3-030-02707-0_39
14. Cogut, A.: Open burning of waste: A global health disaster. R20 Regions of Climate Action. Geneva; R20
15. Banias, G., Batsioula, M., Achillas, C., Patsios, S.I., Kontogiannopoulos, K.N., Bochtis, D., Moussiopoulos, N.: A life cycle analysis approach for the evaluation of municipal solid waste management practices: the case study of the Region of Central Macedonia Greece. Sustainability **12**(8221), 2–17 (2020)
16. Seidel, C.: The application of life cycle assessment to public policy development. Int. J. Life Cycle Assess. **21**, 337–348 (2016). https://doi.org/10.1007/s11367-015-1024-2
17. European Commission Joint Research Centre: Environmental impacts along the supply chain. Raw Materials Information System (2021). https://rmis.jrc.ec.europa.eu/?page=environmental-impacts-along-the-supply-chain-3dfccf, Retrieved 13 May 2021
18. Norris, C., Norris, G., Aulisio, D.: Efficient assessment of social hotspots in the supply chains of 100 product categories using the social hotspots database. Sustainability **6**, 6973–6984 (2014)
19. Fauzi, R.T., Lavoie, P., Sorelli, L., Heidari, M.D., Amor, B.: Exploring the current challenges and opportunities of life cycle sustainability assessment. Sustainability **11**, 636 (2019)
20. United States Environmental Protection Agency: Life cycle analysis: its place in waste management. In: Conference Proceedings: 11th Annual Virginia Waste Management Conference, Richmond Virginia, pp 1–7 (1993)
21. Mancini, L., Sala, S.: Social impact assessment in the mining sector: Review and comparison of indicators frameworks. Resour. Policy **57**, 98–111 (2018)
22. Eynard, U., Mancini, L., Eisfeldt, F., Ciroth, A., Pennington, D.: Social risk in raw materials extraction: a macro-scale assessment. In: Social LCA People and Places for Partnership, 10–12 Sept 2018, Pescara, Italy, Pre-proceeding 6th Social LCA Conference; 2018, ISBN 978-2-9562141-1-3 (online), ISSN 2426-9654 (online), pp. 221–225, JRC110585.
23. African Development Bank: New report by the African Circular Economy Alliance Shines Light on 'Five Big Bets' for Circular Economy (2021)
24. Morales, C.M.B and Sossa, J.W.Z.: Circular economy in Latin America: A systematic literature review. Business Strategy and the Environment 2020: 1–19.

25. Niekerk, S.V., Weghmann, V.: Municipal solid waste management services in Africa. In: Working Paper, Public Service International, pp. 1–68 (2019)
26. Rajput, J., Potgieter, J., Aigbokhan, G., Felgenhauer, K., Smit, T.A.B., Hemkhaus, M., Ahlers, J., Van Hummelen, S., Chewpreecha, U., Smith, A., McGovern, M.: Circular economy in the Africa-EU cooperation—Country report for Nigeria. Country report under EC Contract ENV.F.2./ETU/2018/004 Project: "Circular Economy in Africa-Eu cooperation", Trinomics, B.V., Tomorrow Matters Now Ltd., adelphi Consult GmbH and Cambridge Econometrics Ltd.
27. World Economic Forum: Five Big Bets for the Circular Economy in Africa. Insight Report, p. 4, 23 (2021)
28. Hettiarachchi, H., Ryu, S., Caucci, S., Silva, R.: Municipal solid waste management in Latin America and the Caribbean: issues and potential solutions from the governance perspective. Recycling **3**, 19 (2020). https://doi.org/10.3390/recycling3020019
29. Guerra, C.: Urban solid waste and circular economy in Latin America. Oxford urbanists (2019). https://www.oxfordurbanists.com/oxford-urbanists-monthly/2019/8/29/8yhey82otwagoj1z5e2xr7y7634lah
30. Schröder, P., Albaladejo, M., Ribas, P.A., MacEwen, M., Tilkanen, J.: The Circular Economy In Latin America and the Caribbean. The Royal Institute of International Affairs, Chatham House, London, pp. 4, 23 (2020)
31. Médoc, J., Veenhuizen, R.V.: WABEF: Western Africa bio-wastes for energy and fertilisers. Urban Agriculture Magazine, 32 (2017)
32. Attia, Y.H., Tsabet, E., Mohaddespour, A., Munir, M.J., Farag, S.: Valorisation of waste tire by pyrolysis and hydrothermal liquefaction: a mini-review. J. Mater. Cycles Waste Manag. (2021). https://doi.org/10.1007/s10163-021-01252-1
33. Arabiourrutia, M., Lopez, G., Artetxe, M., Alvarez, J., Bilbao, J., Olazar, M.: Waste tyre valorisation by catalytic pyrolysis—A review. Renew. Sustain. Energy Rev. **129**, 109932 (2020)
34. Mmereki, D., Machola, B., Mokokwe, K.: Status of waste tyres and management practice in Botswana. J. Air Waste Manag. Assoc. **69**(10), 1230–1246 (2019). https://doi.org/10.1080/10962247.2017.1279696
35. Taverne, J.: End of Life Tyres, A Valuable Source with Growing Potential. European Tyre and Rubber Manufacturing Association (ETRMA), Brussels (2010)
36. Transparency Market Research: Pyrolysis Oil Market (2020). https://www.transparencymarketresearch.com/pyrolysis-oil-market.html. Retrieved May 26 May 2021
37. Fiksel, J., Bakshi, B.R., Baral, A., Guerra, E., DeQuervain, E.: Comparative life cycle assessment of beneficial applications for scrap tyres. Clean Technol. Environ. Policy **2011**(13), 19–35 (2011)
38. Bhutta, M.K.S., Omar, A., Yang, X.: Electronic waste: a growing concern in today's environment. Econ. Res. Int. 2021, (Article ID 474230) 8 pages (2011). https://doi.org/10.1155/2011/474230
39. Nekouei, R.K., Saba, M., Golmohammadzadeh, R., Faraji, F.: Valorisation of Electronic Waste (E-waste): Value-Added Products Derived from Processing and Recycling of E-waste. Special Issue 2021, https://www.mdpi.com/journal/sustainability/special_issues/e_waste; Retrieved 11 May 2021
40. Xavier, L.H., Ottoni, M.: A circular approach to the E-Waste Valorisation through urban mining in Rio De Janeiro, Brazil (June 20, 2019). In: Abstract Proceedings of 2019 International Conference on Resource Sustainability—Cities (icRS Cities) 2019. Available at SSRN: https://ssrn.com/abstract=3407254
41. Maphosa, V., Maphosa, M., Tan, A.W.K.: E-waste management in Sub-Saharan Africa: a systematic literature review. Cogent Bus. Manage. **7**, 1 (2020). https://doi.org/10.1080/23311975.2020.1814503
42. Amechi, E.P., Oni, B.A.: Import of electronic waste into Nigeria: the imperative of a regulatory policy shift. Chin. J. Environ. Law **3**(2), 141–166 (2019). https://doi.org/10.1163/24686042
43. Bakhiyi, B., Gravel, S., Ceballos, D., Flynn, M.A., Zayed, J.: Has the question of e-waste opened a Pandora's box? An overview of unpredictable issues and challenges. Environ. Int. **110**, 173–192 (2018). https://doi.org/10.1016/j.envint.2017.10.021

44. Tran, C.D., Salhofer, S.P.: Processes in informal end-processing of e-waste generated from personal computers in Vietnam. J. Mater. Cycles Waste Manag. **20**, 1154–1178 (2018). https://doi.org/10.1007/s10163-017-0678-1
45. Álvarez-de-los-Mozos, E., Rentería-Bilbao, A., Díaz-Martín, F.: WEEE recycling and circular economy assisted by collaborative robots. Appl. Sci. Appl. Sci. **10**, 14–4800 (2020). https://doi.org/10.3390/app10144800
46. Chatterjee, S.: Sustainable electronic waste management and recycling process. Am. J. Environ. Eng. **2**(1), 23–33 (2012). https://doi.org/10.5923/j.ajee.20120201.05

Printed in the United States
by Baker & Taylor Publisher Services